PROBABILITY INEQUALITIES
IN
MULTIVARIATE DISTRIBUTIONS

PROBABILITY INEQUALITIES IN MULTIVARIATE DISTRIBUTIONS

Y. L. TONG
Department of Mathematics and Statistics
University of Nebraska
Lincoln, Nebraska

1980

ACADEMIC PRESS
A Subsidiary of Harcourt Brace Jovanovich, Publishers
New York London Toronto Sydney San Francisco

ACADEMIC PRESS, INC.
111 Fifth Avenue, New York, New York 10003

United Kingdom Edition published by
ACADEMIC PRESS, INC. (LONDON) LTD.
24/28 Oval Road, London NW1 7DX

Library of Congress Cataloging in Publication Data

Tong, Yung Liang, Date
Probability inequalities in multivariate distributions.

(Probability and mathematical statistics series)
Bibliography: p.
Includes index.
1. Distribution (Probability theory)
2. Inequalities (Mathematics) I. Title.
QA273.6.T66 519.2 79–27077
ISBN 0–12–694950–6

1980 Mathematics Subject Classifications: 60E15, 62H99

PRINTED IN THE UNITED STATES OF AMERICA

80 81 82 83 9 8 7 6 5 4 3 2 1

Contents

Bibliography

Preface

This book began as lecture notes for a course given at the University of Nebraska. The course was first taught in 1974. The notes were prepared in the fall of 1977 when this topic was offered for the second time. During my visit to the University of California at Santa Barbara in the spring of 1979, the notes were used again when a similar course was taught. The classes consisted of students in statistics, mathematics, and other fields, and the course was taught with a balanced treatment between theory and applications.

The area of probability inequalities in multivariate distributions is certainly not new. However, it has experienced a remarkable growth and development during the past two decades or so. Today the subject plays an important role in many areas of statistics and probability, and it presents a very challenging and attractive field of research. On the one hand the theory is beautiful and elegant, and on the other hand the applications of the inequalities seem unlimited. I developed my own interest in this area more than ten years ago as a "user" of inequalities. Later I found the theory extremely fascinating in itself. Consequently, I felt the need for a comprehensive treatment of probability inequalities that would make them easily accessible. It is my hope that this book will serve that purpose.

This book consists of eight chapters, a bibliography, and a number of problems which can be used as exercises. References are listed at the end of each chapter. In presenting the inequalities, I have attempted to preserve the direction and the order of developments and to indicate the interrelationships and the differences among the different approaches. Whenever convenient, I have paid attention to conditions under which the inequalities are strict. I have tried to make the book self-contained; but if the proof of a theorem is easily accessible, then I have occasionally given the source of the proof without duplicating the effort. Although I have done my best to be as accurate as possible and have cited most of the major works related to this area (which is a matter of personal judgment), I do wish to apologize for any errors or serious omissions.

The book has been designed both as a reference book and as a text. In addition to those who are primarily interested in inequalities per se, I hope it may also be useful to statisticians and others who need to apply probability inequalities in their work in various areas of statistics. It is my hope that making the inequalities more accessible will encourage a wider application of the theory. As a textbook, most of the material can be covered in one quarter or in one semester following a one-quarter or one-semester course in multivariate analysis. Except possibly for certain parts in Chapters 4 and 6, an ordinary student with such a background should be able to follow the material without difficulty.

Many people have seen an earlier version of the draft, and have provided comments and suggestions. To these people I am deeply grateful. In particular, I wish to express my sincere appreciation to Professors T. W. Anderson, K. L. Chung, E. L. Lehmann, A. W. Marshall, G. S. Mudholkar, I. Olkin, M. D. Perlman, F. Proschan, J. Sethuraman, M. Sobel, and Dr. Z. Šidák for their communications, inspiration, and encouragement; but, of course, I am solely responsible for any errors or omissions. I wish to thank my students and colleagues at Nebraska and UCSB and Ms. Linda Lowe, who typed the manuscript patiently and skillfully. Finally, I thank my wife Aichuan and our children Frank and Betty for their understanding and support.

Acknowledgments

I wish to acknowledge the kind cooperation and permission granted to me by the authors and publishers of research papers and books cited in the text. In particular, I thank the following publishers for their permission: (a) Academic Press, Inc. (for including the statements and the proofs of several theorems which appeared in *Optimizing Methods in Statistics* edited by J. S. Rustagi (1971), pp. 89–113, used in Section 6.4; in *Statistical Decision Theory and Related Topics II* edited by S. S. Gupta and D. S. Moore (1977), pp. 281–296, used in Section 6.3; and in *Inequalities: Theory of Majorization and Its Applications* by A. W. Marshall and I. Olkin (1979), p. 67, p. 73, and p. 482, used as Problems 1, 2, and 16, Chapter 6), (b) Addison-Wesley Publishing Co. (for including an example which appeared in *Limit Distributions for Sums of Independent Random Variables* by B. V. Gnedenko and A. N. Kolmogorov, translated by K. L. Chung (1968), p. 254, used as Problem 5, Chapter 4), (c) Holt, Rinehart and Winston, Inc. (for including a problem which appeared in *Statistical Theory of Reliability and Life Testing* by R. E. Barlow and F. Proschan (1975), p. 104, used as Problem 4, Chapter 4), (d) John Wiley and Sons, Inc. (for citing a result which appeared in *Testing Statistical Hypotheses* by E. L. Lehmann (1959), p. 75, used in the proof of Theorem 5.1.2, and for including a problem which appeared in *Mathematical Statistics* by S. S. Wilks (1962), p. 71, used as Problem 1, Chapter 7), (e) Marcel Dekker, Inc. (for using copyrighted material which appeared in *Communications in Statistics-Theory and Methods*, Vol. ***A6*** (1977), pp. 1105–1120 and

Vol. ***A8*** (1979), pp. 1197–1204, used in Sections 8.1 and 8.3), (f) the American Statistical Association (for including results from several articles which appeared in *Journal of the American Statistical Association*, used in Sections 2.3 and 8.3), (g) the Institute of Mathematical Statistics (for including results from several articles which appeared in *The Annals of Mathematical Statistics*, *The Annals of Statistics*, and *The Annals of Probability*, used in many sections in the text), (h) University of California Press (for including results which appeared in *Proceedings of the Sixth Berkeley Symposium on Mathematical Statistics and Probability*, Vol. II, edited by L. M. LeCam, J. Neyman and E. L. Scott (1972), pp. 241–265, used in Sections 2.2 and 4.3).

I also wish to thank the staff at Academic Press, Inc. for their cooperation in producing this book and for their superb typesetting work.

Notation

In this book vectors and matrices will appear in boldface type. Among the uppercase boldfaced letters, $\boldsymbol{\Sigma}$, **C**, **D**, **H**, **I**, **M**, **P**, **R**, **S**, **T** denote matrices, and **U**, **V**, **W**, **X**, **Y**, **Z** denote multidimensional random variables. Lowercase boldfaced letters such as $\boldsymbol{\alpha}$, $\boldsymbol{\beta}$, $\boldsymbol{\lambda}$, **a**, **b**, **x**, **y**, **z** denote real vectors. Univariate random variables are denoted by uppercase letters in italic type, such as U, V, W, X, Y, Z. All vectors are column vectors; hence transposed vectors such as $\mathbf{X}'$, $\mathbf{x}'$ are row vectors, and $\mathbf{X}$, $\mathbf{x}$, $|\mathbf{x}|$ denote, respectively, $(X_1, \ldots, X_k)'$, $(x_1, \ldots, x_k)'$, and $(|x_1|, \ldots, |x_k|)'$. Among the standard notations, the following will be adopted in various parts of the book without specification: $\text{cov}(X, Y)$ (covariance of X and Y), $\text{cov}(\mathbf{X}, \mathbf{Y})$ (covariance matrix of $\mathbf{X}$ and $\mathbf{Y}$), $\text{var}\, X$ (variance of X), max (maximum), min (minimum), inf (infimum), sup (supremum), lim (limit), a.s. (almost surely). Finally, $\mathcal{R}^k$ denotes the k-dimensional Euclidean space, and the symbol ■ indicates the end of a proof for a theorem, a lemma, or the end of an example.

CHAPTER

1

Introduction

It may be safe to assert that during the course of development of statistics and probability theory, the problem of finding bounds for a joint probability in multivariate distributions is not a new one. For instance, Bonferroni's inequality (of degree one), which says that

$$P\left[\bigcap_{i=1}^{k} A_i\right] \geqslant 1-\sum_{i=1}^{k}\left[1-P(A_i)\right]$$

for all events $A_1,\ldots,A_k$, has been in the literature for some time. However, sharper and more useful inequalities (under suitable conditions) have become available only recently. Some of them were developed because the need arose in certain applied problems, others were discovered for their own sake. During the past two decades, the rate of growth seems to have been increasing rapidly. Today the subject has developed into a self-contained branch of knowledge with both mathematical elegance and a great variety of statistical applications.

1.1. CLASSIFICATION OF PROBABILITY INEQUALITIES

In order to treat the existing probability inequalities systematically, we shall first make an attempt to classify them into several types. For a fixed positive integer k let us consider a random variable $\mathbf{X}=(X_1,\ldots,X_k)'$ with probability density function $f_{\boldsymbol{\theta}}$, which is assumed to exist and which may depend on a parameter vector $\boldsymbol{\theta}$ for $\boldsymbol{\theta}\in\Omega$ (the parameter space). Let A denote a Borel-measurable subset of $\mathcal{R}^k$, and define

$$\alpha(A,f_{\boldsymbol{\theta}}) = P[\mathbf{X}\in A] = \int\cdots\int_A f_{\boldsymbol{\theta}}(\mathbf{x}) \prod_{i=1}^{k} dx_i. \tag{1.1.1}$$

This probability in general depends on the subset A, the functional form of f, and for a given family of probability density functions the parameter $\boldsymbol{\theta}$. Existing inequalities in the literature may generally be classified into one (or several) of the following types:

T1. As a Function of the Subset A. Let both A and A^* be Borel measurable subsets of $\mathcal{R}^k$. If they are comparable through a partial ordering when a relation "$>$" is suitably chosen, then under reasonable conditions on $f_{\boldsymbol{\theta}}$ the corresponding probabilities can be ordered. In symbols, let "$>$" define a relation of partial ordering; then A"$>$"A^* (or A^*"$>$"A) implies that $\alpha(A,f_{\boldsymbol{\theta}})\geqslant\alpha(A^*,f_{\boldsymbol{\theta}})$ for all A,A^*. A simple example is, of course, that A^* is a proper subset of A (in fact this is very well known and sometimes is too trivial to be useful). An important nontrivial relation depends on the location of the subset when the "sizes" of A and A^* are identical. This will be illustrated in Example 1.1.1.

T2. As a Function of the Parameter $\boldsymbol{\theta}$. Assume that $f_{\boldsymbol{\theta}}$ belongs to a certain family of probability density functions, and let $\boldsymbol{\theta}$ and $\boldsymbol{\theta}^*$ be elements of Ω. Then under reasonable conditions on the subset A and for a suitably chosen relation, $\boldsymbol{\theta}$"$>$"$\boldsymbol{\theta}^*$ (or $\boldsymbol{\theta}^*$"$>$"$\boldsymbol{\theta}$) implies that $\alpha(A,f_{\boldsymbol{\theta}})\geqslant\alpha(A,f_{\boldsymbol{\theta}^*})$ for all $\boldsymbol{\theta}$, $\boldsymbol{\theta}^*$ in Ω.

T3. As a Function of the Marginal Probabilities. Suppose that in the special case the subset A is of the form $A=\times_{i=1}^{k}A_i$ and

$$\begin{aligned}\alpha(A,f_{\boldsymbol{\theta}}) &= P_{\boldsymbol{\theta}}\left[\bigcap_{i=1}^{k}\{X_i\in A_i\}\right]\\ &= P_{\boldsymbol{\theta}}[X_i\in A_i,\ i=1,\ldots,k],\end{aligned} \tag{1.1.2}$$

where $A_1,\ldots,A_k$ are Borel-measurable subsets of $\mathcal{R}^1$. Then under suitable conditions on $f_{\boldsymbol{\theta}}$ the joint probability can be bounded below or above by the product of the marginal probabilities. Hence the inequality

$$\alpha(A,f_{\boldsymbol{\theta}}) \geqslant (\leqslant) P_{\boldsymbol{\theta}}\left[\bigcap_{i\in C}\{X_i\in A_i\}\right] P_{\boldsymbol{\theta}}\left[\bigcap_{i\notin C}\{X_i\in A_i\}\right]^{\dagger} \tag{1.1.3}$$

holds for a proper subset C of $\{1,\ldots,k\}$. In most cases one can obtain

$$\alpha(A,f_{\boldsymbol{\theta}}) \geqslant (\leqslant) \prod_{i=1}^{k} P_{\boldsymbol{\theta}}[X_i\in A_i] \tag{1.1.4}$$

by induction. Note that the right-hand side of (1.1.3) is the joint probability when the two subsets of random variables are actually independent. When such inequalities apply, one may take advantage of a reduction in dimensionality, and usually joint probabilities with a lower dimension are more workable.

T4. As a Function of the Moments of the Random Variables. This mainly includes distribution-free inequalities (e.g., the multivariate version of Chebyshev-type inequalities).

T5. Combinations of the Above.

It may be helpful to see an example in which these types of inequality apply. This is shown in Example 1.1.1. For convenience we consider normal variables only.

Example 1.1.1. Let $\mathbf{X}=(X_1,\ldots,X_k)'$ be distributed according to a multivariate normal distribution with mean vector $\mathbf{0}$, variances one, and a common correlation coefficient $\rho\in(-1/(k-1),1)$ (the reason for imposing this condition is to make the covariance matrix positive definite). For fixed $a_i>0$ let us define $A_i=[-a_i,a_i]$ $(i=1,\ldots,k)$. Let $A=\times_{i=1}^{k}A_i$, which is a rectangular subset of $\mathfrak{R}^k$ and is symmetric about the origin. Let $\mathbf{y}=(y_1,\ldots,y_k)'$ denote a given real vector, $\mathbf{y}\neq\mathbf{0}$, and for a real number $\lambda>0$ consider a set

$$A+\lambda\mathbf{y} = \mathop{\times}_{i=1}^{k}\left[-a_i+\lambda y_i,a_i+\lambda y_i\right]. \tag{1.1.5}$$

†Throughout, $\bigcap_{i\notin C}$ means that the intersection is taken over the subset of integers that are not in C but are in $\{1,\ldots,k\}$.

Here $A+\lambda\mathbf{y}$ is a shift in location of A with a distance of λ units along the direction specified by $\mathbf{y}$.

(a) For $\lambda_2>\lambda_1\geqslant 0$, $A+\lambda_1\mathbf{y}$ and $A+\lambda_2\mathbf{y}$ are partially ordered, and the inequality

$$P[\mathbf{X}\in(A+\lambda_1\mathbf{y})] > P[\mathbf{X}\in(A+\lambda_2\mathbf{y})] \tag{1.1.6}$$

holds for all ρ. This follows from Theorem 4.1.1 (Chapter 4). Note that the volumes of $A+\lambda_1\mathbf{y}$ and $A+\lambda_2\mathbf{y}$ are identical, but the set $A+\lambda_2\mathbf{y}$ is farther away from the origin. This inequality is intuitively clear because the functional value of a multivariate normal density is decreasing when a point is moving away from the origin along a given direction (type T1).

(b) The probability $P_\rho[\mathbf{X}\in A]$ as a function of ρ is monotonically decreasing in ρ for $\rho<0$ and monotonically increasing in ρ for $\rho>0$. As a special consequence

$$P_\rho[\mathbf{X}\in A] > \prod_{i=1}^{k} P[-a_i\leqslant X_i\leqslant a_i] \tag{1.1.7}$$

holds for all $\rho\neq 0$. This follows from Theorem 2.2.5 or Theorem 4.3.1. Therefore only the univariate normal probabilities are involved when a lower bound is to be obtained for the probability content of A from the right-hand side of (1.1.7) (type T2 or T3).

(c) For all ρ and all proper subsets C of $\{1,\ldots,k\}$ we have from Theorem 2.2.2 or 2.2.3

$$P[\mathbf{X}\in A] > P\left[\bigcap_{i\in C}\{X_i\in A_i\}\right]P\left[\bigcap_{i\notin C}\{X_i\in A_i\}\right]. \tag{1.1.8}$$

A repeated application of this inequality yields the same inequality given in (1.1.7) (type T2 or T3).

(d) In the special case for which

$$a_1=\cdots=a_k=a \tag{1.1.9}$$

holds we have

$$P[\mathbf{X}\in A] \geqslant 1-\left[(k-1)(1-\rho)^{1/2}+\{1+(k-1)\rho\}^{1/2}\right]^2/ka^2 \tag{1.1.10}$$

for all ρ (Corollary 2 of Theorem 7.2.2). Here the lower bound depends only on the value of a, the variances (which are one), and the common covariance (correlation) ρ (type T4).

(e) Assume that (1.1.9) holds. Let us now consider a more general case in which the variances are one, the correlations are ρ, but the means are not necessarily zero. Consider two different configurations for the mean vector:

$$\boldsymbol{\theta} = (\theta_1,\ldots,\theta_k)', \tag{1.1.11}$$

$$\boldsymbol{\theta}^* = (\theta_1^*,\ldots,\theta_k^*)', \tag{1.1.12}$$

and assume that $\theta_i \geqslant \theta_{i+1}$, $\theta_i^* \geqslant \theta_{i+1}^*$ for $i=1,\ldots,k-1$. If $\boldsymbol{\theta}$ majorizes $\boldsymbol{\theta}^*$, i.e., if

$$\sum_{i=1}^{k} \theta_i = \sum_{i=1}^{k} \theta_i^* \quad \text{and} \quad \sum_{i=1}^{r} \theta_i \geqslant \sum_{i=1}^{r} \theta_i^* \quad \text{for all} \quad r<k, \tag{1.1.13}$$

then the inequality

$$P_{\boldsymbol{\theta}}[\mathbf{X}\in A] \leqslant P_{\boldsymbol{\theta}^*}[\mathbf{X}\in A] \tag{1.1.14}$$

holds (Theorem 6.2.1 or 6.2.2). It therefore follows as a special case that for any $\boldsymbol{\theta}$ satisfying $\Sigma_{i=1}^{k}\theta_i=0$ the inequality

$$P_{\boldsymbol{\theta}}[\mathbf{X}\in A] \leqslant P_{\mathbf{0}}[\mathbf{X}\in A] \tag{1.1.15}$$

holds, where $\mathbf{0}$ is the zero vector (type T2). Note that this particular inequality also follows from (1.1.6) by letting $\lambda_1=0$, $\lambda_2=1$, and $\boldsymbol{\theta}=-\mathbf{y}$ (type T1).

1.2. SCOPE AND ORGANIZATION

In this book we shall be concerned only with those inequalities that are of types T1–T5. The conditions for such inequalities range from very specific to very general. In Chapter 2 we study inequalities for multivariate normal distribution, and see how they depend on the correlation coefficients. Chapter 3 concerns inequalities for other well-known distributions, including the multivariate distributions of *t*, chi-square, and *F*. Basic tools for proving those inequalities include an identity due to Plackett, an application of Anderson's theorem, and a conditional argument.

The chapters are organized according to the order of increasing generality. Chapters 2 and 3 deal with specific distributions. Chapters 4 and 5 contain inequalities for a class of symmetric unimodal distributions and for a certain class of random variables that are positively dependent by association or by mixture. Basic tools involved are a volume inequality and a similar conditional argument.

Inequalities obtainable through the mathematical tool of majorization and weak majorization are given in Chapter 6. The role played by majorization and weak majorization is this: If two parameter vectors or two subsets are comparable through majorization or weak majorization, then under certain conditions the corresponding probabilities can be ordered. Particular applications of these basic inequalities give bounds for location parameter families (hence the normal family) and for the distributions of order statistics when observations are taken from heterogeneous populations. Chapter 7 contains some distribution-free inequalities. Those include Bonferroni-, Chebyshev-, and Kolmogorov-type inequalities, and they are developed under less restrictive conditions.

In Chapter 8 we give some applications of these inequalities to the areas of simultaneous confidence regions, hypothesis testing, multiple decision problems, and reliability theory. This is done to demonstrate the main point: that these inequalities, although discovered and proved on a theoretical basis, are important and powerful tools for solving many real-life problems in statistical applications.

A bibliography on probability inequalities in multivariate distributions, their applications, and related statistical tables is included. It contains mostly references published during the past two decades. They have been divided into nine sections. Some of them are relevant to more than one section, and hence have been cross-listed accordingly.

CHAPTER

2

Inequalities for Multivariate Normal Distribution

Probability inequalities for multivariate normal distribution have received a considerable amount of attention in the statistical literature, especially during the early stage of the development. This may be partially attributed to the fact that the assumption of normality was usually imposed in the applied problems, and partially due to the mathematical simplicity of the functional form of the multivariate normal density function. Existing inequalities for the normal distribution concern mainly the quadrant and rectangular probability contents as functions of either the correlation coefficients or the mean vector. This chapter contains inequalities that depend on the correlation coefficients only. Certain additional inequalities will also be given in Chapter 4, following some results for elliptically contoured distributions. Inequalities that depend on means follow from more general results, and they will be discussed in Chapters 4 and 6.

For an arbitrary but fixed positive integer k let $\mathbf{R}$ be a $k\times k$ matrix such that $\mathbf{R}$ is symmetric and positive semidefinite. Let the random variable $\mathbf{X}=(X_1,\ldots,X_k)'$ be distributed according to a multivariate normal distribution with mean vector $\mathbf{0}$ and a covariance matrix $\mathbf{R}$.

This is to be interpreted as the following: If $\mathbf{R}$ is positive definite (i.e., if $\mathbf{R}$ is nonsingular), then the density of $\mathbf{X}$ is given by

$$\varphi_k(\mathbf{x},\mathbf{R}) = (2\pi)^{-k/2}|\mathbf{R}|^{-1/2}\exp\left(-\tfrac{1}{2}\mathbf{x}'\mathbf{R}^{-1}\mathbf{x}\right) \tag{2.0.1}$$

for all $\mathbf{x}=(x_1,\ldots,x_k)'$. If $\mathbf{R}$ is singular, then $X_1,\ldots,X_k$ are linearly dependent, and the distribution of $\mathbf{X}$ is defined according to Section 24.3 in Cramér (1946) or Definition 2.4.1 in Anderson (1958). We consider the probability content of a subset A, which is assumed to be Lebesgue measurable. In particular, for an arbitrary but fixed real vector $\mathbf{a}=(a_1,\ldots,a_k)'$, we consider the probabilities

$$\alpha_1(k,\mathbf{a},\mathbf{R}) = P_{\mathbf{R}}\left[\bigcap_{i=1}^{k}\{X_i \leqslant a_i\}\right], \tag{2.0.2}$$

$$\alpha_2(k,\mathbf{a},\mathbf{R}) = P_{\mathbf{R}}\left[\bigcap_{i=1}^{k}\{|X_i| \leqslant a_i\}\right], \qquad a_i > 0, \quad i = 1,\ldots,k, \tag{2.0.3}$$

and

$$\alpha_3(k,\mathbf{a},\mathbf{R}) = P_{\mathbf{R}}\left[\bigcap_{i=1}^{k}\{|X_i| \geqslant a_i\}\right], \qquad a_i > 0, \quad i = 1,\ldots,k, \tag{2.0.4}$$

as functions of the correlation coefficients. Since $\mathbf{a}$ is arbitrary, we may therefore assume that the variances of the X_is are one (i.e., $\mathbf{R}=(\rho_{ij})$ is a correlation matrix) in most discussions (otherwise we can replace a_i by a_i/σ_i and the problem remains unchanged). Most inequalities contained in this chapter concern the monotonocity properties of these probabilities as functions of the correlations.

2.1. SLEPIAN'S INEQUALITY

This section concerns an inequality for the quadrant probability $\alpha_1(k,\mathbf{a},\mathbf{R})$ as a function of the elements of $\mathbf{R}=(\rho_{ij})$. The probability in question is that for which the random variables simultaneously take smaller values. As pointed out by Slepian (1962), the correlation matrix $\mathbf{R}$ may generally be regarded as an indicator of how much the random variables $X_1,\ldots,X_k$ "hang together." Therefore it is intuitively obvious that when the variables are highly correlated (with large ρ_{ij}s), they should hang together more and are more likely to maintain the same magnitude. This view is confirmed by an inequality of Slepian which says that the quadrant probability is a monotonically increasing function of the ρ_{ij}s.

Before seeing this theorem we first observe a lemma of Plackett (1954) concerning a basic property of the multivariate normal density function. For $k=2$ this property is an old result and it follows from straightforward calculation. In his paper on the reduction in dimensionality for multivariate normal probability integrals, Plackett gave an elegant proof of this property for general k.

Lemma 2.1.1. Assume that $\mathbf{R}=(\rho_{lm})$ is a positive definite correlation matrix. Then for $\varphi_k(\mathbf{x},\mathbf{R})$ defined as in (2.0.1)

$$\frac{\partial}{\partial\rho_{lm}}\varphi_k(\mathbf{x},\mathbf{R}) = \frac{\partial^2}{\partial x_l\,\partial x_m}\varphi_k(\mathbf{x},\mathbf{R}) \tag{2.1.1}$$

holds for all $l\neq m$.

Proof. The normal density $\varphi_k(\mathbf{x},\mathbf{R})$ can be written as the transform of its characteristic function (see, e.g., Cramér (1946, Section 10.6)), which with $i^2=-1$ is

$$\varphi_k(\mathbf{x},\mathbf{R}) = (2\pi)^{-k}\int\cdots\int\exp\left(-i\mathbf{t}'\mathbf{x}-\tfrac{1}{2}\mathbf{t}'\mathbf{R}\mathbf{t}\right)\prod_{j=1}^{k}dt_j. \tag{2.1.2}$$

After interchanging the order of differentiation and integration in (2.1.2) (which is obviously permissible), both sides of (2.1.1) become

$$-(2\pi)^{-k}\int\cdots\int t_l t_m\exp\left(-i\mathbf{t}'\mathbf{x}-\tfrac{1}{2}\mathbf{t}'\mathbf{R}\mathbf{t}\right)\prod_{j=1}^{k}dt_j.$$

Hence the identity follows. ■

Now let us observe the following fact: If we keep the other elements of $\mathbf{R}$ fixed and change only the value of ρ_{ij}, then the set of admissible points of ρ_{ij} (in which $\mathbf{R}$ remains positive definite or positive semidefinite) is an interval (see Problem 2). This fact together with the next lemma, which was also due to Plackett (1954), yields an important monotonicity property.

Lemma 2.1.2. Let **X** be distributed according to $N(\mathbf{0},\mathbf{R})$,† where $\mathbf{R}=(\rho_{ij})$ is a positive definite correlation matrix. Then for every fixed **a**

$$\frac{\partial}{\partial\rho_{ij}}\alpha_1(k,\mathbf{a},\mathbf{R}) > 0 \tag{2.1.3}$$

holds for all $i \neq j$.

Proof. For $k=2$ this result is immediate. Hence the proof will be given for $k>2$ only and without loss of generality we may assume that $(i,j)=(1,2)$. This is so because for any other (i,j) the problem remains unchanged and the same proof applies after the coordinates have been properly renumbered.

After interchanging the order of differentiation and integration and applying Lemma 2.1.1, we have

$$\frac{\partial}{\partial\rho_{12}}\alpha_1(k,\mathbf{a},\mathbf{R})$$

$$= \int_{-\infty}^{a_3}\cdots\int_{-\infty}^{a_k}\left[\int_{-\infty}^{a_1}\int_{-\infty}^{a_2}\frac{\partial^2}{\partial x_1\,\partial x_2}\varphi_k(\mathbf{x},\mathbf{R})\,dx_1\,dx_2\right]\prod_{i=3}^{k}dx_i$$

$$= \int_{-\infty}^{a_3}\cdots\int_{-\infty}^{a_k}\varphi_k(a_1,a_2,x_3,\ldots,x_k,\mathbf{R})\prod_{i=3}^{k}dx_i > 0$$

as desired. ■

We now state and prove the inequality due to Slepian (1962).

Theorem 2.1.1. Let **X** be distributed according to $N(\mathbf{0},\boldsymbol{\Sigma})$, where $\boldsymbol{\Sigma}$ is a correlation matrix. Let $\mathbf{R}=(\rho_{ij})$, $\mathbf{T}=(\tau_{ij})$ be two positive semidefinite correlation matrices. If $\rho_{ij} \geqslant \tau_{ij}$ holds for all i,j, then

$$P_{\boldsymbol{\Sigma}=\mathbf{R}}\left[\bigcap_{i=1}^{k}\{X_i \leqslant a_i\}\right] \geqslant P_{\boldsymbol{\Sigma}=\mathbf{T}}\left[\bigcap_{i=1}^{k}\{X_i \leqslant a_i\}\right] \tag{2.1.4}$$

holds for all $\mathbf{a}=(a_1,\ldots,a_k)'$. Furthermore, the inequality is strict if $\mathbf{R},\mathbf{T}$ are positive definite and if the strict inequality $\rho_{ij}>\tau_{ij}$ holds for some i,j.

†We shall use $N(\boldsymbol{\mu},\boldsymbol{\Sigma})$ to denote a multivariate normal distribution with mean $\boldsymbol{\mu}$ and covariance matrix $\boldsymbol{\Sigma}$.

Proof. First let us assume that both $\mathbf{R},\mathbf{T}$ are positive definite. For $\lambda \in [0,1]$ let us define

$$\mathbf{S}(\lambda) = (s_{ij}) = [\lambda\mathbf{R} + (1-\lambda)\mathbf{T}]. \tag{2.1.5}$$

Then $\mathbf{S}(\lambda)$ is also positive definite. For fixed $\mathbf{a}$ $\alpha_1(k,\mathbf{a},\mathbf{S}(\lambda))$ is differentiable with respect to λ, and the directional derivative is

$$\frac{d}{d\lambda}\alpha_1(k,\mathbf{a},\mathbf{S}(\lambda)) = \sum_{i<j}\left(\frac{d}{d\lambda}s_{ij}\right)\frac{\partial}{\partial s_{ij}}\alpha_1(k,\mathbf{a},\mathbf{S}(\lambda)).$$

But for $i<j$

$$\frac{d}{d\lambda}s_{ij} = \rho_{ij} - \tau_{ij}$$

and by Lemma 2.1.2 we have

$$\frac{\partial}{\partial s_{ij}}\alpha_1(k,\mathbf{a},\mathbf{S}(\lambda)) > 0.$$

Hence if $\mathbf{R},\mathbf{T}$ are positive definite and $\rho_{ij} > \tau_{ij}$ holds for some i,j, then $(d/d\lambda)\alpha_1(k,\mathbf{a},\mathbf{S}(\lambda))$ is strictly positive. This implies the strict inequality $\alpha_1(k,\mathbf{a},\mathbf{R}) > \alpha_1(k,\mathbf{a},\mathbf{T})$ because $\mathbf{T}=\mathbf{S}(0)$ and $\mathbf{R}=\mathbf{S}(1)$.

If $\mathbf{R}$ or $\mathbf{T}$ is singular, then for $\varepsilon > 0$ and $\lambda \in [0,1]$ we define

$$\mathbf{S}(\lambda,\varepsilon) = \frac{1}{1+\varepsilon}[\lambda\mathbf{R} + (1-\lambda)\mathbf{T} + \varepsilon\mathbf{I}],$$

where $\mathbf{I}$ is the identity matrix. Clearly $\mathbf{S}(\lambda,\varepsilon)$ is positive definite. Hence for every fixed $\varepsilon > 0$ the directional derivative $(d/d\lambda)$ $\alpha_1(k,\mathbf{a},\mathbf{S}(\lambda,\varepsilon))$ is nonnegative, and it remains nonnegative as ε approaches zero. ∎

Remark. In the above proof the correlations in $\mathbf{S}(\lambda)$ are raised *proportionally* by increasing λ from zero to one. It might be tempting to claim that the proof follows immediately from a repeated application of Lemma 2.1.2, by increasing one correlation coefficient at a time. This is not so, however, because of the difficulty that the matrices so obtained in the steps in between may not be positive semidefinite (see Problem 3).

An alternative proof of this theorem, based on a geometrical argument, was later provided by Chartres (1963).

We now observe some immediate consequences of Theorem 2.1.1. For convenience we write $P_{\mathbf{R}}(A)$ instead of $P_{\Sigma=\mathbf{R}}(A)$.

Corollary 1. Under the same conditions imposed in Theorem 2.1.1

$$P_{\mathbf{R}}\left[\bigcap_{i=1}^{k}\{X_i \geqslant a_i\}\right] \geqslant P_{\mathbf{T}}\left[\bigcap_{i=1}^{k}\{X_i \geqslant a_i\}\right] \tag{2.1.6}$$

holds for all $\mathbf{a}$.

Corollary 2. Let $\mathbf{X}$ be distributed according to $N(\mathbf{0},\mathbf{R})$, where $\mathbf{R}=(\rho_{ij})$ is a positive semidefinite correlation matrix. Let C be a proper subset of $\{1,\ldots,k\}$. If $\rho_{ij} \geqslant (\leqslant) 0$ for all $i \in C$ and $j \notin C$, then

$$P_{\mathbf{R}}\left[\bigcap_{i=1}^{k}\{X_i \leqslant a_i\}\right] \geqslant (\leqslant) P\left[\bigcap_{i \in C}\{X_i \leqslant a_i\}\right] P\left[\bigcap_{j \notin C}\{X_j \leqslant a_j\}\right] \tag{2.1.7}$$

holds for all $\mathbf{a}$. In particular, if $\rho_{ij} \geqslant (\leqslant) 0$ for all $i \neq j$, then

$$P_{\mathbf{R}}\left[\bigcap_{i=1}^{k}\{X_i \leqslant a_i\}\right] \geqslant (\leqslant) \prod_{i=1}^{k} P[X_i \leqslant a_i] \tag{2.1.8}$$

holds for all $\mathbf{a}$.

Corollary 1 is equivalent to Theorem 2.1.1. It follows from the fact that when $\mathbf{a}$ and $\mathbf{X}$ are replaced by $-\mathbf{a}$ and $-\mathbf{X}$, respectively, the correlations remain unchanged. Corollary 2 is immediate, and it has been found useful in certain applications.

2.2. MULTIVARIATE NORMAL PROBABILITIES OF RECTANGLES

In view of Slepian's elegant result it is certainly reasonable to expect that rectangular multivariate normal probabilities also possess some monotonicity properties as functions of the correlations. This problem was first considered by Dunn (1958) in a paper concerning simultaneous confidence regions for the means of correlated normal variables. In order to obtain a conservative solution based on univariate normal probabilities only, she needed to apply the inequality

$$\alpha_2(k,\mathbf{a},\mathbf{R}) = P_{\mathbf{R}}\left[\bigcap_{i=1}^{k}\{|X_i| \leqslant a_i\}\right] \geqslant \prod_{i=1}^{k} P[|X_i| \leqslant a_i] \tag{2.2.1}$$

and was able to prove this for certain special cases. (Her results were given in the special form of $a_1 = \cdots = a_k$ only, but it is clear that the proof she provided also applies even when the a_is are different.)

Before stating her theorem we define a particular structure of a covariance matrix $\mathbf{R}$. It was first discussed by Dunnett and Sobel (1955) and has been found useful in certain multiple comparison problems. (This will be illustrated in Chapter 8.)

Definition 2.2.1. A positive definite covariance matrix $\mathbf{R}=(\rho_{ij})$ is said to have the structure l if there exist real numbers $\lambda_1,\ldots,\lambda_k$ in $(-1,1)$ and $\sigma_1,\ldots,\sigma_k$ in $(0,\infty)$ such that $\rho_{ii}=\sigma_i^2$ for all i and $\rho_{ij}=\lambda_i\lambda_j\sigma_i\sigma_j$ for all $i\neq j$. If $\mathbf{R}$ is already a correlation matrix (i.e., if $\rho_{ii}=1$ for all i), then $\mathbf{R}$ is said to have the structure l if $\rho_{ij}=\lambda_i\lambda_j$ for all $i\neq j$.

Now let $\mathbf{X}$ be distributed according to $N(\mathbf{0},\mathbf{R})$. If $\mathbf{R}$ has the structure l, then $\mathbf{X}$ can be obtained through a linear transformation of independent standard normal variables $U, U_1,\ldots,U_k$; i.e., we can define

$$X_i = \sigma_i\left[\left(1-\lambda_i^2\right)^{1/2}U_i - \lambda_i U\right], \qquad i = 1,\ldots,k. \tag{2.2.2}$$

If $\mathbf{R}$ is a correlation matrix (i.e., the variances of X_i are one), then $\sigma_1,\ldots,\sigma_k$ are chosen to be one. Applying such a transformation, the probability contents of $\mathbf{X}$ can be expressed in the form of a single integral. This integral depends only on the standard normal distribution function.

We also observe a basic lemma that has become a useful tool in obtaining certain types of probability inequalities. This was originally due to Chebyshev (see Hardy, Littlewood, and Pólya (1959, p. 43)), and it was also proved later by Kimball (1951) and Khatri (1967).

Lemma 2.2.1. Let $\mathbf{U}$ be a s-dimensional random variable and $g_1, g_2\colon \mathcal{R}^s \to \mathcal{R}^1$ be Borel-measurable real-valued functions. If g_1, g_2 satisfy

$$\left[g_1(\mathbf{u}_2)-g_1(\mathbf{u}_1)\right]\left[g_2(\mathbf{u}_2)-g_2(\mathbf{u}_1)\right] \geqslant 0\,(\leqslant 0) \tag{2.2.3}$$

for all $\mathbf{u}_1, \mathbf{u}_2$ in the support of the distribution of $\mathbf{U}$, then

$$E\left[g_1(\mathbf{U})g_2(\mathbf{U})\right] \geqslant (\leqslant)\left[Eg_1(\mathbf{U})\right]\left[Eg_2(\mathbf{U})\right] \tag{2.2.4}$$

holds provided the expectations exist, or equivalently,

$$\operatorname{cov}(g_1(\mathbf{U}), g_2(\mathbf{U})) \geqslant 0 (\leqslant 0). \tag{2.2.5}$$

The inequality is strict unless $g_1(\mathbf{U})$ or $g_2(\mathbf{U})$ is singular.

Remark. Let U be a one-dimensional random variable. It follows that if $g_1, \ldots, g_k$ are bounded, nonnegative, and monotone in the same direction, then

$$E \prod_{i=1}^{k} g_i(U) \geqslant \left[E \prod_{i \in C} g_i(U) \right] \left[E \prod_{i \notin C} g_i(U) \right] \geqslant \cdots \geqslant \prod_{i=1}^{k} E g_i(U) \tag{2.2.6}$$

holds for every subset C of $\{1, \ldots, k\}$. The inequalities are strict if the random variables $g_i(U)$ $(i=1,\ldots,k)$ are nonsingular.

The inequality in (2.2.6) was proved independently by Kimball (1951), and quite frequently it is called Kimball's inequality in the literature.

We now state and prove Dunn's result.

Theorem 2.2.1. Let $\mathbf{X}$ be distributed according to $N(\mathbf{0}, \mathbf{R})$, where $\mathbf{R}$ is positive definite. If (a) $k \leqslant 3$ or (b) $k > 3$ and $\mathbf{R}$ has the structure l with $\lambda_i \in [0,1)$ $(i=1,\ldots,k)$, then the inequality in (2.2.1) holds for all $\mathbf{a}$.

Outline of the Proof. Without loss of generality it may be assumed that $\mathbf{R}$ is a correlation matrix.

It follows from elementary calculation that for all k we have

$$\frac{\partial}{\partial \rho_{ij}} P_{\mathbf{R}} \left[\bigcap_{i=1}^{k} \{ |X_i| \leqslant a_i \} \right] \Bigg|_{\mathbf{R}=\mathbf{I}} = 0,$$

$$\frac{\partial^2}{\partial \rho_{ij}\, \partial \rho_{lm}} P_{\mathbf{R}} \left[\bigcap_{i=1}^{k} \{ |X_i| \leqslant a_i \} \right] \Bigg|_{\mathbf{R}=\mathbf{I}} \begin{matrix} > 0 & \text{for} \quad (l,m)=(i,j) \\ = 0 & \text{otherwise,} \end{matrix} \tag{2.2.7}$$

where $\mathbf{I}$ is the identity matrix (with $\rho_{ij}=0$, $i \neq j$). Therefore in the expansion of the probability $P_{\mathbf{R}}[\bigcap_{i=1}^{k} \{|X_i| \leqslant a_i\}]$ about $\mathbf{R}=\mathbf{I}$, the first degree terms vanish and the second degree terms give a positive definite quadratic form, so that, as a function of the ρ_{ij}s, this probability has a relative minimum at the origin (with $\rho_{ij}=0$ for all $i \neq j$) for all k. Now suppose that $k=2$ or $k=3$. Then from the functional form of the left-hand side of (2.2.7) at any point other than the origin at least one of the first derivatives is different from zero. Therefore the absolute minimum must occur at either the origin or at a boundary point of the set of admissible points. It follows that for both $k=2$ and $k=3$ it cannot be at a boundary point. Hence (2.2.1) follows for $k \leqslant 3$ for any $\mathbf{R}$ which is positive definite.

The technique used by Dunn for proving (b) depends on a different argument. It depends on an integral form, and it was first adopted by Dunnett and Sobel (1955). Suppose that $\mathbf{R}$ has the structure l with $\lambda_i \in [0,1)$, and let φ and Φ denote the univariate $N(0,1)$ density and distribution functions, respectively. Then from the transformation given in (2.2.2) we can write

$$\begin{aligned}
\alpha_2(k,\mathbf{a},\mathbf{R}) &= P_{\mathbf{R}}\left[\bigcap_{i=1}^{k}\left\{\frac{(\lambda_i U - a_i)}{(1-\lambda_i^2)^{1/2}} \leqslant U_i \leqslant \frac{(\lambda_i U + a_i)}{(1-\lambda_i^2)^{1/2}}\right\}\right] \\
&= \int_{-\infty}^{\infty}\prod_{i=1}^{k}\left[\Phi\left(\frac{(\lambda_i u + a_i)}{(1-\lambda_i^2)^{1/2}}\right) - \Phi\left(\frac{(\lambda_i u - a_i)}{(1-\lambda_i^2)^{1/2}}\right)\right]\varphi(u)\,du \\
&\equiv \int_{-\infty}^{\infty}\prod_{i=1}^{k} g(u;\lambda_i,a_i)\varphi(u)\,du. \qquad (2.2.8)
\end{aligned}$$

After changing variables in the integral over the interval $(-\infty,0)$, using the symmetry property of Φ, and applying (2.2.6), we have

$$\begin{aligned}
\alpha_2(k,\mathbf{a},\mathbf{R}) &= E^{|U|}\prod_{i=1}^{k} g(|U|,\lambda_i,a_i) \\
&\geqslant \prod_{i=1}^{k} E^{|U|} g(|U|,\lambda_i,a_i) = \prod_{i=1}^{k} P[|X_i| \leqslant a_i],
\end{aligned}$$

as was to be shown. Here the expectations are taken over the distribution of $|U|$, which has a density $2\varphi(u)$ for $u \geqslant 0$, and the inequality follows from the fact that $g(|u|,\lambda_i,a_i)$ is monotonically nonincreasing in $|u|$ for all $\lambda_i \geqslant 0$ and $a_i > 0$. ■

Although not able to obtain a general proof, Dunn conjectured that (2.2.1) should hold for an arbitrary positive definite correlation matrix $\mathbf{R}$ when $k>3$. The proofs of her conjecture were given almost simultaneously by Scott (1967), Khatri (1967), and Šidák (1967, 1968). Scott's proof was later found to be in error (see Das Gupta, Eaton, Olkin, Perlman, Savage and Sobel (1972) and Šidák (1975)).

The proofs given by Khatri and Šidák follow from more general results under different approaches. Since their results depend on a special application of Anderson's theorem to normal distribution, we first state this application in the following lemma. Its proof follows immediately from Anderson's theorem, and the theorem itself will be seen in Chapter 4 (Theorem 4.1.1).

Lemma 2.2.2. Let $\varphi_k(\mathbf{x},\mathbf{R})$ be the multivariate normal density function given in (2.0.1), where $\mathbf{R}$ is a positive definite covariance matrix. Let $A \subset \mathfrak{R}^k$ be a convex set symmetric in $\mathbf{x}$ about the origin, and let $\mathbf{y} \in \mathfrak{R}^k$, $\mathbf{y} \neq \mathbf{0}$, be arbitrary but fixed. Then as a function of $u \in \mathfrak{R}^1$

$$g(u,\mathbf{y},A) = \int_A \cdots \int \varphi_k(\mathbf{x} - u\mathbf{y},\mathbf{R}) \prod_{i=1}^{k} dx_i; \qquad (2.2.9)$$

is monotonically nonincreasing in $|u|$. If A is a bounded set with positive probability content, then it is strictly decreasing in $|u|$.

Note that for every given u $g(u,\mathbf{y},A)$ is the probability content under a multivariate normal distribution with mean $u\mathbf{y}$ and covariance matrix $\mathbf{R}$ over the symmetric convex region A. If the mean of the distribution moves away from the origin along the direction specified by the vector $\mathbf{y}$ (with a larger $|u|$ value), then the probability content becomes smaller. We also observe the symmetry property of g, i.e., $g(u,\mathbf{y},A)=g(-u,\mathbf{y},A)$, which follows immediately from the symmetry of φ_k and A.

Before stating the result of Khatri let us first introduce some notation. Assume that $\mathbf{X}$ is distributed according to $N(\mathbf{0},\mathbf{R})$, where $\mathbf{R}$ is $k \times k$ and is positive semidefinite. For a positive integer $q<k$ let us consider a partition of the components of $\mathbf{X}$ and the elements of $\mathbf{R}$ as follows:

$$\mathbf{X} = \begin{pmatrix} \mathbf{X}^{(1)} \\ \mathbf{X}^{(2)} \end{pmatrix}, \qquad \mathbf{R} = \begin{pmatrix} \mathbf{R}_{11} & \mathbf{R}_{12} \\ \mathbf{R}_{21} & \mathbf{R}_{22} \end{pmatrix}, \qquad (2.2.10)$$

where $\mathbf{X}^{(1)}$, $\mathbf{X}^{(2)}$, $\mathbf{R}_{11}$, and $\mathbf{R}_{22}$ are $q \times 1$, $(k-q)\times 1$, $q \times q$, and $(k-q)\times(k-q)$, respectively. It is well known that the marginal distributions of $\mathbf{X}^{(1)}$ and $\mathbf{X}^{(2)}$ are $N(\mathbf{0},\mathbf{R}_{11})$ and $N(\mathbf{0},\mathbf{R}_{22})$, respectively.

Theorem 2.2.2. Let $\mathbf{X}=(\mathbf{X}^{(1)},\mathbf{X}^{(2)})'$ be distributed according to $N(\mathbf{0},\mathbf{R})$, where $\mathbf{R}$ is a positive semidefinite covariance matrix of the form given in (2.2.10). Let A_1, A_2 be two convex regions symmetric about the origin in $\mathcal{R}^q$ and $\mathcal{R}^{k-q}$, respectively. If the rank of $\mathbf{R}_{12}$ is either zero† or one, then

$$P_{\mathbf{R}}\left[\bigcap_{i=1}^{2}\{\mathbf{X}^{(i)}\in A_i\}\right] \geqslant \prod_{i=1}^{2} P_{\mathbf{R}_{ii}}\left[\mathbf{X}^{(i)}\in A_i\right]. \tag{2.2.11}$$

If $\mathbf{R}$ is positive definite, $\mathbf{R}_{12}$ has rank one, and if the sets A_1, A_2 are bounded and have positive probability contents, then the inequality is strict.

Proof. If the rank of $\mathbf{R}_{12}$ is zero, then $\mathbf{X}^{(1)}$ and $\mathbf{X}^{(2)}$ are independent and (2.2.11) is immediate. Now let us assume that the rank of $\mathbf{R}_{12}$ is one. If $\mathbf{R}$ is positive definite and $\mathbf{R}_{12}$ has rank one, then we can always find a new random variable U with variance one such that from the joint distribution of $\mathbf{V}=(\mathbf{X}^{(1)},\mathbf{X}^{(2)},U)'$ (which is $N(\mathbf{0},\boldsymbol{\Sigma}_{\mathbf{V}})$) $\mathbf{X}^{(1)}$ and $\mathbf{X}^{(2)}$ are conditionally independent for given $U=u$. This is so for the reason that the covariance matrix of $\mathbf{V}$ is of the form

$$\boldsymbol{\Sigma}_{\mathbf{V}} = \begin{pmatrix} \mathbf{R}_{11} & \mathbf{R}_{12} & \mathbf{b} \\ \mathbf{R}_{21} & \mathbf{R}_{22} & \mathbf{c} \\ \mathbf{b}' & \mathbf{c}' & 1 \end{pmatrix},$$

where $\mathbf{b}$ is $q\times 1$, $\mathbf{c}$ is $(k-q)\times 1$, and they are both to be determined. For given $U=u$ the conditional covariance matrix of $\mathbf{X}_1$ and $\mathbf{X}_2$ is $\mathbf{0}$, provided that $\mathbf{b}$ and $\mathbf{c}$ are chosen to satisfy $\mathbf{bc}'=\mathbf{R}_{12}$, which is possible if the rank of $\mathbf{R}_{12}$ is one. For given $U=u$ the random variables $\mathbf{X}^{(1)}$ and $\mathbf{X}^{(2)}$ are then independent with conditional distributions $N(u\mathbf{b},\mathbf{R}_{11}-\mathbf{bb}')$ and $N(u\mathbf{c},\mathbf{R}_{22}-\mathbf{cc}')$, respectively. (Note that when the rank of $\mathbf{R}_{12}$ is one, then (from $\mathbf{bc}'=\mathbf{R}_{12}$) $\mathbf{b}$ and $\mathbf{c}$ cannot be $\mathbf{0}$.) Clearly the probability content depends on $\boldsymbol{\Sigma}_{\mathbf{V}}$ only through $\mathbf{R}$. Therefore from the distribution of $\mathbf{V}$ we can write

$$P_{\mathbf{R}}\left[\bigcap_{i=1}^{2}\{\mathbf{X}^{(i)}\in A_i\}\right] = \int_{-\infty}^{\infty} P_{\mathbf{R}}\left[\bigcap_{i=1}^{2}\{\mathbf{X}^{(i)}\in A_i\}\,|\,U=u\right]\varphi(u)\,du,$$

where $\varphi(u)$ is the univariate standard normal density. From the conditional distributions of $\mathbf{X}^{(1)}$ and $\mathbf{X}^{(2)}$, given $U=u$, we have

$$P_{\mathbf{R}}\left[\bigcap_{i=1}^{2}\{\mathbf{X}^{(i)}\in A_i\}\,|\,U=u\right] = g(u,\mathbf{b},A_1)\,g(u,\mathbf{c},A_2),$$

†The rank of $\mathbf{R}_{12}$ is zero means that $\mathbf{R}_{12}=\mathbf{0}$.

where g is defined as in (2.2.9) under the covariance matrices in the conditional distributions. Therefore it follows from the symmetry property of g (in u) and Lemmas 2.2.1 and 2.2.2 that

$$\begin{aligned} P_{\mathbf{R}}\left[\bigcap_{i=1}^{2}\{\mathbf{X}^{(i)}\in A_i\}\right] &= E^{U}[g(U,\mathbf{b},A_1)g(U,\mathbf{c},A_2)] \\ &= E^{|U|}[g(|U|,\mathbf{b},A_1)g(|U|,\mathbf{c},A_2)] \\ &\geqslant [E^{|U|}g(|U|,\mathbf{b},A_1)][E^{|U|}g(|U|,\mathbf{c},A_2)] \\ &= \prod_{i=1}^{2} P_{\mathbf{R}_{ii}}[\mathbf{X}^{(i)}\in A_i]. \end{aligned}$$

Moreover, the inequality is strict if the sets A_i $(i=1,2)$ are bounded with positive probability contents. This completes the proof for the case in which $\mathbf{R}$ is positive definite (i.e., $\mathbf{R}$ is nonsingular). If $\mathbf{R}$ is singular, consider the probability content under a new covariance matrix

$$\mathbf{S} = \mathbf{S}(\varepsilon) = \frac{1}{(1+\varepsilon)}(\mathbf{R}+\varepsilon\mathbf{I}),$$

where $\mathbf{I}$ is the identity matrix. Clearly $\mathbf{S}$ is positive definite for every $\varepsilon>0$. Hence

$$P_{\mathbf{S}}\left[\bigcap_{i=1}^{2}\{\mathbf{X}^{(i)}\in A_i\}\right] \geqslant \prod_{i=1}^{2} P[\mathbf{X}^{(i)}\in A_i]$$

holds for every $\varepsilon>0$. The inequality is preserved as ε approaches zero. ■

We note that the proof of this theorem is quite similar to that of Theorem 2.2.1, except for the greater generality. The key point is to introduce the new random variable U so that, given $U=u$, $\mathbf{X}^{(1)}$ and $\mathbf{X}^{(2)}$ are independent. Also, the condition on the rank of $\mathbf{R}_{12}$ is immediately satisfied when $\mathbf{R}$ has the structure l, or when either $\mathbf{X}^{(1)}$ or $\mathbf{X}^{(2)}$ contains only one component (i.e., $q=1$ or $q=k-1$). To see that Theorem 2.2.2 gives a proof for Dunn's conjecture, we observe the following result which follows from Theorem 2.2.2.

Corollary. Let $\mathbf{X}=(X_1,\mathbf{X}^{(2)})'$ be distributed according to $N(\mathbf{0},\mathbf{R})$, where $\mathbf{X}^{(2)}$ is $(k-1)\times 1$ and $\mathbf{R}$ is a positive semidefinite covariance matrix. Let A_2 be a convex region symmetric about the origin in $\mathcal{R}^{k-1}$. Then

$$P_{\mathbf{R}}\left[|X_1| \leqslant a_1, \mathbf{X}^{(2)} \in A_2\right] \geqslant P\left[|X_1| \leqslant a_1\right] P_{\mathbf{R}_{22}}\left[\mathbf{X}^{(2)} \in A_2\right] \quad (2.2.12)$$

holds for all $a_1 > 0$. The inequality is strict if $\mathbf{R}$ is positive definite, X_1 and $\mathbf{X}^{(2)}$ are dependent, and A_2 is a bounded set with positive probability content.

Now a repeated application of (2.2.12), with one component being excluded from $\mathbf{X}^{(2)}$ at a time, yields the inequality in (2.2.1) (which is strict if $\mathbf{R}$ is positive definite, $\mathbf{R}$ is not a diagonal matrix, and all the a_is are positive).

Theorem 2.2.2 was later generalized by Das Gupta, Eaton, Olkin, Perlman, Savage, and Sobel (1972) to the case in which the mean vector is not necessarily zero. This appears to be the first inequality available for nonzero means.

Theorem 2.2.3. Let $\mathbf{X} = (\mathbf{X}^{(1)}, \mathbf{X}^{(2)})'$ be distributed according to $N(\boldsymbol{\mu}, \mathbf{R})$, where $\boldsymbol{\mu} = (\boldsymbol{\mu}^{(1)}, \boldsymbol{\mu}^{(2)})'$ and $\mathbf{R}$ is a positive semidefinite matrix of the form given in (2.2.10). Let A_1, A_2 be two convex regions symmetric about the origin in $\mathcal{R}^q$, $\mathcal{R}^{k-q}$, respectively. If the rank of R_{12} is one and if there exists a scalar η such that $\eta^2 \mathbf{R}_{ii} - \boldsymbol{\mu}^{(i)}(\boldsymbol{\mu}^{(i)})'$ is positive semidefinite for $i = 1, 2$ and $\eta^2 \mathbf{R}_{12} = \boldsymbol{\mu}^{(1)}(\boldsymbol{\mu}^{(2)})'$, then

$$P\left[\bigcap_{i=1}^{2} \{\mathbf{X}^{(i)} \in A_i\}\right] \geqslant \prod_{i=1}^{2} P\left[\mathbf{X}^{(i)} \in A_i\right]. \quad (2.2.13)$$

Proof. From the proof of Theorem 2.2.2 we may assume that $\mathbf{R}$ is positive definite. If $\eta = 0$, we choose $\mathbf{b}, \mathbf{c}$ satisfying $\mathbf{bc}' = \mathbf{R}_{12}$. Otherwise let $\mathbf{b} = \boldsymbol{\mu}^{(1)}/\eta$ and $\mathbf{c} = \boldsymbol{\mu}^{(2)}/\eta$. Let us consider the random variable $\mathbf{V} = (\mathbf{X}^{(1)}, \mathbf{X}^{(2)}, U)'$ that is normally distributed with mean vector $(\boldsymbol{\mu}^{(1)}, \boldsymbol{\mu}^{(2)}, \eta)'$ and a covariance matrix satisfying

$$\operatorname{var} U = 1, \qquad \operatorname{cov}(\mathbf{X}^{(1)}, U) = \mathbf{b}, \qquad \operatorname{cov}(\mathbf{X}^{(2)}, U) = \mathbf{c},$$

and

$$\operatorname{cov}(\mathbf{X}^{(1)}, \mathbf{X}^{(2)}) = \mathbf{bc}'.$$

Then for given $U = u$, $\mathbf{X}^{(1)}$ and $\mathbf{X}^{(2)}$ are independent with conditional means $u\mathbf{b}, u\mathbf{c}$ and conditional covariance matrices $\boldsymbol{\Sigma}_{11} - \mathbf{bb}', \boldsymbol{\Sigma}_{11} - \mathbf{cc}'$, respectively. The proof follows from a similar argument after applying Lemmas 2.2.1 and 2.2.2. ■

Remarks. (1) It should be noted that in Lemma 2.2.2 and Theorems 2.2.2 and 2.2.3 $\mathbf{R}$ is in general a covariance matrix, and it is not necessarily a correlation matrix. However, without loss of generality one can always assume for convenience that $\mathbf{R}$ is already a correlation matrix (i.e., one can assume that the variances are one). (Otherwise the set A can be standardized and the problem remains unchanged.)

(2) In Theorems 2.2.2 and 2.2.3 the condition that $\mathbf{R}_{12}$ have rank zero or one is crucial. When the rank of $\mathbf{R}_{12}$ is greater than one, this problem still remains open.

If $\mathbf{R}$ has the structure l, then a stronger result can be obtained. Suppose that for a fixed positive integer r we consider the partition of $\mathbf{X}$ into r subvectors:

$$\mathbf{X} = (\mathbf{X}^{(1)},\dots,\mathbf{X}^{(r)})',$$

where $\mathbf{X}^{(i)}$ is $k_i \times 1$, $k_i > 0$, and $\Sigma_{i=1}^r k_i = k$.

Theorem 2.2.4. Let $\mathbf{X}$ be distributed according to $N(\mathbf{0},\mathbf{R})$. For $i=1,\dots,r$ let the A_is be convex regions symmetric about the origin in $\mathfrak{R}^{k_i}$. If $\mathbf{R}$ has the structure l, then

$$P_{\mathbf{R}}\left[\bigcap_{i=1}^{r}\{\mathbf{X}^{(i)}\in A_i\}\right] \geqslant P\left[\bigcap_{i\in C}\{\mathbf{X}^{(i)}\in A_i\}\right]P\left[\bigcap_{i\notin C}\{\mathbf{X}^{(i)}\in A_i\}\right]$$

$$\geqslant \prod_{i=1}^{r} P[\mathbf{X}^{(i)}\in A_i]$$

holds for every subset C of $\{1,\dots,r\}$. The inequalities are strict if the A_is are bounded sets with positive probability contents and if the λ_is are different from zero.

Proof. Without loss of generality assume that $\mathbf{R}$ is a correlation matrix. Since $\mathbf{R}$ has the structure l, we can form the random variable $\mathbf{X}$ as given in (2.2.2). Then for given $U=u$ the components of $\mathbf{X}$ are independent. The probability in question is $E^{|U|}\prod_{i=1}^r g(|U|,\mathbf{b}_i,A_i)$, where $\mathbf{b}_i$ is a given real vector (depending on the λ_is) and g is defined in (2.2.9). The proof now follows similarly from Lemma 2.2.2 and the inequality in (2.2.6). ■

Khatri's approach to the problem begins with any partition of the random variable **X**, and the inequality concerns the probability contents of all symmetric convex regions. His theorem says that when the covariance matrix of $\mathbf{X}^{(1)}, \mathbf{X}^{(2)}$ has rank one, the probability that the components $\mathbf{X}^{(i)}$ will simultaneously take values in two symmetric and convex regions can be bounded below, and the lower bound is attainable when $\mathbf{X}^{(1)}$ and $\mathbf{X}^{(2)}$ are independent with $\mathbf{R}_{12}=\mathbf{0}$. However the theorem does not give a monotonicity property of the probability content as a function of the correlation coefficients. This problem was studied by Šidák (1968) for a particular partition of the components of **X** and for rectangular regions only. His theorem is given below.

Theorem 2.2.5. Let **X** be distributed according to $N(\mathbf{0}, \mathbf{R}(\lambda))$, where $\mathbf{R}(\lambda)=(\rho_{ij}(\lambda))$ is a correlation matrix depending on $\lambda \in [0,1]$ in the following way: For a fixed positive semidefinite correlation matrix $\mathbf{T}=(\tau_{ij})$, we define $\rho_{ij}(\lambda)=\tau_{ij}$ for all $i,j=2,\ldots,k$ and $\rho_{1j}(\lambda)=\rho_{j1}(\lambda)=\lambda\tau_{1j}$ for $j=2,\ldots,k$. Then

$$\alpha_2(k, \mathbf{a}, \mathbf{R}(\lambda)) = P_{\mathbf{R}(\lambda)}\left[\bigcap_{i=1}^{k} \{|X_i| \leqslant a_i\}\right]$$

is monotonically nondecreasing in $\lambda \in [0,1]$ for every **a**. If **T** is positive definite, $\tau_{1j} \neq 0$ for some $j>1$, and all the a_is are positive, then it is strictly increasing in λ.

Remarks. (1) If **T** is positive semidefinite, so is $\mathbf{R}(\lambda)$ for $\lambda \in [0,1]$. This follows from the identity

$$\mathbf{R}(\lambda) = \lambda\mathbf{R}(1) + (1-\lambda)\mathbf{R}(0) = \lambda\mathbf{T} + (1-\lambda)\mathbf{R}(0)$$

and the fact that

$$\mathbf{x}'\mathbf{R}(\lambda)\mathbf{x} = \lambda\mathbf{x}'\mathbf{T}\mathbf{x} + (1-\lambda)\mathbf{x}'\mathbf{R}(0)\mathbf{x} \geqslant 0$$

holds for all **x**. It is also immediate that $\mathbf{R}(\lambda)$ is positive definite if **T** is.

(2) The result given in the theorem has the following meaning: Consider a partition $\mathbf{X}=(X_1, \mathbf{X}^{(2)})'$, where X_1 is a univariate $N(0,1)$ variable and the correlation structure of $\mathbf{X}^{(2)}$ remains fixed. X_1 and $\mathbf{X}^{(2)}$ then become more correlated (along the direction specified by the vector $\boldsymbol{\tau}=(\tau_{12},\ldots,\tau_{1k})'$) when λ approaches one. Hence the rectangular probability content is monotonically nondecreasing in

the correlations in this fashion, and the theorem represents a two-sided analogue to Slepian's result.

Proof of Theorem 2.2.5. The original proof given by Šidák (1968) also makes use of the transformation given in (2.2.2). Since it is quite lengthy, we shall instead give here the shorter proof obtained by Jogdeo (1970). Since the case $\tau_{1j}=0$ for all $j>1$ is trivial, it will be assumed that $\tau_{1j}\neq 0$ for some $j>1$.

First assume that $\mathbf{T}$ is positive definite. For fixed $\mathbf{a}$, since $\alpha_2(k,\mathbf{a},\mathbf{R}(\lambda))$ is differentiable with respect to λ, it suffices to show that the directional derivative is positive, i.e., the inequality

$$\frac{d}{d\lambda}\alpha_2(k,a_1,\mathbf{a}_2,\mathbf{R}(\lambda)) > 0 \tag{2.2.14}$$

holds for all a_1 and $\mathbf{a}_2=(a_2,\ldots,a_k)'$. Since interchanging the order of differentiation and integration is permissible, simple calculation yields for $\mathbf{x}_2=(x_2,\ldots,x_k)'$ and $A_2=\times_{i=2}^{k}[-a_i\leqslant x_i\leqslant a_i]$

$$\begin{aligned}
\lambda&\left[\frac{d}{d\lambda}\alpha_2(k,a_1,\mathbf{a}_2,\mathbf{R}(\lambda))\right]\\
&=\lambda\int\cdots\int_{A_2}\int_{-a_1}^{a_1}\left[\frac{d}{d\lambda}\varphi_k(x_1,\mathbf{x}_2,\mathbf{R}(\lambda))\right]dx_1\prod_{i=2}^{k}dx_i\\
&=\int\cdots\int_{A_2}\int_{-a_1}^{a_1}\left[\sum_{j=2}^{k}\tau_{1j}\left\{\frac{\partial}{\partial\tau_{1j}}\varphi_k(x_1,\mathbf{x}_2,\mathbf{T})\right\}\right]dx_1\prod_{i=2}^{k}dx_i\\
&=2\int\cdots\int_{A_2}\int_{0}^{a_1}\left[\sum_{j=2}^{k}\tau_{1j}\left\{\frac{\partial^2}{\partial x_1\,\partial x_j}\varphi_k(x_1,\mathbf{x}_2,\mathbf{T})\right\}\right]dx_1\prod_{i=2}^{k}dx_i\\
&=2\int\cdots\int_{A_2}\left[\sum_{j=2}^{k}\tau_{1j}\left\{\frac{\partial}{\partial x_j}\varphi_k(a_1,\mathbf{x}_2,\mathbf{T})\right\}\right]\prod_{i=2}^{k}dx_i\\
&\quad-2\int\cdots\int_{A_2}\left[\sum_{j=2}^{k}\tau_{1j}\left\{\frac{\partial}{\partial x_j}\varphi_k(0,\mathbf{x}_2,\mathbf{T})\right\}\right]\prod_{i=2}^{k}dx_i\\
&\equiv I_1+I_2.
\end{aligned}$$

Note that the third equality here follows from the identity established by Plackett (see Lemma 2.1.1). Using the facts that φ_k and the intervals $[-a_j,a_j]$ are symmetric about the origin, it follows from elementary calculation that $I_2=0$. Clearly we can rewrite (as in

Plackett (1954)) $\varphi_k(a_1, \mathbf{x}_2, \mathbf{T})$ as the product of a marginal density and the conditional density of $\mathbf{X}^{(2)}$, given $X_1 = a_1$, in the form

$$\varphi_k(a_1, \mathbf{x}_2, \mathbf{T}) = b\varphi_{k-1}(\mathbf{x}_2 - a_1\boldsymbol{\tau}, \mathbf{R}_{2\cdot 1}),$$

where $b > 0$ (which involves the marginal density of X_1 at a_1), $\boldsymbol{\tau} = (\tau_{12}, \ldots, \tau_{1k})'$, and $\mathbf{R}_{2\cdot 1}$ is the conditional covariance matrix, which is positive definite. It then suffices to show that

$$\int \cdots \int_{A_2} \left[\sum_{j=2}^{k} \tau_{1j} \left\{ \frac{\partial}{\partial x_j} \varphi_{k-1}(\mathbf{x}_2 - a_1\boldsymbol{\tau}, \mathbf{R}_{2\cdot 1}) \right\} \right] \prod_{i=2}^{k} dx_i > 0,$$

or equivalently,

$$\Delta = \int \cdots \int_{A_2} \left[\sum_{j=2}^{k} c_j \left\{ \frac{\partial}{\partial x_j} \varphi_{k-1}(\mathbf{x}_2 + \mathbf{c}, \mathbf{R}_{2\cdot 1}) \right\} \right] \prod_{i=2}^{k} dx_i < 0,$$

where $\mathbf{c} = (c_2, \ldots, c_k)'$ and $c_j = -a_1\tau_{1j}$ $(j = 2, \ldots, k)$. But Δ is the directional derivative of the function

$$g(u) = \int \cdots \int_{A_2} \varphi_{k-1}(\mathbf{x}_2 + u\mathbf{c}, \mathbf{R}_{2\cdot 1}) \prod_{i=2}^{k} dx_i$$

at $u = 1$. By Lemma 2.2.2 $g(u)$ is strictly decreasing in u because $\mathbf{R}_{2\cdot 1}$ is nonsingular and $\mathbf{c} \neq \mathbf{0}$. Hence Δ is strictly negative, and hence (2.2.14) follows when $\mathbf{T}$ is positive definite and $a_i > 0$ $(i = 1, \ldots, k)$.

If $\mathbf{T}$ is singular, then we can construct a sequence of nonsingular matrices converging to $\mathbf{T}$ (as we did in the proof of Theorem 2.2.2), and the inequality will be preserved when passing to the limit. ■

The following corollaries are immediate consequences of Theorem 2.2.5.

Corollary 1. Let $\mathbf{X}$ be as stated in Theorem 2.2.5, except that its correlation matrix $\mathbf{R}(\boldsymbol{\lambda}) = (\rho_{ij}(\boldsymbol{\lambda}))$ depends on $\boldsymbol{\lambda} = (\lambda_1, \ldots, \lambda_k)'$ in the following way: $\rho_{ij} = \lambda_i\lambda_j\tau_{ij}$ for all $i \neq j$, where $\lambda_i \in [0, 1]$ for $i = 1, \ldots, k$. Then

$$\alpha_2(k, \mathbf{a}, \mathbf{R}(\boldsymbol{\lambda})) = P_{\mathbf{R}(\boldsymbol{\lambda})}\left[\bigcap_{i=1}^{k} \{|X_i| \leqslant a_i\} \right]$$

is monotonically nondecreasing in each $\lambda_i \in [0, 1]$ $(i = 1, \ldots, k)$. It is strictly increasing in λ_i if $\mathbf{T} = (\tau_{ij})$ is positive definite, $\lambda_i\tau_{ij} \neq 0$ for some $j \neq i$, and all the a_is are positive.

Corollary 2. Let $\mathbf{X}$ be as stated in Theorem 2.2.5, except that $\mathbf{R}(\lambda)=(\rho_{ij}(\lambda))$ depends on a real number $\lambda\in[0,1]$ such that $\rho_{ij}(\lambda)=\lambda\tau_{ij}$ for all $i\neq j$. Then

$$\alpha_2(k,\mathbf{a},\mathbf{R}(\lambda)) = P_{\mathbf{R}(\lambda)}\left[\bigcap_{i=1}^{k}\{|X_i|\leqslant a_i\}\right]$$

is monotonically nondecreasing in $\lambda\in[0,1]$. It is strictly increasing in λ if $\mathbf{T}=(\tau_{ij})$ is positive definite, is not the identity matrix, and all the a_is are positive.

Now the proof of Dunn's conjecture follows immediately from Corollary 2 by taking two extreme values $\lambda=1$ and $\lambda=0$, respectively.

We note that both corollaries may be regarded as partial two-sided analogues to the one-sided Slepian result. A natural question to ask is then this: Does a complete analogue exist? That is, is the rectangular probability content monotonically nondecreasing in ρ_{ij} for all $i\neq j$ when the other correlation coefficients remain fixed? The answer to this question is negative, and the following is one of the two counterexamples offered by Šidák (1968).

Example 2.2.1. Consider $\mathbf{X}=(X_1,X_2,X_3)'$ which is distributed according to $N(\mathbf{O},\mathbf{R})$. If the partial correlation coefficient $\rho_{13\cdot 2}$ is negative (positive) (i.e., if $\rho_{13}-\rho_{12}\rho_{23}<0$ (>0)), then for small a_2

$$\frac{\partial}{\partial\rho_{13}}P_{\mathbf{R}}\left[\bigcap_{i=1}^{3}\{|X_i|\leqslant a_i\}\right]<0\ (>0).$$

Thus in the first case the probability in question decreases when ρ_{13} increases.

Proof. We again apply Lemma 2.1.1 and take the derivative under the integral sign. This yields

$$\frac{\partial}{\partial\rho_{13}}P_{\mathbf{R}}\left[\bigcap_{i=1}^{3}\{|X_i|\leqslant a_i\}\right]$$
$$=2\int_{-a_2}^{a_2}\left[\varphi_3(a_1,x_2,a_3,\mathbf{R})-\varphi_3(a_1,x_2,-a_3,\mathbf{R})\right]dx_2, \tag{2.2.15}$$

which is equal to zero when $a_2=0$. At the point $a_2=0$ we have

$$\frac{d}{da_2}\int_{-a_2}^{a_2}\left[\varphi_3(a_1,x_2,a_3,\mathbf{R})-\varphi_3(a_1,x_2,-a_3,\mathbf{R})\right]dx_2$$

$$=2\left[\varphi_3(a_1,0,a_3,\mathbf{R})-\varphi_3(a_1,0,-a_3,\mathbf{R})\right]. \qquad (2.2.16)$$

Now write the two joint densities on the right-hand side of (2.2.16) as products of the marginal density of X_2 at $x_2=0$ and the conditional density of (X_1,X_3), given $X_2=0$, at (a_1,a_3) and $(a_1,-a_3)$, respectively. Clearly the correlation in the conditional density function is $\rho_{13\cdot 2}$, and the right-hand side of (2.2.16) can be made negative (or positive) by choosing $\rho_{13\cdot 2}<0$ (or >0). The proof of the assertion now follows from the fact that the quantity given in (2.2.15) is a continuous function of a_2. ■

The inequalities stated above concern only events of the form $[\mathbf{X}^{(1)}\in A_1$ and $\mathbf{X}^{(2)}\in A_2]$. A natural question one may wish to ask is this: If $\mathbf{X}$ is a normal variable with means zero and a correlation matrix $\mathbf{R}$ and if A is convex and symmetric about the origin in $\mathfrak{R}^k$, is $P_{\mathbf{R}}[\mathbf{X}\in A]$ also monotonically increasing in the correlations in a certain fashion? This problem was considered by Das Gupta, Eaton, Olkin, Perlman, Savage, and Sobel (1972), and the answer is negative. To illustrate this let us consider the simple case with

$$A=\left\{\mathbf{x}\,\middle|\,\sum_{i=1}^{k}x_i^2\leqslant a\right\}$$

and assume that the variances of the X_is are one. When $\mathbf{R}=\mathbf{I}$ (the identity matrix), the probability content is a chi-square probability with k degrees of freedom. But when the correlations are close to one, then the probability content is approximately $P[\chi_1^2\leqslant a/k]$, where χ_1^2 is a chi-square variable with one degree of freedom. In this case the direction of the inequality depends on the value of a, and no general results can be obtained.

2.3. OTHER INEQUALITIES FOR MULTIVARIATE NORMAL DISTRIBUTION

We first consider the behavior of the probability $\alpha_3(k,\mathbf{a},\mathbf{R})$ as a function of the correlations. This is the probability that a multivariate normal variable **X** will take values over "the corners." In view of the results in the previous sections it may be expected that this probability will be larger when the components of **X** are correlated in a certain fashion. This view is confirmed by the following theorem due to Khatri (1967).

Theorem 2.3.1. Let **X** be distributed according to $N(\mathbf{0},\mathbf{R})$. If **R** has the structure *l*, then

$$P_{\mathbf{R}}\left[\bigcap_{i=1}^{k}\{|X_i|\geqslant a_i\}\right]\geqslant\prod_{i=1}^{k}P[|X_i|\geqslant a_i] \tag{2.3.1}$$

holds.

Instead of showing the proof of this theorem, we give the proof for the next theorem, which is in a more general form.

Theorem 2.3.2. Let $\mathbf{X},A_1,\ldots,A_r$ be as stated in Theorem 2.2.4. If **R** has the structure *l*, then

$$P_{\mathbf{R}}\left[\bigcap_{i=1}^{r}\{\mathbf{X}^{(i)}\notin A_i\}\right]\geqslant P\left[\bigcap_{i\in C}\{\mathbf{X}^{(i)}\notin A_i\}\right]P\left[\bigcap_{i\notin C}\{\mathbf{X}^{(i)}\notin A_i\}\right]$$

$$\geqslant\prod_{i=1}^{r}P[\mathbf{X}^{(i)}\notin A_i] \tag{2.3.2}$$

holds for every subset C of $\{1,\ldots,r\}$. Moreover, the inequalities are strict if the A_is are bounded sets with positive probability contents and if the λ_is are different from zero.

Outline of the Proof. The proof follows essentially the same type of arguments as those given in the proof of Theorem 2.2.4, except that the probability here under consideration is

$$E^{|U|}\prod_{i=1}^{r}\left[1-g(|U|,\mathbf{b}_i,A_i)\right]. \quad ■$$

After successfully proving Theorems 2.2.2 and 2.3.1 in his 1967 paper, Khatri (1970) tried to generalize those results by making an attempt to prove the following two conjectures.

Conjecture 2.3.1. For $r \geqslant 2$ let $\mathbf{X}, A_1, \ldots, A_r$ be as stated in Theorem 2.2.4, where $\mathbf{R}$ is any positive semidefinite covariance matrix. Then

$$P_{\mathbf{R}}\left[\bigcap_{i=1}^{r}\{\mathbf{X}^{(i)} \in A_i\}\right] \geqslant P[\mathbf{X}^{(1)} \in A_1] P\left[\bigcap_{i=2}^{r}\{\mathbf{X}^{(i)} \in A_i\}\right] \geqslant \prod_{i=1}^{r} P[\mathbf{X}^{(i)} \in A_i].$$

Conjecture 2.3.2. Using the same notation as that in Conjecture 2.3.1, we have

$$P_{\mathbf{R}}\left[\bigcap_{i=1}^{r}\{\mathbf{X}^{(i)} \notin A_i\}\right] \geqslant P[\mathbf{X}^{(1)} \notin A_1] P\left[\bigcap_{i=2}^{r}\{\mathbf{X}^{(i)} \notin A_i\}\right] \geqslant \prod_{i=1}^{r} P[\mathbf{X}^{(i)} \notin A_i].$$

Conjectures 2.3.1 and 2.3.2 would be, if true, generalizations of Theorems 2.2.4 and 2.3.2, respectively, by removing the conditions on the covariance matrix. But unfortunately Khatri's proofs depend on a generalization of the argument used by Scott (1967) which is in error. Therefore as pointed out by Das Gupta, Eaton, Olkin, Perlman, Savage, and Sobel (1972) and Šidák (1975), the proofs are incorrect. The correctness of the statement in Conjecture 2.3.1 remains open, and Conjecture 2.3.2 was found to be false (Šidák, 1971). The following is a counterexample Šidák obtained.

Example 2.3.1. Let X_1, X_2 be independent $N(0,1)$ variables and let $X_3 = (X_1 + X_2)/\sqrt{2}$. Then for any $a > 0$ and for sufficiently small $a_3 > 0$, we have

$$P[|X_1| \geqslant a, |X_2| \geqslant a, |X_3| \geqslant a_3] < P[|X_1| \geqslant a] P[|X_2| \geqslant a] P[|X_3| \geqslant a_3]. \quad (2.3.3)$$

Proof. To prove this statement one first introduces two independent $N(0,1)$ variables U_1, U_2 and defines

$$X_1 = (U_1 - U_2)/\sqrt{2}, \qquad X_2 = (U_1 + U_2)/\sqrt{2}, \qquad X_3 = U_1.$$

Then after rewriting the probabilities in terms of the U variables, the difference between the right-hand side and left-hand side of (2.3.3) is bounded below by

$$\Delta = c\left[\Phi(-\sqrt{2}\,a - a_3) - 2\Phi^2(-a)\right],$$

where $c > 0$ and Φ is the $N(0,1)$ distribution function. It follows from the properties of Φ that $\Delta > 0$ holds for all a and for small a_3. Since the distribution of $(X_1, X_2, X_3)'$ (which is singular) can be approximated by a sequence of nonsingular ones and the inequality in (2.3.3) is strict, there exists a nonsingular distribution such that (2.3.3) holds. ■

Although the statement in Conjecture 2.3.2 is false, the probability content in question does depend on the correlation coefficients in a certain fashion. A positive result, depending on the correlations, was proved by Šidák (1971) for the special case in which the sets A_i are intervals.

Theorem 2.3.3. Let $\mathbf{X}$ be distributed according to $N(\mathbf{0}, \mathbf{R})$. If $\mathbf{R}$ has the structure l, then as a function of the λ_is the probability $P\left[\cap_{i=1}^{k}\{|X_i| \geqslant a_i\}\right]$ is strictly increasing in $|\lambda_i|$ for every $|\lambda_i| \in [0,1]$ when the a_is are positive. In particular,

$$P\left[\bigcap_{i=1}^{k}\{|X_i| \geqslant a_i\}\right] > \prod_{i=1}^{k} P\left[|X_i| \geqslant a_i\right].$$

Proof. Without loss of generality assume that the variances are one and consider the probability as a function of λ_1, denoted by $H(\lambda_1)$, where $\lambda_2, \ldots, \lambda_k$ are arbitrary but fixed. Employing the transformation given in (2.2.2), $H(\lambda_1)$ can be written in the form

$$H(\lambda_1) = \int_{-\infty}^{\infty} h(\lambda_1, u)\varphi(u)\,du,$$

where

$$h(\lambda_1, u) = \prod_{i=1}^{k}\left[1 - \Phi\left((\lambda_i u + a_i)/(1-\lambda_i^2)^{1/2}\right) + \Phi\left((\lambda_i u - a_i)/(1-\lambda_i^2)^{1/2}\right)\right]$$

and φ and Φ are the standard normal density function and distribution function, respectively. Since φ is symmetric about the origin, $h(\lambda_1, u)$ depends on $\lambda_1, \ldots, \lambda_k$ only through $|\lambda_1|, \ldots, |\lambda_k|$ for every fixed u. Therefore it suffices to show that

$$\frac{\partial}{\partial \lambda_1} H(\lambda_1) \geqslant 0 \qquad \text{for} \quad \lambda_i > 0, \quad i = 1, \ldots, k.$$

After differentiating under the integral sign, this follows from the fact that

$$\frac{\partial}{\partial \lambda_1} h(\lambda_1, u) \geqslant 0 \quad \text{for all} \quad u,$$

which in turn follows from rearranging the terms and making use of the condition that $\lambda_i > 0$. ∎

In the following we give a stronger result for the special case in which the events are exchangeable. The proof of this result depends on a moment inequality for nonnegative random variables. Let $W \geqslant 0$ a.s., and for $k \geqslant 1$ consider the kth moment μ_k of W. Since w^k is a convex function of w for $w \geqslant 0$, one has immediately from Jensen's inequality $EW^k \geqslant (EW)^k$. This inequality is well known; it has been used quite frequently in the literature, and the equality is attainable either when $k = 1$ or when W is a singular random variable. However, for $k \geqslant 2$ if the variance of W (var W) is large, then this bound is not very sharp (for $k = 2$ this is obvious because $EW^2 = (EW)^2 + \text{var}\, W$). In proving certain probability inequalities Tong (1970) obtained the following lemma which offers a sharper bound on EW^k.

Lemma 2.3.1. Let W satisfy $P[W \geqslant 0] = 1$. If $EW^k < \infty$, then

$$EW^k \geqslant (EW^{k/r})^r \geqslant (EW)^k + \left(EW^{k/r} - (EW)^{k/r}\right)^r \tag{2.3.4}$$

holds for all $k > r \geqslant 2$; a special form of this inequality is

$$EW^k \geqslant (EW)^k + (\text{var}\, W)^{k/2}. \tag{2.3.5}$$

The inequalities are strict unless W is singular.

Proof. The first inequality in (2.3.4) follows from the known result that as a function of r $(EW^{1/r})^r$ is monotonically nonincreasing in r. To prove the second inequality in (2.3.4) it suffices to show that for any a,b satisfying $a>b>0$ we have

$$a^r - b^r \geqslant (a-b)^r \qquad \text{for every} \quad r \geqslant 1. \tag{2.3.6}$$

From $1=(1/a)[b+(a-b)]$ the inequality

$$1 = (1/a^r)[b+(a-b)]^r \geqslant (1/a^r)[b^r+(a-b)^r]$$

holds for every $r \geqslant 1$. This then implies $a^r \geqslant b^r+(a-b)^r$ and (2.3.6); hence (2.3.4) follows. Inequality (2.3.5) immediately follows from (2.3.4) with $r=k/2$. By Jensen's inequality all inequalities are strict when W is nonsingular. ■

Let us now consider the special case in which $\mathbf{X}$ is distributed according to $N(\mathbf{0},\mathbf{R})$, where the components of $\mathbf{X}$ have a common variance and they are equally correlated with a correlation coefficient $\rho \geqslant 0$. This additional condition does not appear to be restrictive in many applications, and it is equivalent to the condition that the covariance matrix $\mathbf{R}$ have the structure l with equal variances and equal λ_is ($\geqslant 0$). Now for a fixed real number a let us define

$$\beta_1(k) = P\left[\bigcap_{i=1}^{k} \{X_i \leqslant a\}\right], \tag{2.3.7}$$

$$\beta_2(k) = P\left[\bigcap_{i=1}^{k} \{|X_i| \leqslant a\}\right], \qquad a>0, \tag{2.3.8}$$

$$\beta_3(k) = P\left[\bigcap_{i=1}^{k} \{|X_i| \geqslant a\}\right], \qquad a>0, \tag{2.3.9}$$

and observe the following theorem obtained by Tong (1970).

Theorem 2.3.4. Let $\mathbf{X}$ be distributed according to $N(\mathbf{0},\mathbf{R})$. If the elements of $\mathbf{R}$ satisfy

$$\sigma_{ij} = \begin{cases} \sigma^2 & \text{for} \quad i=j, \\ \rho\sigma^2 & \text{for} \quad i \neq j \end{cases}$$

for some $\rho \in [0,1)$, then for $m=1,2,3$

$$\beta_m(k) \geqslant [\beta_m(r)]^{k/r} \geqslant [\beta_m(1)]^k \tag{2.3.10}$$

holds for $k>r\geqslant 2$. Furthermore, the inequalities are strict if $\rho>0$.

Proof. Only the proof for $m=1$ will be given because the proofs for the other cases are similar. Employing the transformation in (2.2.2) with $\lambda_i = \sqrt{\rho}$ for $i = 1, \ldots, k$, it follows that $\beta_1(k) = EW^k$, where

$$W = \Phi\big((\sqrt{\rho}\ U + a/\sigma)/(1-\rho)^{1/2}\big) \tag{2.3.11}$$

is nonnegative, U is a standard normal variable, and Φ is the standard normal distribution function. Therefore (2.3.10) follows from Lemma 2.3.1. If $\rho > 0$, then W is nonsingular; hence by Lemma 2.3.1 the inequalities in (2.3.10) are strict. ■

If the conditions in the above theorem are satisfied, then the events defined by the components of **X** are exchangeable in a positive fashion. In this theorem we take advantage of this positive exchangeability condition, and the lower bound obtained is simply the (k/r)th power of the corresponding marginal probability. The bound is fairly sharp so long as the ratio k/r is close to one (although the difference between k and r may be large). In Chapters 3, 5, and 6 this concept will be investigated further, and more general inequalities will be given there.

PROBLEMS

1. Let $\mathbf{X} = (X_1, X_2)'$ be a bivariate normal variable with means zero, variances one, and correlation coefficient $\rho \in (-1, 1)$. Find the values of a_1, a_2 such that the derivative

$$\frac{\partial}{\partial \rho} P_\rho[X_1 \leqslant a_1, X_2 \leqslant a_2]$$

   is maximized.
2. Let **R** be a positive definite (semidefinite) matrix. Show that if we keep the other elements of **R** fixed and change the value of ρ_{ij}, then the set of admissible points of ρ_{ij} (in which **R** remains positive definite (semidefinite)) is an interval.
3. Let $\mathbf{R} = (\rho_{ij})$ and $\mathbf{T} = (\tau_{ij})$ be two positive definite correlation matrices such that $\rho_{ij} \geqslant \tau_{ij}$ and strict inequalities hold for at least two pairs of (i,j). Let $\mathbf{S} = (s_{ij})$ be such that s_{ij} is either ρ_{ij} or τ_{ij} but $\mathbf{S} \neq \mathbf{R}$ and $\mathbf{S} \neq \mathbf{T}$. Give an example to show that **S** may not be positive semidefinite.
4. Let **X** be distributed according to $N(\mathbf{0}, \mathbf{R})$, where **X** and **R** are of the forms given in (2.2.10) with $q = k/2$ (k is even) and

$\mathbf{R}_{11}=\mathbf{R}_{22}$. For fixed $\gamma\in(0,1)$ let a, a_1, a_2 satisfy, respectively,

$$P_{\mathbf{R}}\left[\bigcap_{i=1}^{k}\{X_i\leqslant a\}\right]=\gamma,\qquad P_{\mathbf{R}_{11}}\left[\bigcap_{i=1}^{q}\{X_i\leqslant a_1\}\right]=(\gamma)^{1/2},$$

$$P_{\mathbf{R}_{11}}\left[\bigcap_{i=1}^{q}\{X_i\leqslant a_2\}\right]=(1+\gamma)/2.$$

Show that if all the elements in $\mathbf{R}_{12}$ are $\leqslant 0$, then we have $a_1\leqslant a<a_2$.

5. Let $\mathbf{X}$ be distributed according to $N(\mathbf{0},\mathbf{R})$. Apply Lemma 2.2.1 to show that if $\mathbf{R}$ has the structure l with $\lambda_i\in[0,1]$, $i=1,\ldots,k$, then

$$P_{\mathbf{R}}\left[\bigcap_{i=1}^{k}\{X_i\leqslant a_i\}\right]\geqslant P\left[\bigcap_{i\in C}\{X_i\leqslant a_i\}\right]P\left[\bigcap_{i\notin C}\{X_i\leqslant a_i\}\right]$$

holds for all subsets C of $\{1,\ldots,k\}$.

6. Let $\mathbf{X}$ be a multivariate normal variable with mean vector $\mathbf{0}$, variances σ_i^2, and correlations ρ_{ij} for $1\leqslant i,j\leqslant k$. For fixed $\gamma\in(0,1)$ denote

$$W=\max_{1\leqslant i\leqslant k}X_i,\qquad V=\min_{1\leqslant i\leqslant k}X_i;$$
$$P[W\leqslant w]=\gamma,\quad P[V\leqslant v]=\gamma.$$

Show that w, EW are nonincreasing in ρ_{ij} and v, EV are nondecreasing in ρ_{ij} for all $i\neq j$. If $\rho_{ij}=\rho\in(-1/(k-1),1)$ for all $i\neq j$, then w, EW are strictly decreasing in ρ and v, EV are strictly increasing in ρ. (See, e.g., the tables in Gupta (1963) and Owen and Steck (1962).)

7. Show that if the variables X_i are equally correlated with correlation $\rho\geqslant 0$, then the statement in Problem 6 remains true when W, V are defined by

$$W=\max_{1\leqslant i\leqslant k}|X_i|,\qquad V=\min_{1\leqslant i\leqslant k}|X_i|.$$

8. Give a proof for Lemma 2.2.1. (Hint (Khatri, 1967): Define $\mathbf{U}_1$, $\mathbf{U}_2$, $\mathbf{U}$ independent and identically distributed random variables and consider the expectation of $\prod_{i=1}^{2}\{g_i(\mathbf{U}_2)-g_i(\mathbf{U}_1)\}$.)

9. Let $\mathbf{X}$ be a bivariate normal variable with mean vector $\boldsymbol{\mu}=(b,-b)'$ and covariance matrix $\mathbf{R}$. Let $A\subset\mathfrak{R}^2$ be a convex set that is symmetric about the origin. Show that if $P_{\boldsymbol{\mu}}[\mathbf{X}\in A]$ is

in (0, 1), then it is strictly decreasing in b. Hence, in particular,

$$P_{\mu=0}\left[\bigcap_{i=1}^{2}\{|X_i|\leqslant a_i\}\right] > P_{\mu=(b,-b)'}\left[\bigcap_{i=1}^{2}\{|X_i|\leqslant a_i\}\right]$$

holds for all $a_i > 0$ and $b \neq 0$. (When the variances are equal, this also follows from the inequalities via majorization to be discussed in Chapter 6.)

10. Let $\mathbf{X}$ be distributed according to $N(\mathbf{0},\mathbf{R})$, where $\mathbf{R}$ is a correlation matrix. If $\mathbf{R}$ has the structure l such that

$$\rho_{ij} = \lambda_i\lambda_j, \qquad |\lambda_i| < 1 \quad \text{for all} \quad i \neq j,$$

then

$$P_{\mathbf{R}}\left[\bigcap_{i=1}^{k}\{|X_i|\leqslant a_i\}\right] \leqslant \prod_{i=1}^{k} P\left[|X_i| \leqslant \frac{a_i}{(1-\lambda_i^2)^{1/2}}\right],$$

and strict inequality holds if $a_i > 0$ and $\lambda_i \neq 0$ $(i = 1,\ldots,k)$.

11. Let $\mathbf{R}$ be a positive definite correlation matrix that is of the form given in (2.2.10). Show that if $\mathbf{R}$ has the structure l, then the rank of $\mathbf{R}_{12}$ is either zero or one.

12. Let $(X_1,X_2)'$ be a bivariate normal variable with means $\mu_1,\mu_2 \neq 0$, variances one, and correlation $\rho \neq 0$. Show that if

$$0 < \rho \leqslant (\mu_1/\mu_2) \leqslant 1/\rho \qquad \text{or} \qquad 1/\rho \leqslant (\mu_1/\mu_2) \leqslant \rho < 0,$$

then

$$P\left[\bigcap_{i=1}^{2}\{|X_i|\leqslant a_i\}\right] \geqslant \prod_{i=1}^{2} P[|X_i|\leqslant a_i]$$

(Das Gupta, Eaton, Olkin, Perlman, Savage, and Sobel, 1972).

13. Show that in the proof of Theorem 2.2.5 the value of I_2 is zero.

14. Let W be a nonnegative random variable and let us denote $\mu_k = EW^k$ (assume that it exists). It is known that (see, e.g., Loève (1963, p. 156)) as a function of $k \in (0,\infty)$ $\log \mu_k$ is convex. (It is strictly convex if W is nonsingular.) Use this result to show that for fixed k

$$h(r) = \mu_r\mu_{k-r}$$

is monotonically nondecreasing in r for $r \in [\tfrac{1}{2}k, k]$, and it is strictly increasing in r if W is nonsingular. Hence the inequality in (2.3.10) is sharper when r is closer to k.

REFERENCES

Anderson, T. W. (1958). *An Introduction to Multivariate Statistical Analysis*. Wiley, New York.

Chartres, B. (1963). A geometrical proof of a theorem due to D. Slepian. *SIAM Rev.* **5**, 335–341.

Cramér, H. (1946). *Mathematical Methods of Statistics*. Princeton Univ. Press, Princeton, New Jersey.

Das Gupta, S., Eaton, M. L., Olkin, I., Perlman, M. D., Savage, L. J., and Sobel, M. (1972). Inequalities on the probability content of convex regions for elliptically contoured distributions. *Proc. Sixth Berkeley Symp. Math. Statist. Probab.* **2**, eds. L. M. LeCam, J. Neyman and E. L. Scott, 241–265. Univ. of California Press, Berkeley, California.

Dunn, O. J. (1958). Estimation of the means of dependent variables. *Ann. Math. Statist.* **29**, 1095–1111.

Dunnett, C. W., and Sobel, M. (1955). Approximations to the probability integral and certain percentage points to a multivariate analogue of Student's *t*-distribution. *Biometrika* **42**, 258–260.

Gupta, S. S. (1963). Probability integrals of multivariate normal and multivariate *t*. *Ann. Math. Statist.* **34**, 792–828.

Hardy, G. H., Littlewood, J. E., and Pólya, G. (1959). *Inequalities*, 2nd ed. Cambridge Univ. Press, London and New York.

Jogdeo, K. (1970). A simple proof of an inequality for multivariate normal probabilities of rectangles. *Ann. Math. Statist.* **41**, 1357–1359.

Khatri, C. G. (1967). On certain inequalities for normal distributions and their applications to simultaneous confidence bounds. *Ann. Math. Statist.* **38**, 1853–1867.

Khatri, C. G. (1970). Further contributions to some inequalities for normal distributions and their applications to simultaneous confidence bounds. *Ann. Inst. Statist. Math.* **22**, 451–458.

Kimball, A. W. (1951). On dependent tests of significance in the analysis of variance. *Ann. Math. Statist.* **22**, 600–602.

Loève, M. (1963). *Probability Theory*, 3rd ed. Van Nostrand-Reinhold, New York.

Owen, D. B., and Steck, G. P. (1962). Moments of order statistics from the equicorrelated multivariate normal distribution. *Ann. Math. Statist.* **33**, 1286–1291.

Plackett, R. L. (1954). A reduction formula for normal multivariate integrals. *Biometrika* **41**, 351–360.

Scott, A. (1967). A note on conservative confidence regions for the mean of a multivariate normal. *Ann. Math. Statist.* **38**, 278–280. [Corrigenda (1968). *Ann. Math. Statist.* **39**, 2161.]

Šidák, Z. (1967). Rectangular confidence regions for the means of multivariate normal distributions. *J. Amer. Statist. Assoc.* **62**, 626–633.

Šidák, Z. (1968). On multivariate normal probabilities of rectangles: Their dependence on correlations. *Ann. Math. Statist.* **39**, 1425–1434.

Šidák, Z. (1971). On probabilities of rectangles in multivariate Student distributions: Their dependence on correlations. *Ann. Math. Statist.* **42**, 169–175.

Šidák, Z. (1975). A note on C. G. Khatri and A. Scott's papers on multivariate normal distributions. *Ann. Inst. Statist. Math.* **27**, 181–184.

Slepian, D. (1962). The one-sided barrier problem for Gaussian noise. *Bell System Tech. J.* **41**, 463–501.

Tong, Y. L. (1970). Some probability inequalities of multivariate normal and multivariate *t*. *J. Amer. Statist. Assoc.* **65**, 1243–1247.

CHAPTER

3

Inequalities for Other Well-known Distributions

In this chapter we shall see some useful inequalities for the multivariate distributions of t, chi-square, F, and a few others. Most of them were obtained on an individual basis when the needs in applications arose, and their proofs depend on a similar type of conditional argument. A more general inequality, given in Section 3.3, can be applied to all such multivariate distributions when the random variables are exchangeable in a certain fashion.

3.1. MULTIVARIATE t DISTRIBUTION

The generalizations of the Student t distribution appear to have been made along two different directions. One generalization is, of course, Hotelling's T^2 distribution, which is again one-dimensional. The other type of generalization is the multivariate t. When applying them to obtain confidence regions for the mean vector of a normal distribution, the regions are elliptical and rectangular, respectively. This section concerns some inequalities for multivariate t distribution.

Let $\mathbf{X}=(X_1,\ldots,X_k)'$ be distributed according to $N(\mathbf{0},\sigma^2\mathbf{R})$, where $\mathbf{R}$ is a positive semidefinite correlation matrix, and let S be independent of $\mathbf{X}$ such that nS^2/σ^2 is a chi-square variable with n degrees of freedom. Let $\mathbf{t}=(t_1,\ldots,t_k)'$ denote the multivariate t variable with components

$$t_i = \frac{X_i}{S}, \qquad i = 1,\ldots,k. \tag{3.1.1}$$

The density function of $\mathbf{t}$ given below, which depends on the parameters $(n,\mathbf{R})$, was obtained by Dunnett and Sobel (1954) for $\mathbf{R}$ positive definite.

$$f(\mathbf{t}) = \frac{\Gamma((k+n)/2)}{(n\pi)^{k/2}\Gamma(n/2)|\mathbf{R}|^{1/2}}\left[1+\frac{1}{n}\mathbf{t}'\mathbf{R}^{-1}\mathbf{t}\right]^{-(k+n)/2},$$
$$-\infty < t_i < \infty, \qquad i = 1,\ldots,k. \tag{3.1.2}$$

This distribution is useful for making inferences on the normal means when the k populations have a common unknown variance σ^2 (which is estimated by S^2) and the marginal distributions are Student's t. The following theorem says that inequalities that depend on correlations can be preserved when passing from a normal distribution to such a t distribution. Therefore most inequalities given in Chapter 2 also apply here. This approach was first adopted by Dunnett and Sobel (1955).

Theorem 3.1.1. Let $\mathbf{t}$ be a multivariate t variable as defined in (3.1.1), where $\mathbf{X}$ is an $N(\mathbf{0},\sigma^2\mathbf{R})$ variable, nS^2/σ^2 is a chi-square variable with n degrees of freedom, and $\mathbf{X}$ and S^2 are independent. Let $\mathbf{R}_1$, $\mathbf{R}_2$ be two positive semidefinite correlation matrices and $A\subset\Re^k$ be Borel-measurable. If

$$P_{\mathbf{R}=\mathbf{R}_1}[\mathbf{X}\in A] \geqslant (\text{or} >) P_{\mathbf{R}=\mathbf{R}_2}[\mathbf{X}\in A]$$

holds for all $\sigma^2>0$, then

$$P_{\mathbf{R}=\mathbf{R}_1}[\mathbf{t}\in A] \geqslant (\text{or} >) P_{\mathbf{R}=\mathbf{R}_2}[\mathbf{t}\in A] \tag{3.1.3}$$

holds for all n.

Proof. Clearly we have

$$P_{\mathbf{R}}[\mathbf{t}\in A] = EP_{\mathbf{R}}[(\mathbf{X}/S)\in A|S=s].$$

By independence, the conditional distribution of $\mathbf{X}/S$, given $S=s$ is $N(\mathbf{0},\sigma^2\mathbf{R}/s^2)$, and the inequality (the strict inequality) follows by taking expectations on both sides. ∎

Remark. A special case of interest is that in which $\mathbf{R}$ is the identity matrix $\mathbf{I}$. In this case the components of $\mathbf{t}$ are not independent because they all depend on the same S, and from Lemma 2.2.1 we immediately obtain

$$P_{\mathbf{R}=\mathbf{I}}\left[\bigcap_{i=1}^{k}\{t_i \leqslant a_i\}\right] \geqslant \prod_{i=1}^{k} P[t_i \leqslant a_i], \tag{3.1.4}$$

$$P_{\mathbf{R}=\mathbf{I}}\left[\bigcap_{i=1}^{k}\{|t_i| \leqslant a_i\}\right] \geqslant \prod_{i=1}^{k} P[|t_i| \leqslant a_i] \tag{3.1.5}$$

for all $a_i \geqslant 0$, $i=1,\ldots,k$. We note that the inequality given in Theorem 2.3.4 also holds for this multivariate t distribution. This is so because the proof given there can easily be modified by taking expectations over the distribution of S.

Next we consider a more general version of the multivariate t distribution. Let $\mathbf{X}$ again denote a normal variable with mean vector $\mathbf{0}$ and correlation matrix $\mathbf{R}$; without loss of generality assume that the variances are already one. Let $\mathbf{V}_1,\ldots,\mathbf{V}_n$ be independent and independent of $\mathbf{X}$ so that $\mathbf{V}_m$ is normally distributed with mean vector $\mathbf{0}$, variances one, and correlations

$$\operatorname{cor}(V_{im}, V_{jm}) = \lambda_i(m)\lambda_j(m), \qquad |\lambda_i(m)| < 1, \tag{3.1.6}$$

for $i \neq j$ and for $m=1,\ldots,n$ (i.e., each correlation matrix has the structure l). Let us define

$$S_i = \left(\frac{1}{n}\sum_{m=1}^{n} V_{im}^2\right)^{1/2}, \qquad t_i = \frac{X_i}{S_i} \quad i = 1,\ldots,k. \tag{3.1.7}$$

For the t_i variables defined in (3.1.7) inequalities for the distribution of $\mathbf{t}=(t_1,\ldots,t_k)'$ have been given by several authors. Halperin (1967) proved for $k=2$ only that the inequality

$$P\left[\bigcap_{i=1}^{2}\{|t_i| \leqslant a_i\}\right] \geqslant \prod_{i=1}^{2} P[|t_i| \leqslant a_i]$$

holds. His proof also depends on a conditional argument with the aid of the inequality given in (2.2.6) (the $\mathbf{t}$ variable he defined is slightly different). Scott (1967) and Khatri (1970) made attempts to prove an inequality of this type for the general case when $\mathbf{X}, \mathbf{V}_1, \ldots, \mathbf{V}_n$ have a common correlation matrix with a general form, but unfortunately their proofs were later found to be in error. The correctness of such an inequality for general k is still an open problem. Šidák (1971) attacked this problem by studying the monotonicity property of the rectangular probability content as a function of the correlations; he obtained the following theorem.

Theorem 3.1.2. Let $\mathbf{t}$ be a multivariate t variable as defined in (3.1.7), where $\mathbf{X}, \mathbf{V}_1, \ldots, \mathbf{V}_n$ are independent normal variables with means $\mathbf{0}$ and variances one. Assume that the correlations of the components of the $\mathbf{V}_m$s are of the form given in (3.1.6) and that $\mathrm{cor}(X_i, X_j) = \kappa_i \kappa_j \tau_{ij}$ $(i \neq j)$, where $|\kappa_i| \leqslant 1$ and $\mathbf{T} = (\tau_{ij})$ is a positive definite correlation matrix. Then for arbitrary but fixed $a_i > 0$ $(i = 1, \ldots, k)$, the probability $P[\bigcap_{i=1}^{k} \{|t_i| \leqslant a_i\}]$ is (a) strictly increasing (strictly decreasing) in κ_i for $\kappa_i \geqslant 0$ $(\leqslant 0)$ for each i and (b) strictly increasing in $|\lambda_i(m)|$ for each i and each m.

Proof. The proof of (a) is immediate by first applying Corollary 1 of Theorem 2.2.5 to the conditional distribution of $\mathbf{X}$, given $\mathbf{V}_m = \mathbf{v}_m$ $(m = 1, \ldots, n)$ (which is normal) and then taking expectations. The proof of (b) is again based on a conditional argument. For given $\mathbf{X} = \mathbf{x}$ and $\mathbf{V}_m = \mathbf{v}_m$ for $m = 2, \ldots, n$, this amounts to proving that the probability

$$P\left[\bigcap_{i=1}^{k} \left\{ |V_{i1}| \geqslant \left(n\left(\frac{x_i}{a_i}\right)^2 - \sum_{m=2}^{n} v_{im}^2 \right)^{1/2} \right\}\right]$$

is strictly increasing in $|\lambda_i(1)|$ for $i = 1, \ldots, k$, which follows from Theorem 2.3.3. ■

The above theorem establishes a monotonicity property for the probability content as a function of the correlations of the components of both $\mathbf{X}$ and the $\mathbf{V}_m$s. As an immediate consequence, if the correlations of the $\mathbf{V}_m$s are fixed, the rectangular probability content of $\mathbf{t}$ is minimized when the components of $\mathbf{X}$ are independent. This result, of course, is also an immediate consequence of Corollary 1 of Theorem 2.2.5.

3.2. MULTIVARIATE CHI-SQUARE AND F DISTRIBUTIONS

The study of inequalities for multivariate chi-square distribution was motivated mainly by the problem of constructing joint confidence regions for normal variances when the random variables are correlated. In the following we give a theorem which says that if each of the correlation matrices of the variables has the structure l, then a multivariate chi-square probability is bounded below by the product of the marginal probabilities. Again, without loss of generality the common variance may be assumed to be one.

Theorem 3.2.1. For $m=1,\ldots,n$ let $\mathbf{V}_m=(V_{1m},\ldots,V_{km})'$ be independent $N(\mathbf{0},\mathbf{R}_m)$ variables. If each of the correlation matrices $\mathbf{R}_m$ is of the form given in (3.1.6), then

$$P\left[\bigcap_{i=1}^{k}\{\chi_i^2\leqslant a_i\}\right]\geqslant P\left[\bigcap_{i\in C}\{\chi_i^2\leqslant a_i\}\right]P\left[\bigcap_{i\notin C}\{\chi_i^2\leqslant a_i\}\right], \tag{3.2.1}$$

$$P\left[\bigcap_{i=1}^{k}\{\chi_i^2>a_i\}\right]\geqslant P\left[\bigcap_{i\in C}\{\chi_i^2>a_i\}\right]P\left[\bigcap_{i\notin C}\{\chi_i^2>a_i\}\right] \tag{3.2.2}$$

hold for every subset C of $\{1,\ldots,k\}$, where $\chi_i^2=\Sigma_{m=1}^n V_{im}^2$ are chi-square variables with n degrees of freedom each for $i=1,\ldots,k$.

It should be noted that this result is not an immediate consequence of Theorem 2.2.2 or 2.2.4. Its proof, given here, depends on an application of the following lemma concerning the expectations of monotone functions of independent random variables. This lemma represents an extension of the inequality in (2.2.6) for which $\mathbf{U}$ is multidimensional. Also, it is a special case of more general results given by Lehmann (1966) and Esary, Proschan, and Walkup (1967) (see Theorems 5.1.4 and 5.2.2).

Lemma 3.2.1. Let $\mathbf{U}=(U_1,\ldots,U_k)'$ be a random variable with independent components and let $g_1(\mathbf{u}),g_2(\mathbf{u})$ be Borel-measurable real-valued functions. If $E|g_1(\mathbf{U})g_2(\mathbf{U})|<\infty$ and if in each argument g_1, g_2 are either nondecreasing together or nonincreasing together in the same direction, then

$$E[g_1(\mathbf{U})g_2(\mathbf{U})] \geqslant [Eg_1(\mathbf{U})][Eg_2(\mathbf{U})];$$

or, equivalently,

$$\operatorname{cov}(g_1(\mathbf{U}), g_2(\mathbf{U})) \geqslant 0.$$

The inequality is strict unless $g_1(\mathbf{U})$ or $g_2(\mathbf{U})$ is singular.

Proof. For $i=1,\ldots,k$ let $F_i(u_i)$ denote the distribution of U_i. Then by Lemma 2.2.1 we have

$$\begin{aligned}
E\prod_{j=1}^{2} g_j(\mathbf{U}) &= \int\cdots\int\left[\int\prod_{j=1}^{2} g_j(u_1,\ldots,u_k)\,dF_1(u_1)\right]\prod_{i=2}^{k} dF_i(u_i)\\
&\geqslant \int\cdots\int\left[\prod_{j=1}^{2}\int g_j(u_1,\ldots,u_k)\,dF_1(u_1)\right]\prod_{i=2}^{k} dF_i(u_i)\\
&\geqslant\cdots\geqslant \prod_{j=1}^{2}\int\cdots\int g_j(u_1,\ldots,u_k)\prod_{i=1}^{k} dF_i(u_i)\\
&= \prod_{j=1}^{2} Eg_j(\mathbf{U});
\end{aligned}$$

and unless $g_1(\mathbf{U})$ or $g_2(\mathbf{U})$ is singular, the inequality is strict. ∎

Remark. It should be noted that for $k>1$ the condition imposed in this lemma is different from that stated in (2.2.3). To illustrate this point let us consider the following example.

Example 3.2.1. Let the distribution of $\mathbf{U}$ and g_1, g_2 be given by

$$P[\mathbf{U}=(3,0)'] = P[\mathbf{U}=(4,0)'] = P[\mathbf{U}=(3,3)'] = P[\mathbf{U}=(4,3)'] = \tfrac{1}{4},$$

$$g_1(u_1,u_2) = \max(u_1,u_2), \qquad g_2(u_1,u_2) = \tfrac{1}{2}(u_1+u_2).$$

Then from Lemma 3.2.1 we can conclude that $E\prod_{j=1}^{2} g_j(\mathbf{U}) > \prod_{j=1}^{2} Eg_j(\mathbf{U})$, but Lemma 2.2.1 fails to apply.

Proof of Theorem 3.2.1. Without loss of generality it will be assumed that $C=\{1,\ldots,q\}$ for $q<k$, and that $\lambda_i(m) \geqslant 0$ for all i, m (because otherwise we can replace V_{im} by $-V_{im}$ and the probability under consideration remains unchanged). Let $U_1,\ldots,U_n$ be independent standard normal variables such that, given $U_m=u_m$, $V_{1m},\ldots,V_{km}$ are

independent normal variables with means $\lambda_i(m)u_m$ and variances $(1-\lambda_i^2(m))$. Let φ denote the $N(0,1)$ density and let

$$A_1 = \left\{(v_{11},\ldots,v_{qn})' \mid \sum_{m=1}^{n} v_{im}^2 \leqslant a_i,\ i=1,\ldots,q\right\}, \tag{3.2.3}$$

$$A_2 = \left\{(v_{q+1,1},\ldots,v_{kn})' \mid \sum_{m=1}^{n} v_{im}^2 \leqslant a_i,\ i=q+1,\ldots,k\right\}. \tag{3.2.4}$$

Then clearly we have

$$P\left[\bigcap_{i=1}^{k} \{\chi_i^2 \leqslant a_i\}\right] = E\prod_{j=1}^{2} g_j(\mathbf{U}),$$

where

$$g_1(\mathbf{u}) = \int_{A_1}\cdots\int \prod_{i=1}^{q}\prod_{m=1}^{n} \varphi\left(\frac{(v_{im}-\lambda_i(m)u_m)}{(1-\lambda_i^2(m))^{1/2}}\right) dv_{11}\cdots dv_{qn},$$

$$g_2(\mathbf{u}) = \int_{A_2}\cdots\int \prod_{i=q+1}^{k}\prod_{m=1}^{n} \varphi\left(\frac{(v_{im}-\lambda_i(m)u_m)}{(1-\lambda_i^2(m))^{1/2}}\right) dv_{q+1,1}\cdots dv_{kn}.$$

It then follows from Theorem 4.1.4 of Chapter 4 that g_1, g_2 are symmetric in u_m and monotonically nonincreasing in $|u_m|$ for each m. Therefore from Lemma 3.2.1 we conclude that

$$\begin{aligned} E\prod_{j=1}^{2} g_j(\mathbf{U}) &= E\prod_{j=1}^{2} g_j(|U_1|,\ldots,|U_n|) \\ &\geqslant \prod_{j=1}^{2} Eg_j(|U_1|,\ldots,|U_n|) = \prod_{j=1}^{2} Eg_j(\mathbf{U}). \end{aligned}$$

This establishes (3.2.1). The proof of (3.2.2) is similar. ■

As an immediate consequence of this result, we obtain

$$P\left[\bigcap_{i=1}^{k} \{\chi_i^2 \leqslant a_i\}\right] \geqslant \prod_{i=1}^{k} P[\chi_i^2 \leqslant a_i] \tag{3.2.5}$$

and

$$P\left[\bigcap_{i=1}^{k} \{\chi_i^2 > a_i\}\right] \geqslant \prod_{i=1}^{k} P[\chi_i^2 > a_i]. \tag{3.2.6}$$

In fact the inequality in (3.2.5) also holds when the correlation matrices $\mathbf{\Sigma}_m$ are equal but do not have the structure l. This follows as a special case of a theorem given by Das Gupta, Eaton, Olkin, Perlman, Savage, and Sobel (1972) for a class of elliptically contoured distributions (see Corollary 1 of Theorem 4.3.5).

An inequality for the bivariate chi-square distribution was obtained by Jensen (1969). He proved that under more general conditions on the correlation matrix

$$P\left[\bigcap_{i=1}^{2}\{b\leqslant\chi_i^2\leqslant a\}\right]\geqslant\prod_{i=1}^{2}P\left[b\leqslant\chi_i^2\leqslant a\right] \tag{3.2.7}$$

holds for all $a>b\geqslant 0$. The proof he gave depends on a specific expression for the joint density of $(\chi_1^2,\chi_2^2)'$. Hence an extension of this inequality to the general case of $k>2$ is difficult.

A multivariate version of the F distribution was formulated by Kimball (1951) in connection with an analysis-of-variance problem. For $k\geqslant 2$ let $\chi_0^2,\chi_1^2,\ldots,\chi_k^2$ be independent chi-square variables with degrees of freedom $n_0,n_1,\ldots,n_k$, respectively. In the analysis-of-variance problem χ_0^2 depends on the sum of the squares of the residuals, and $\chi_1^2,\ldots,\chi_k^2$ represent the sums of the squares of the treatment effect, block effect, etc. When testing the significance of the treatment effect and block effect *simultaneously*, one needs to consider the joint distribution of two F variables. In general let us define a random variable $(F_1,\ldots,F_k)'$, where

$$F_i=\left(\chi_i^2/n_i\right)/\left(\chi_0^2/n_0\right),\qquad i=1,\ldots,k. \tag{3.2.8}$$

Kimball's result says

Theorem 3.2.2. Let $\chi_0^2,\chi_1^2,\ldots,\chi_k^2$ be independent chi-square variables with degrees of freedom $n_0,n_1,\ldots,n_k$, respectively, and let $F_1,\ldots,F_k$ be defined according to (3.2.8). Then for $a_i>0$ $(i=1,\ldots,k)$

$$P\left[\bigcap_{i=1}^{k}\{F_i\leqslant a_i\}\right]>P\left[\bigcap_{i\in C}\{F_i\leqslant a_i\}\right]P\left[\bigcap_{i\notin C}\{F_i\leqslant a_i\}\right], \tag{3.2.9}$$

$$P\left[\bigcap_{i=1}^{k}\{F_i>a_i\}\right]>P\left[\bigcap_{i\in C}\{F_i>a_i\}\right]P\left[\bigcap_{i\notin C}\{F_i>a_i\}\right] \tag{3.2.10}$$

hold for all subsets C of $\{1,\ldots,k\}$. Hence, in particular,

$$P\left[\bigcap_{i=1}^{k}\{F_i \leqslant a_i\}\right] > \prod_{i=1}^{k} P[F_i \leqslant a_i], \tag{3.2.11}$$

$$P\left[\bigcap_{i=1}^{k}\{F_i > a_i\}\right] > \prod_{i=1}^{k} P[F_i > a_i]. \tag{3.2.12}$$

Proof. By independence we can write

$$P\left[\bigcap_{i=1}^{k}\{F_i \leqslant a_i\}\right] = E\prod_{i=1}^{k} P\left[\chi_i^2 \leqslant (a_i n_i/n_0)\chi_0^2\right],$$

$$P\left[\bigcap_{i=1}^{k}\{F_i > a_i\}\right] = E\prod_{i=1}^{k} P\left[\chi_i^2 > (a_i n_i/n_0)\chi_0^2\right],$$

and the proof follows immediately from Lemma 2.2.1. ■

We note that the inequalities given in Section 3.1 may also be regarded as inequalities for the multivariate F distribution in the special case with one degree of freedom in the numerator. Also, another inequality for the multivariate F distribution was given by Olkin (1972). Since that result was obtained through majorization, its discussion is postponed here and will be given in Section 6.2.

3.3. AN INEQUALITY VIA EXCHANGEABILITY

In this section we give an inequality obtained by Šidák (1973) for the case in which the events defined by the random variables are exchangeable in a certain fashion, and we discuss some applications. He found that in the proof of the result given by Tong (1970) (see Theorem 2.3.4) the same type of argument applies to a much larger class of random variables. He then proceeded to extract the core of that argument for the general case and obtained the following theorem.

Theorem 3.3.1. Let $\mathbf{Y}_1,\ldots,\mathbf{Y}_k$ denote independent and identically distributed $p\times 1$ random variables, and let $\mathbf{U}$ be a $q\times 1$ random variable that is independent of the $\mathbf{Y}_i$s. Let $\psi:\mathfrak{R}^{p+q}\to\mathfrak{R}^s$, $A\subset\mathfrak{R}^s$

($1 \leqslant s \leqslant p+q$) be Borel-measurable. For $r=1,\ldots,k$ define

$$\beta(r) = P\left[\bigcap_{i=1}^{r} \{\psi(\mathbf{Y}_i, \mathbf{U}) \in A\}\right]. \tag{3.3.1}$$

Then the inequalities

$$\beta(k) \geqslant [\beta(r)]^{k/r} \geqslant \cdots \geqslant [\beta(2)]^{k/2} \geqslant [\beta(1)]^{k} \tag{3.3.2}$$

hold for all $k > r \geqslant 2$. Moreover, the inequalities are strict unless the conditional probability $P[\psi(\mathbf{Y}_1, \mathbf{U}) \in A | \mathbf{U}=\mathbf{u}]$ is a constant with probability one.

Proof. The proof is similar to that of Theorem 2.3.4 with W being the conditional probability $P[\psi(\mathbf{Y}_1, \mathbf{U}) \in A | \mathbf{U}=\mathbf{u}]$. ∎

Although the proof of this theorem does not involve new difficulty, the theorem covers more ground when the exchangeability condition is satisfied. In the following we give some examples of applications, most of which were originally provided by Šidák ($s=1$ in Examples 3.3.2–3.3.7).

Example 3.3.1 (multivariate normal). Let $\mathbf{X}=(\mathbf{X}^{(1)},\ldots,\mathbf{X}^{(k)})'$ be a $sk \times 1$ $N(\boldsymbol{\mu}, \mathbf{R})$ random variable, where each $\mathbf{X}^{(i)}$ is $s \times 1$ ($s \geqslant 1$), all the elements in $\boldsymbol{\mu}$ are equal, $\mathbf{R}$ is a correlation matrix, and all the correlations in $\mathbf{R}$ are equal to $\rho \geqslant 0$. Then for every Borel-measurable set $A \subset \mathfrak{R}^s$ we have

$$P\left[\bigcap_{i=1}^{k} \{\mathbf{X}^{(i)} \in A\}\right] \geqslant \left\{P\left[\bigcap_{i=1}^{r} \{\mathbf{X}^{(i)} \in A\}\right]\right\}^{k/r}$$
$$\geqslant \{P[\mathbf{X}^{(1)} \in A]\}^{k}$$

for all $k > r \geqslant 2$. If the probabilities are in $(0,1)$, then the inequalities are strict unless $\rho = 0$.

Example 3.3.2 (multivariate t). Let $t_1,\ldots,t_k$ be as defined in 3.1.1. If the X_is have means zero and a common correlation coefficient $\rho \geqslant 0$, then

$$P\left[\bigcap_{i=1}^{k} \{b \leqslant t_i \leqslant a\}\right] > \left\{P\left[\bigcap_{i=1}^{r} \{b \leqslant t_i \leqslant a\}\right]\right\}^{k/r}$$
$$> \{P[b \leqslant t_1 \leqslant a]\}^{k}$$

for all $k > r \geqslant 2$ and $a > b$.

Example 3.3.3 (multivariate chi-square). In Theorem 3.2.1 if $\mathbf{R}_1 = \cdots = \mathbf{R}_n = \mathbf{R}$ and the correlations in $\mathbf{R}$ are equal to $\rho \geqslant 0$, then

$$P\left[\bigcap_{i=1}^{k} \{b \leqslant \chi_i^2 \leqslant a\}\right] \geqslant \left\{P\left[\bigcap_{i=1}^{r} \{b \leqslant \chi_i^2 \leqslant a\}\right]\right\}^{k/r}$$
$$\geqslant \{P[b \leqslant \chi_1^2 \leqslant a]\}^k$$

for all $k > r \geqslant 2$ and $a > b \geqslant 0$. The inequalities are strict unless $\rho = 0$.

Example 3.3.4 (multivariate F). In Theorem 3.2.2 if $n_1 = \cdots = n_k$, then

$$P\left[\bigcap_{i=1}^{k} \{b \leqslant F_i \leqslant a\}\right] > \left\{P\left[\bigcap_{i=1}^{r} \{b \leqslant F_i \leqslant a\}\right]\right\}^{k/r}$$
$$> \{P[b \leqslant F_1 \leqslant a]\}^k$$

for all $k > r \geqslant 2$ and $a > b \geqslant 0$.

Example 3.3.5 (multivariate beta). If the Y_is and U are independent chi-square variables with degrees of freedom n_y and n_u, respectively, and if we define

$$X_i = Y_i \Big/ \left(\sum_{i=1}^{k} Y_i + U\right), \qquad i = 1, \ldots, k,$$

then from Olkin and Rubin (1964) $(X_1, \ldots, X_k)'$ is a multivariate beta variable with density

$$f(\mathbf{x}) = c\left(\prod_{i=1}^{k} x_i^{r-1}\right)\left(1 - \sum_{j=1}^{k} x_j\right)^{s-1}, \qquad x_i \geqslant 0, \quad \sum_{i=1}^{k} x_i \leqslant 1, \tag{3.3.3}$$

for some c, r, and s, and Theorem 3.3.1 applies.

Example 3.3.6 (multivariate exponential). A special case of the multivariate exponential variable $(X_1, \ldots, X_k)'$ may be defined by the model $X_i = \min(Y_i, U)$, where the Y_is and U are independent exponential variables with parameters θ_y and θ_u, respectively. This model was considered by Marshall and Olkin (1967), and Theorem 3.3.1 applies to this class of distributions with $\psi(y, u) = \min(y, u)$.

Example 3.3.7 (multivariate Poisson). If the Y_is and U are independent Poisson variables with parameters θ_y and θ_u, respectively, then defining $X_i = Y_i + U$, $(X_1,\ldots,X_k)'$ is a multivariate Poisson variable (the bivariate case was considered by Holgate (1964)), and Theorem 3.3.1 applies.

In general Theorem 3.3.1 may be applied to any distribution when the random variables $\mathbf{Y}_i$ are independent and identically distributed and when the transformation ψ and the set A do not depend on i. Therefore the random variables $\psi(\mathbf{Y}_i,\mathbf{U})$ here are conditionally independent and identically distributed, which by the definition of exchangeability implies that they are exchangeable. This concept will be explored further in Section 5.3, and a refinement of Theorem 3.3.1 will be given in Theorem 6.2.11.

PROBLEMS

1. Let $t_1,\ldots,t_k$ be defined according to (3.1.1), and assume that the covariance matrix of the X_is is the identity matrix. Show that if $a_i \neq 0$ $(i=1,\ldots,k)$, then

$$P\left[\bigcap_{i=1}^{k}\{t_i \leqslant a_i\}\right] < P\left[\bigcap_{i\in C}\{t_i \leqslant a_i\}\right]P\left[\bigcap_{i\notin C}\{t_i \leqslant a_i\}\right],$$

where $C=\{i|1\leqslant i\leqslant k, a_i<0\}$. In particular, if $a_1a_2<0$, then

$$P[t_1\leqslant a_1, t_2\leqslant a_2] < P[t_1\leqslant a_1]P[t_2\leqslant a_2].$$

2. Show that the second inequality in Problem 1 holds in general so long as $a_1a_2<0$ and the correlation between X_1 and X_2 is nonpositive.
3. Show that the assertions in Problems 6 and 7 of Chapter 2 remain true when $\mathbf{X}$ is a multivariate t variable with density function as given in (3.1.2) (see, e.g., the tables in Krishnaiah and Armitage (1966) and Hahn and Hendrickson (1971)).
4. Give an example to show that Lemma 3.2.1 fails if $U_1,\ldots,U_k$ are not independent.
5. Verify the statement in Example 3.2.1.
6. Show that the inequalities in (3.2.9)–(3.2.12) all approach equalities as $n_0\to\infty$.

7. Verify the inequality in Example 3.3.3 by identifying the random variables $\mathbf{Y}_i$ and $\mathbf{U}$ and the transformation ψ in Theorem 3.3.1.

8. For $i=0,1,\ldots,k$ let $\{Y_{im}\}_{m=1}^{n}$ be a sequence of independent and identically distributed random variables with continuous distributions F_i $(i=0,\ldots,k)$, where $F_1=\cdots=F_k$. Let $R_{im}=$rank of Y_{im} in the combined sample $\{Y_{im},Y_{0m}\}_{m=1}^{n}$, and define $S_i=\sum_{m=1}^{n}R_{im}$, $i=1,\ldots,k$. The rank sums $S_1,\ldots,S_k$ may be used to compare k experimental populations simultaneously with a control under certain nonparametric procedures (see Steel (1959) or Miller (1966, Section 4.3)). Show that the inequalities

$$P\left[\bigcap_{i=1}^{k}\{S_i\in A\}\right]\geqslant\left\{P\left[\bigcap_{i=1}^{r}\{S_i\in A\}\right]\right\}^{k/r}\geqslant\{P[S_1\in A]\}^{k}$$

hold for every Borel-measurable subset A of $\mathfrak{R}^1$ and every $k>r\geqslant 2$.

REFERENCES

Das Gupta, S., Eaton, M. L., Olkin, I., Perlman, M. D., Savage, L. J., and Sobel, M. (1972). Inequalities on the probability content of convex regions for elliptically contoured distributions. *Proc. Sixth Berkeley Symp. Math. Statist. Probab.* **2** (L. M. LeCam, J. Neyman, and E. L. Scott, eds.), 241–265. Univ. of California Press, Berkeley, California.

Dunnett, C. W., and Sobel, M. (1954). A bivariate generalization of Student's *t*-distribution with tables for special cases. *Biometrika* **41**, 153–169.

Dunnett, C. W., and Sobel, M. (1955). Approximations to the probability integral and certain percentage points to a multivariate analogue of Student's *t*-distribution. *Biometrika* **42**, 258–260.

Esary, J. D., Proschan, F., and Walkup, D. W. (1967). Association of random variables, with applications. *Ann. Math. Statist.* **38**, 1466–1474.

Hahn, G. J., and Hendrickson, R. W. (1971). A table of percentage points of the distribution of the largest absolute value of *k* Student *t* variates and its applications. *Biometrika* **58**, 323–332.

Halperin, M. (1967). An inequality on a bivariate Student's "*t*" distribution. *J. Amer. Statist. Assoc.* **62**, 603–606.

Holgate, P. (1964). Estimation for the bivariate Poisson distribution. *Biometrika* **51**, 241–245.

Jensen, D. R. (1969). An inequality for a class of bivariate chi-square distributions. *J. Amer. Statist. Assoc.* **64**, 333–336.

Kimball, A. W. (1951). On dependent tests of significance in the analysis of variance. *Ann. Math. Statist.* **22**, 600–602.

Khatri, C. G. (1970). Further contributions to some inequalities for normal distributions and their applications to simultaneous confidence bounds. *Ann. Inst. Statist. Math.* **22**, 451–458.

Krishnaiah, P. R., and Armitage, J. V. (1966). Tables for multivariate *t* distribution. *Sankhyā Ser. B*, **28**, 31–56.

Lehmann, E. L. (1966). Some concepts of dependence. *Ann. Math. Statist.* **37**, 1137–1153.

Marshall, A. W., and Olkin, I. (1967). A generalized bivariate exponential distribution. *J. Appl. Probab.* **4**, 291–302.

Miller, R. G., Jr. (1966). *Simultaneous Statistical Inference*. McGraw-Hill, New York.

Olkin, I. (1972). Monotonicity properties of Dirichlet integrals with applications to the multinomial distribution and the analysis of variance. *Biometrika* **59**, 303–307.

Olkin, I., and Rubin, H. (1964). Multivariate beta distributions and independence properties of the Wishart distribution. *Ann. Math. Statist.* **35**, 261–269.

Scott, A. (1967). A note on conservative confidence regions for the mean of a multivariate normal. *Ann. Math. Statist.* **38**, 278–280. [Corrigenda (1968). *Ann. Math. Statist.* **39**, 2161].

Šidák, Z. (1971). On probabilities of rectangles in multivariate Student distributions: Their dependence on correlations. *Ann. Math. Statist.* **42**, 169–175.

Šidák, Z. (1973). A chain of inequalities for some types of multivariate distributions, with nine special cases. *Apl. Mat.* **18**, 110–118.

Steel, R. G. D. (1959). A multiple comparison rank sum test: Treatments versus control. *Biometrics* **15**, 560–572.

Tong, Y. L. (1970). Some probability inequalities of multivariate normal and multivariate *t*. *J. Amer. Statist. Assoc.* **65**, 1243–1247.

CHAPTER

4

Integral Inequalities Over A Symmetric Convex Set

Two fundamental inequalities with greater generality will be studied in this chapter. Basically they involve the probability content of a convex set that is symmetric about the origin. The first inequality was obtained by Anderson (1955), and we have already seen a special case of it (Lemma 2.2.2) when applied to multivariate normal distribution. Generalizations of this inequality will be discussed in Section 4.2. The second inequality, due to Das Gupta, Eaton, Olkin, Perlman, Savage, and Sobel (1972), represents a generalization of an inequality of Šidák (see Theorem 2.2.5) and yields a number of new interesting results for the normal, t, chi-square, and F distributions.

4.1. ANDERSON'S THEOREM AND RELATED RESULTS

Let $f(\mathbf{x})$ denote the density function of a k-dimensional random variable $\mathbf{X}=(X_1,\ldots,X_k)'$. It is assumed that the probability measure under consideration is absolutely continuous with respect to

Lebesgue measure. In the univariate case ($k=1$) we say that f is unimodal if $f(x)$ is nondecreasing (nonincreasing) in x for $x<c$ ($x>c$) for some real number c, which is equivalent to saying that the set $D_u=\{x|f(x)\geqslant u\}$ is convex for all $u\geqslant 0$. Anderson extended this concept of unimodality to the multivariate case and gave the following definition:

Definition 4.1.1. $f(\mathbf{x})$: $\mathcal{R}^k\to[0,\infty)$ is said to be unimodal if the set $D_u=\{\mathbf{x}|f(\mathbf{x})\geqslant u\}$ is convex for all $u\geqslant 0$.

We also define

Definition 4.1.2. A function $f(\mathbf{x})$: $\mathcal{R}^k\to[0,\infty)$ is said to be symmetric about the origin if $f(\mathbf{x})=f(-\mathbf{x})$ for all $\mathbf{x}$. A set $A\subset\mathcal{R}^k$ is said to be symmetric about the origin if $\mathbf{x}\in A$ implies $-\mathbf{x}\in A$.

As Anderson (1955) noted, if one has a unimodal and symmetric function on the real line (with $k=1$), then "it is obvious that the integral of $f(x)$ over an interval of fixed length is maximized if the interval is centered at the origin; in fact, the integral is a nonincreasing function of the distance of the midpoint from the origin" (p. 170). He showed that this remains true in the multivariate case in the following sense: The probability integral is nonincreasing if the center of a symmetric convex set A is moving away from the origin along a given direction.

Before stating his theorem we shall introduce another notation. Let A and B be subsets of $\mathcal{R}^k$. For real numbers λ_1 and λ_2, $\lambda_1 A+\lambda_2 B$ denotes the set given by

$$\lambda_1 A+\lambda_2 B = \{\mathbf{z}|\mathbf{z}=\lambda_1\mathbf{x}+\lambda_2\mathbf{y},\mathbf{x}\in A,\mathbf{y}\in B\}. \tag{4.1.1}$$

In the special case in which B contains a single point $\mathbf{y}$ and in which $\lambda_1=1,\lambda_2=\lambda$, one may write $A+\lambda\mathbf{y}$ instead of $A+\lambda B$.

We now observe Anderson's theorem. To avoid the trivial case the vector $\mathbf{y}$ is to be interpreted as different from $\mathbf{0}$ in the theorem.

Theorem 4.1.1. Let $f(\mathbf{x})$: $\mathcal{R}^k\to[0,\infty)$ be symmetric about the origin and unimodal. Let $A\subset\mathcal{R}^k$ by symmetric about the origin and convex. If $\int_A f(\mathbf{x})\,d\mathbf{x} < \infty$[†] (in the Lebesgue sense), then

$$\int_A f(\mathbf{x}+\lambda\mathbf{y})\,d\mathbf{x} \geqslant \int_A f(\mathbf{x}+\mathbf{y})\,d\mathbf{x} \tag{4.1.2}$$

holds for all $\mathbf{y}$ and all $\lambda \in [0,1]$, or equivalently,

$$\int_{A+\lambda\mathbf{y}} f(\mathbf{x})\,d\mathbf{x} \geqslant \int_{A+\mathbf{y}} f(\mathbf{x})\,d\mathbf{x}. \tag{4.1.3}$$

The proof of this theorem depends on an application of a volume inequality known as the Brunn–Minkowski theorem, which is stated below as a lemma.

Lemma 4.1.1 (Brunn–Minkowski Theorem).[‡] Let A, B be non-empty, bounded, and convex subsets of $\mathfrak{R}^k$, and let $V\{A\}$ denote the volume of A. Then for all $\alpha \in [0,1]$

$$\left[V\{\alpha A+(1-\alpha)B\}\right]^{1/k} \geqslant \alpha\left[V\{A\}\right]^{1/k}+(1-\alpha)\left[V\{B\}\right]^{1/k}. \tag{4.1.4}$$

Proof of Theorem 4.1.1. For every fixed $u>0$, let D_u denote the set

$$D_u = \{\mathbf{x} \mid f(\mathbf{x}) \geqslant u\}. \tag{4.1.5}$$

First note that since $(A-\mathbf{y})\cap D_u$ is the mirror image of $(A+\mathbf{y})\cap D_u$ through the origin, one must have

$$V\{(A-\mathbf{y})\cap D_u\} = V\{(A+\mathbf{y})\cap D_u\}. \tag{4.1.6}$$

[†]For notational convenience the multiple integral

$$\int \cdots_A \int f(\mathbf{x}) \prod_{i=1}^{k} dx_i$$

will be denoted by $\int_A f(\mathbf{x})\,d\mathbf{x}$.

[‡]Borell (1975) studied a class of convex measures P with the following property: For some $r \in [-\infty, 1/k)$

$$P[\alpha A+(1-\alpha)B] \geqslant \{\alpha[P(A)]^r+(1-\alpha)[P(B)]^r\}^{1/r}$$

holds for all Borel-measurable convex sets A, B in $\mathfrak{R}^k$ and all $\alpha \in [0,1]$. When $r=0$, by continuity this is interpreted to mean that P is log-concave, i.e.,

$$P[\alpha A+(1-\alpha)B] \geqslant [P(A)]^{\alpha}[P(B)]^{1-\alpha},$$

which was shown to be true for certain measures (including the measures defined by multivariate normal distribution) by Prékopa (1971). The Brunn–Minkowski theorem says that the Lebesgue measure is a member of this class with $r=1/k$.

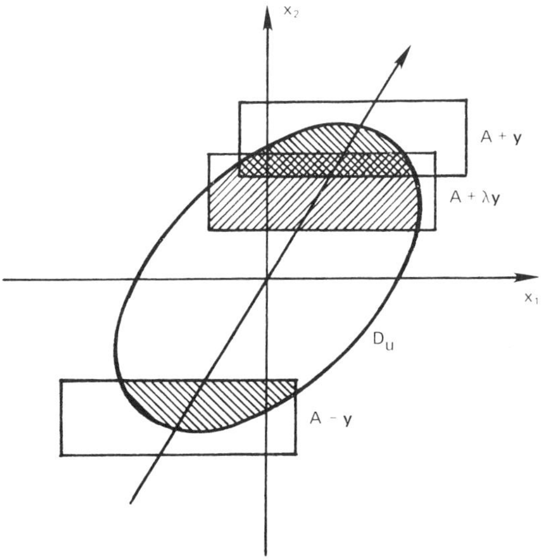

Fig. 4.1.1. Illustration for the proof of Theorem 4.1.1.

By letting $\alpha=(1+\lambda)/2$, one has

$$(A+\lambda\mathbf{y}) \supset [\{\alpha A+(1-\alpha)A\}+\lambda\mathbf{y}]^{\dagger}$$
$$=[\alpha(A+\mathbf{y})+(1-\alpha)(A-\mathbf{y})]. \tag{4.1.7}$$

Since A and D_u are convex, this implies that

$$[(A+\lambda\mathbf{y})\cap D_u] \supset [\alpha\{(A+\mathbf{y})\cap D_u\}+(1-\alpha)\{(A-\mathbf{y})\cap D_u\}]. \tag{4.1.8}$$

Combining (4.1.6) and (4.1.8) and applying Lemma 4.1.1 yield

$$H(u)\equiv V\{(A+\lambda\mathbf{y})\cap D_u\} \geqslant V\{(A+\mathbf{y})\cap D_u\} \equiv H^*(u) \tag{4.1.9}$$

for all $\lambda\in[0,1]$.

Now by the definitions of the Lebesgue and the Lebesgue–Stieltjes integrals one can write

$$\delta=\int_{A+\lambda\mathbf{y}} f(\mathbf{x})\,d\mathbf{x}-\int_{A+\mathbf{y}} f(\mathbf{x})\,d\mathbf{x}$$
$$=-\int_0^\infty u\,dH(u)+\int_0^\infty u\,dH^*(u)$$
$$=\int_0^\infty u\,d[H^*(u)-H(u)].$$

†This implication was given in Anderson's proof. In fact it is also true that $(A+\lambda\mathbf{y})=[\{\alpha A+(1-\alpha)A\}+\lambda\mathbf{y}]$.

Integrating by parts and using the fact that $\int_A f(\mathbf{x})\,d\mathbf{x} < \infty$ and the inequality in (4.1.9), one has

$$\delta = \lim_{b\to\infty} b\left[H^*(b) - H(b)\right] - \lim_{a\to 0} a\left[H^*(a) - H(a)\right] + \int_0^{\infty}\left[H(u) - H^*(u)\right]du \geq 0,$$

as was to be shown. ∎

Remarks. (1) By the symmetry conditions on f and A it is easy to verify that the integrals in (4.1.2) and (4.1.3) are symmetric in $\mathbf{y}$, i.e., for every fixed λ

$$\int_{A+\lambda\mathbf{y}} f(\mathbf{x})\,d\mathbf{x} = \int_{A-\lambda\mathbf{y}} f(\mathbf{x})\,d\mathbf{x}. \tag{4.1.10}$$

Thus since $\mathbf{y}$ is arbitrary, Theorem 4.1.1 can be restated as the following: Under the conditions given the integral $\int_{A+\lambda\mathbf{y}} f(\mathbf{x})\,d\mathbf{x}$ is monotonically nonincreasing in $|\lambda|$ for every fixed $\mathbf{y}$. Therefore for two sets A_1 and A_2 if $A_i = A + \lambda_i\mathbf{y}$ for some $\mathbf{y}$ and if $|\lambda_1| < |\lambda_2|$, then A_1 and A_2 are comparable through a partial ordering, and the probability content of A_2 is smaller than that of A_1.

(2) It can be seen from the proof of the theorem that the inequality in (4.1.3) (hence in (4.1.2)) becomes an equality if and only if

$$\left[(A+\mathbf{y}) \cap D_u\right] = \left[(A \cap D_u) + \mathbf{y}\right]$$

holds for every $u > 0$. If f is a normal density with mean vector $\mathbf{0}$ and a positive definite covariance matrix $\mathbf{\Sigma}$, then f is unimodal and this condition is not satisfied (Anderson, 1955). Hence for normal distribution the inequality is strict.

When applying Theorem 4.1.1, it is required to check that the unimodality condition is satisfied. A sufficient condition for independent random variables was provided by Mudholkar (1969) for this purpose.

Theorem 4.1.2. For $i = 1, \ldots, k$ if each f_i: $\mathcal{R}^1 \to [0, \infty)$ is log-concave, then $f(\mathbf{x}) = \prod_{i=1}^{k} f_i(x_i)$: $\mathcal{R}^k \to [0, \infty)$ is a unimodal function of $\mathbf{x}$.

Proof. For arbitrary but fixed $u>0$ let $\mathbf{x}$, $\mathbf{y}$ satisfy $f(\mathbf{x})\geqslant u$, $f(\mathbf{y})\geqslant u$. Then for $\alpha\in[0,1]$ one has

$$\begin{aligned}\log f(\alpha\mathbf{x}+(1-\alpha)\mathbf{y}) &= \sum_{i=1}^{k}\log f_i(\alpha x_i+(1-\alpha)y_i)\\ &\geqslant \alpha\sum_{i=1}^{k}\log f_i(x_i)\\ &\quad + (1-\alpha)\sum_{i=1}^{k}\log f_i(y_i)\\ &\geqslant \log u,\end{aligned}$$

which implies the convexity of the set D_u given in (4.1.5). ■

Now an application of Theorem 4.1.1 gives the following theorem, which is also due to Anderson (1955).

Theorem 4.1.3. Let $\mathbf{X}$ have density function $f(\mathbf{x})$, and assume that f, A satisfy the conditions imposed in Theorem 4.1.1. Let $\mathbf{Y}$ be a random variable that is independent of $\mathbf{X}$. Then

(a) $P[(\mathbf{X}+\lambda\mathbf{Y})\in A]$ is nonincreasing in $|\lambda|$.

(b) In general, let ψ: $\mathfrak{R}^k\to\mathfrak{R}^1$ be symmetric about the origin such that the set $\{\mathbf{x}|\psi(\mathbf{x})\leqslant v\}$ is convex; then $P[\psi(\mathbf{X}+\lambda\mathbf{Y})\leqslant v]$ is nonincreasing in $|\lambda|$.

Moreover, it is strictly decreasing in $|\lambda|$ unless $[(A+\mathbf{Y})\cap D_u]=[(A\cap D_u)+\mathbf{Y}]$ holds with probability one for all $u>0$.

Proof. Let λ_1, λ_2 satisfy $|\lambda_1|<|\lambda_2|$. Then from Theorem 4.1.1 we have

$$P\big[(\mathbf{X}+\lambda_1\mathbf{Y})\in A|\mathbf{Y}=\mathbf{y}\big] \geqslant P\big[(\mathbf{X}+\lambda_2\mathbf{Y})\in A|\mathbf{Y}=\mathbf{y}\big]$$

for every $\mathbf{y}$, and the inequality is preserved after taking expectations on both sides. This proves (a). The proof for the general case (b) is similar. ■

Anderson (1955) also gave the following special result for multivariate normal distribution, which says that in the normal case inequalities can be obtained through a partial ordering of the covariance matrices.

Corollary. Let $\mathbf{X}$ be distributed according to $N(\mathbf{0},\boldsymbol{\Sigma})$, and assume that $A \subset \mathcal{R}^k$ is symmetric about the origin and convex. Let $\mathbf{\Sigma}_1$, $\mathbf{\Sigma}_2$ be two positive definite covariance matrices. If $\mathbf{\Sigma}_2 - \mathbf{\Sigma}_1$ is positive semidefinite, then

$$P_{\mathbf{\Sigma}=\mathbf{\Sigma}_1}[\mathbf{X} \in A] \geqslant P_{\mathbf{\Sigma}=\mathbf{\Sigma}_2}[\mathbf{X} \in A]. \tag{4.1.11}$$

In general, if $\psi(\mathbf{X})$ is symmetric about the origin such that $\{\mathbf{x} | \psi(\mathbf{x}) \leqslant v\}$ is convex, then

$$P_{\mathbf{\Sigma}=\mathbf{\Sigma}_1}[\psi(\mathbf{X}) \leqslant v] \geqslant P_{\mathbf{\Sigma}=\mathbf{\Sigma}_2}[\psi(\mathbf{X}) \leqslant v]. \tag{4.1.12}$$

If A is a bounded set and $\mathbf{\Sigma}_2 - \mathbf{\Sigma}_1 \neq \mathbf{0}$, then the inequalities in (4.1.11) and (4.1.12) are strict.

Proof. Let $\mathbf{X}_i$ be distributed according to $N(\mathbf{0}, \mathbf{\Sigma}_i)$ for $i = 1, 2$, and let $\mathbf{Y}$ be an $N(\mathbf{0}, \mathbf{\Sigma}_2 - \mathbf{\Sigma}_1)$ variable that is independent of $\mathbf{X}_1$. Then $\mathbf{X}_2$ and $\mathbf{X}_1 + \mathbf{Y}$ are identically distributed, and the proof follows immediately from Theorem 4.1.3. ■

Note that a special case of the above corollary occurs when both $\mathbf{\Sigma}_1$ and $\mathbf{\Sigma}_2$ are diagonal matrices. In this special case the statement is, of course, trivial.

In the following we shall give a theorem that is analogous to but different from Anderson's theorem. Anderson's theorem says basically that if f is unimodal, A is convex, and f, A are symmetric about the origin, then the probability content becomes smaller if the set is moving away from the origin with a larger $|\lambda|$ value *along a given direction* specified by the vector $\mathbf{y}$. Now consider two real vectors $\mathbf{y}$, $\mathbf{y}^*$ in $\mathcal{R}^k$. We define

Definition 4.1.3. Let $|\mathbf{y}|$ denote the vector $(|y_1|, \ldots, |y_k|)'$. $\mathbf{y}$ is said to have larger absolute values than $\mathbf{y}^*$ (in symbols $|\mathbf{y}| > |\mathbf{y}^*|$) if $|y_i| \geqslant |y_i^*|$ holds for each i and strict inequality holds for some i.

Let $A + \mathbf{y}$ and $A + \mathbf{y}^*$ denote two sets centered at $\mathbf{y}$ and $\mathbf{y}^*$, respectively. If $|\mathbf{y}| > |\mathbf{y}^*|$, then the set $A + \mathbf{y}$ is farther away from the origin, but not necessarily proportionally so in each coordinate, i.e., not necessarily along a given direction. A natural question one may wish to ask is this: Does the conclusion in Theorem 4.1.1 remain true when $A + \lambda\mathbf{y}$ is replaced by $A + \mathbf{y}^*$? If this does remain true, then it would imply Anderson's theorem as a special case by simply moving away from the origin along one coordinate at a time. The following example shows that under the same conditions on f and A this is not true even when y_i and y_i^* are of the same sign for all i.

Example 4.1.1. Suppose that $k=2$,

$$f(\mathbf{x}) = \begin{cases} \frac{1}{3} & \text{for} \quad \mathbf{x} \in S, \\ 0 & \text{otherwise,} \end{cases}$$

where

$$S = \{\mathbf{x} \mid |x_1| \leqslant 1, |x_2| \leqslant 1, |x_1 - x_2| \leqslant 1\}.$$

Let A denote the set

$$A = \{\mathbf{x} \mid |x_1| \leqslant \tfrac{1}{2}, |x_2| \leqslant \tfrac{1}{2}\}.$$

Then it is easy to check that f, A satisfy the conditions stated in Theorem 4.1.1; but for $\mathbf{y} = (\frac{1}{2}, \frac{1}{2})'$ and $\mathbf{y}^* = (\frac{1}{4}, \frac{1}{2})'$ we have instead

$$\int_{A+\mathbf{y}^*} f(\mathbf{x})\, d\mathbf{x} < \int_{A+\mathbf{y}} f(\mathbf{x})\, d\mathbf{x}.$$

This example suggests that in order to establish corresponding inequalities under the weaker condition $|\mathbf{y}| > |\mathbf{y}^*|$ stronger conditions on f and A are required. We define

Definition 4.1.4. A real-valued function $f(\mathbf{x})$ is said to be sign invariant if for all $\mathbf{x}$ in the domain

$$f(\mathbf{x}) = f(x_1, \ldots, x_{i-1}, -x_i, x_{i+1}, \ldots, x_k) \tag{4.1.13}$$

holds for all i.

Definition 4.1.5. A set $A \subset \mathfrak{R}^k$ is said to be sign invariant if $\mathbf{x} \in A$ implies

$$(x_1, \ldots, x_{i-1}, -x_i, x_{i+1}, \ldots, x_k) \in A \tag{4.1.14}$$

for all $\mathbf{x}$ and all i, that is, its indicator function is sign invariant.

It is clear that if $f(\mathbf{x})$ (or A) is sign invariant, then it is symmetric about the origin, but not conversely. Also, if $f(\mathbf{x})$ is sign invariant, then the set D_u defined in (4.1.5) is also sign invariant. With this stronger symmetry property we can now prove the following theorem.

Theorem 4.1.4. Let $f(\mathbf{x})$: $\mathcal{R}^k \to [0, \infty)$ be sign invariant and unimodal. Let $A \subset \mathcal{R}^k$ be sign invariant and convex. If $\int_A f(\mathbf{x})\,d\mathbf{x} < \infty$ (in the Lebesgue sense), then

$$\int_{A+\mathbf{y}} f(\mathbf{x})\,d\mathbf{x} = \int_{A+|\mathbf{y}|} f(\mathbf{x})\,d\mathbf{x} \tag{4.1.15}$$

and the inequality

$$\int_{A+\mathbf{y}^*} f(\mathbf{x})\,d\mathbf{x} \geqslant \int_{A+\mathbf{y}} f(\mathbf{x})\,d\mathbf{x} \tag{4.1.16}$$

holds for all $\mathbf{y}, \mathbf{y}^*$ satisfying $|\mathbf{y}| > |\mathbf{y}^*|$.

This theorem was actually applied when proving an inequality for the multivariate chi-square distribution in Theorem 3.2.1, and it follows immediately from Theorem 4.2.2 or 5.2.5. We give below an alternative proof which is analogous to that of Theorem 4.1.1. Note that in this proof the Brunn–Minkowski theorem is not needed.

Proof. For every $\mathbf{x}$ and every given i let $\mathbf{x}_0$ denote the vector

$$(x_1, \ldots, x_{i-1}, -x_i, x_{i+1}, \ldots, x_k)'.$$

It follows that

$$\int_{A+\mathbf{y}} f(\mathbf{x})\,d\mathbf{x} = \int_A f(\mathbf{x}+\mathbf{y})\,d\mathbf{x} = \int_A f(\mathbf{x}_0+\mathbf{y}_0)\,d\mathbf{x}$$

$$= \int_A f(\mathbf{x}+\mathbf{y}_0)\,d\mathbf{x} = \int_{A+\mathbf{y}_0} f(\mathbf{x})\,d\mathbf{x},$$

which establishes (4.1.15). To prove (4.1.16) it suffices to prove only the case $|y_i| > |y_i^*|$ and $|y_j| = |y_j^*|$ for $j \neq i$ for some i. For notational convenience we may then assume without loss of generality that $i = 1$ and all y_j and y_j^* are nonnegative ($j = 1, \ldots, k$).

Let $V\{A\}$ denote the volume of A. The essential part of the proof is to show that under the conditions imposed we have

$$V\{(A+\mathbf{y}^*) \cap D_u\} \geqslant V\{(A+\mathbf{y}) \cap D_u\}$$

for every $u > 0$. After this inequality is established, the rest of the proof follows similarly from the proof of Theorem 4.1.1.

Since A, D_u are convex and $y_j = y_j^*$ for $j=2,\ldots,k$, there exists a set A_0 (which is the projection of $A+\mathbf{y}$ or $A+\mathbf{y}^*$ onto $\mathbf{x}_2$) such that

$$V\{(A+\mathbf{y}^*)\cap D_u\} = \int_{A_0} [a_2(\mathbf{x}_2) - a_1(\mathbf{x}_2)]\, d\mathbf{x}_2,$$

$$V\{(A+\mathbf{y})\cap D_u\} = \int_{A_0} [b_2(\mathbf{x}_2) - b_1(\mathbf{x}_2)]\, d\mathbf{x}_2,$$

where $\mathbf{x}_2 = (x_2,\ldots,x_k)'$. Now for every fixed $\mathbf{x}_2$ let us define

$$\begin{aligned} c_1(\mathbf{x}_2,\mathbf{y}^*) &= \inf\{x_1 | (x_1,\mathbf{x}_2)' \in (A+\mathbf{y}^*)\},\\ c_2(\mathbf{x}_2,\mathbf{y}^*) &= \sup\{x_1 | (x_1,\mathbf{x}_2)' \in (A+\mathbf{y}^*)],\\ d_1(\mathbf{x}_2) &= \inf\{x_1 | (x_1,\mathbf{x}_2)' \in D_u\},\\ d_2(\mathbf{x}_2) &= \sup\{x_1 | (x_1,\mathbf{x}_2)' \in D_u\}. \end{aligned}$$

We consider the following possible cases:

$$\begin{array}{llll} \text{(a)} & c_1(\mathbf{x}_2,\mathbf{y}^*) \leqslant d_1(\mathbf{x}_2) & \text{and} & c_2(\mathbf{x}_2,\mathbf{y}^*) \geqslant d_2(\mathbf{x}_2),\\ \text{(b)} & c_1(\mathbf{x}_2,\mathbf{y}^*) \geqslant d_1(\mathbf{x}_2) & \text{and} & c_2(\mathbf{x}_2,\mathbf{y}^*) \leqslant d_2(\mathbf{x}_2),\\ \text{(c)} & c_1(\mathbf{x}_2,\mathbf{y}^*) \geqslant d_1(\mathbf{x}_2) & \text{and} & c_2(\mathbf{x}_2,\mathbf{y}^*) \geqslant d_2(\mathbf{x}_2),\\ \text{(d)} & c_1(\mathbf{x}_2,\mathbf{y}^*) < d_1(\mathbf{x}_2) & \text{and} & c_2(\mathbf{x}_2,\mathbf{y}^*) < d_2(\mathbf{x}_2). \end{array}$$

Case (d) can be ruled out because both A and D_u are symmetric in x_1 and $y_1^* \geqslant 0$. But in each of the other cases we always have

$$\Delta(\mathbf{x}_2) = [a_2(\mathbf{x}_2) - a_1(\mathbf{x}_2)] - [b_2(\mathbf{x}_2) - b_1(\mathbf{x}_2)] \geqslant 0.$$

This is so because by $y_1 > y_1^* \geqslant 0$ we have

$$c_1(\mathbf{x}_2,\mathbf{y}) > c_1(\mathbf{x}_2,\mathbf{y}^*)$$

and

$$c_2(\mathbf{x}_2,\mathbf{y}) - c_1(\mathbf{x}_2,\mathbf{y}) = c_2(\mathbf{x}_2,\mathbf{y}^*) - c_1(\mathbf{x}_2,\mathbf{y}^*).$$

Therefore

$$V\{(A+\mathbf{y}^*)\cap D_u\} - V\{(A+\mathbf{y})\cap D_u\} = \int_{A_0} \Delta(\mathbf{x}_2)\, d\mathbf{x}_2 \geqslant 0,$$

and the proof of the theorem is thus complete. ■

From Theorem 4.1.4 we can obtain a new inequality for the probability content of independent random variables whose distributions are continuous and symmetric about the origin.

Theorem 4.1.5. Let $X_1,\ldots,X_k$ be independent random variables with continuous density functions $f_i(x_i)$, where $f_i(x_i)=f_i(-x_i)$ and $f_i(x_i)$ is log-concave $(i=1,\ldots,k)$. Let $A\subset\mathfrak{R}^k$ be sign invariant and convex. If $|\mathbf{y}|>|\mathbf{y}^*|$, then the inequality

$$P[\mathbf{X}\in(A+\mathbf{y}^*)] \geqslant P[\mathbf{X}\in(A+\mathbf{y})] \tag{4.1.17}$$

holds.

Proof. By Theorem 4.1.2 the joint density function $f(\mathbf{x})=\prod_{i=1}^k f_i(x_i)$ of $\mathbf{X}$ is unimodal. It is clear that the other conditions in Theorem 4.1.4 are also satisfied; hence the proof follows immediately from Theorem 4.1.4. ■

Corollary. Let $X_1,\ldots,X_k$ be independent random variables with continuous densities $g_{\theta_i}(x_i)=g(x_i-\theta_i)$ (a location parameter family) for $i=1,\ldots,k$ such that $g(x)=g(-x)$ and $g(x)$ is log-concave. Let $A\subset\mathfrak{R}^k$ be sign invariant and convex. If $|\boldsymbol{\xi}|>|\boldsymbol{\theta}|$, then

$$P_{\boldsymbol{\theta}}[\mathbf{X}\in A] \geqslant P_{\boldsymbol{\xi}}[\mathbf{X}\in A]. \tag{4.1.18}$$

As a final note, we observe that this corollary applies to the normal distribution.

4.2. GENERALIZATIONS OF ANDERSON'S THEOREM

We now consider several different versions of generalizations of Anderson's theorem. A result of Sherman (1955) concerns mainly the convolutions of unimodal functions. Applying the properties of convex sets and the Brunn–Minkowski theorem, he first proved

Lemma 4.2.1. Let A and D_u be convex sets in $\mathfrak{R}^k$, and define $V\{A\}$ to be the volume of A. Then the set

$$B_v = \{\mathbf{y}\,|\,V\{(A+\mathbf{y})\cap D_u\}\geqslant v\}$$

is convex for all $v\geqslant 0$. If in addition A and D_u are symmetric about the origin, then B_v is also symmetric about the origin for all $v\geqslant 0$.

Now let $\mathcal{C}_0$, $\mathcal{C}_1$ denote the closed convex cones generated by the indicator functions of all symmetric, compact, and convex sets in $\mathfrak{R}^k$, respectively, in the uniform norm ($\|f\|_\infty = \sup|f(\mathbf{x})|$) and in the L_1 norm. Let $\mathcal{C}_3$ be the closed convex cone generated in the $\|f_3\|$ norm by the same indicator functions, where $\|f\|_3 = \max\{\|f\|_\infty, \|f\|_1\}$ (i.e., $\mathcal{C}_3$ is generated in both the uniform norm and the L_1 norm). For two symmetric, compact, and convex sets A_1, A_2 in $\mathcal{R}^k$ the convolution of their indicator functions is given by

$$\chi_{A_1} * \chi_{A_2}(\mathbf{x}) = \int \chi_{A_1}(\mathbf{x}-\mathbf{t})\chi_{A_2}(\mathbf{t})\, d\mathbf{t}.$$

Applying Lemma 4.2.1 and the continuity property of convolution in the L_1 and $\|f\|_3$ norms, Sherman (1955) proved that

Theorem 4.2.1. For $i=1,3$ if $f_1, f_2 \in \mathcal{C}_i$, then $f_1 * f_2 \in \mathcal{C}_i$, where

$$f_1 * f_2(\mathbf{x}) = \int f_1(\mathbf{x}-\mathbf{t}) f_2(\mathbf{t})\, d\mathbf{t}. \tag{4.2.1}$$

This theorem says basically that the convolution of two symmetric unimodal functions in $\mathcal{R}^k$ is unimodal. (It should be noted that without the symmetry condition even in the univariate case the assertion is false. A counter-example on this (see Problem 5) was given by Chung (1953), and it was later included in Appendix II of Gnedenko and Kolmogorov as translated by Chung (1968, p. 254).) Now for $\lambda \in [0,1]$ the indicator function h of any symmetric convex set satisfies $h(\lambda\mathbf{y}) \geqslant h(\mathbf{y})$, and also holds for any convex combination of them. Since this property is preserved by taking uniform limits, it follows that $h(\lambda\mathbf{y}) \geqslant h(\mathbf{y})$ holds for every $h \in \mathcal{C}_3$ and every $\mathbf{y}$. This implies Anderson's theorem as a special case because for f, A satisfying the conditions specified in Theorem 4.1.1 the function

$$h(\mathbf{y}) = \int_A f(\mathbf{t}+\mathbf{y})\, d\mathbf{t} = \int f(\mathbf{y}-\mathbf{t})\chi_A(\mathbf{t})\, d\mathbf{t}$$

is in $\mathcal{C}_3$.

Another interesting and useful generalization of Anderson's theorem was given by Mudholkar (1966), and the generalization was made to include other symmetry properties. To motivate his approach let us first consider an example that is a special case of his general result. We shall adopt the following notation: Let g: $\mathcal{R}^k \to \mathcal{R}^k$ be a Borel-measurable transformation. Then for $A \subset \mathcal{R}^k$

$$g(A) = \{\mathbf{z} | \mathbf{z} = g(\mathbf{x}), \mathbf{x} \in A\}. \tag{4.2.2}$$

Example 4.2.1. Let $\mathbf{x}\in\mathcal{R}^k$ and $\mathcal{G}=\{g_1,g_2\}$ be a group of transformations such that $g_1(\mathbf{x})=\mathbf{x}, g_2(\mathbf{x})=-\mathbf{x}$ (hence $g_i=g_j^{-1}$, $i\neq j$). It is easy to check that the symmetry conditions on f and A given in Theorem 4.1.1 can be restated as f and A are $\mathcal{G}$-invariant (i.e., $f(\mathbf{x})=f(g(\mathbf{x}))$ and $\mathbf{x}\in A$ implies $g(\mathbf{x})\in A$ for all $g\in\mathcal{G}$). Also, for every $\mathbf{y}$ and $\lambda\in[0,1]$ there exist two real numbers $\alpha_1,\alpha_2\in[0,1]$ such that $\alpha_1+\alpha_2=1$ and $\lambda\mathbf{y}=\Sigma_{i=1}^2\alpha_i g_i(\mathbf{y})=\boldsymbol{\alpha}(\mathbf{y})$ (say). Therefore Theorem 4.1.1 can be restated in the following way: If f is unimodal, A is convex, and f, A are $\mathcal{G}$-invariant, then

$$\int_A f(\mathbf{x}+\boldsymbol{\alpha}(\mathbf{y}))\,d\mathbf{x} \geqslant \int_A f(\mathbf{x}+\mathbf{y})\,d\mathbf{x}.$$

This example illustrates how the symmetry property in Theorem 4.1.1 can be replaced by a more general invariance property. For a finite positive integer N let $\mathcal{G}=\{g_1,\ldots,g_N\}$ denote a finite group of Borel-measurable, Lebesgue-measure-preserving† linear transformations from $\mathcal{R}^k$ onto $\mathcal{R}^k$. We say that f is $\mathcal{G}$-invariant if $f(\mathbf{x})=f(g(\mathbf{x}))$ for all $\mathbf{x}\in\mathcal{R}^k$ and all $g\in\mathcal{G}$, and that A is $\mathcal{G}$-invariant if $\mathbf{x}\in A$ implies $g(\mathbf{x})\in A$ for all $g\in\mathcal{G}$. With this new invariance property Mudholkar (1966) proved

Theorem 4.2.2. Let $\mathcal{G}$ be a finite group of Borel-measurable, Lebesgue-measure-preserving linear transformations from $\mathcal{R}^k$ onto $\mathcal{R}^k$. Let $f(\mathbf{x})$: $\mathcal{R}^k\to[0,\infty)$ be $\mathcal{G}$-invariant and unimodal. Further, let $A\subset\mathcal{R}^k$ be $\mathcal{G}$-invariant‡ and convex. If $\int_A f(\mathbf{x})\,d\mathbf{x}<\infty$ (in the Lebesgue sense), then

$$\int_A f(\mathbf{x}+\boldsymbol{\alpha}(\mathbf{y}))\,d\mathbf{x} \geqslant \int_A f(\mathbf{x}+\mathbf{y})\,d\mathbf{x} \tag{4.2.3}$$

holds for all $\mathbf{y}\in\mathcal{R}^k$ and all $\boldsymbol{\alpha}$ satisfying

$$\boldsymbol{\alpha}(\mathbf{y}) = \sum_{i=1}^{N}\alpha_i g_i(\mathbf{y}), \qquad \alpha_i \geqslant 0, \quad \sum_{i=1}^{N}\alpha_i = 1; \tag{4.2.4}$$

or equivalently,

$$\int_{A+\boldsymbol{\alpha}(\mathbf{y})} f(\mathbf{x})\,d\mathbf{x} \geqslant \int_{A+\mathbf{y}} f(\mathbf{x})\,d\mathbf{x}. \tag{4.2.5}$$

†This means that for every $g\in\mathcal{G}$ and every Lebesgue-measurable set A the volume of A equals that of $g(A)$.

‡Since $\mathcal{G}$ is a group, A is $\mathcal{G}$-invariant implies $g(A)=A$ for all $g\in\mathcal{G}$.

The proof of this theorem is similar to that of Theorem 4.1.1, except that (4.1.8) is replaced by

$$[(A+\boldsymbol{\alpha}(\mathbf{y}))\cap D_u] \supset \sum_{i=1}^{N} \alpha_i[(A+g_i(\mathbf{y}))\cap D_u],$$

which follows from the invariance property and the convexity of A and D_u. The details are omitted.

The following theorem, also due to Mudholkar (1966), follows from Theorem 4.2.2. The proof is similar to that of Theorem 4.1.3; hence it is not given here.

Theorem 4.2.3. Let $\mathbf{X}$ have density function $f(\mathbf{x})$, and assume that $\mathcal{G}$, f, A, and $\boldsymbol{\alpha}$ satisfy the conditions imposed in Theorem 4.2.2. Let $\mathbf{Y}$ be independent of $\mathbf{X}$. Then

$$P[(\mathbf{X}+\boldsymbol{\alpha}(\mathbf{Y}))\in A] \geqslant P[(\mathbf{X}+\mathbf{Y})\in A]. \tag{4.2.6}$$

In general let ψ: $\mathcal{R}^k \to \mathcal{R}^1$ be $\mathcal{G}$-invariant such that $\{\mathbf{x}|\psi(\mathbf{x})\leqslant v\}$ is convex; then

$$P[\psi(\mathbf{X}+\boldsymbol{\alpha}(\mathbf{Y}))\leqslant v] \geqslant P[\psi(\mathbf{X}+\mathbf{Y})\leqslant v]. \tag{4.2.7}$$

This implies

$$P[\psi(\mathbf{X}+\boldsymbol{\alpha}(\mathbf{y}))\leqslant v] \geqslant P[\psi(\mathbf{X}+\mathbf{y})\leqslant v] \tag{4.2.8}$$

for all $\mathbf{y}\in\mathcal{R}^k$.

It is clear that Theorems 4.2.2 and 4.2.3 are generalizations of Theorems 4.1.1 and 4.1.3, respectively, by choosing $\mathcal{G}$ to be a group of sign-change transformations. In addition, new inequalities can be obtained from Theorem 4.2.2 by considering other groups of transformations. An important special case is that in which, as pointed out by Mudholkar (1966), $\mathcal{G}$ is the group of permutations of coordinates. Because of a known result on the linear transformation of a vector through a doubly stochastic matrix, probability inequalities can then be derived via majorization. This deserves special attention, and it will be discussed in Chapter 6.

A multivariate version of Anderson's theorem was obtained by Das Gupta (1976a). Let us consider two finite groups of measurable one-to-one transformations $\mathcal{G}_1$, $\mathcal{G}_2$ from $\mathcal{R}^k$ onto $\mathcal{R}^k$ and from $\mathcal{R}^m$ onto $\mathcal{R}^m$, respectively. Let $\mathcal{G}^*$ denote a subgroup of $\mathcal{G}_1\times\mathcal{G}_2$ such that the following conditions are satisfied:

(1) $\mathcal{G}_1$ is Lebesgue-measure-preserving, and

(2) for every $g_2 \in \mathcal{G}_2$ (which contains N elements) there exists a $g_1 \in \mathcal{G}_1$ such that $(g_1, g_2) \in \mathcal{G}^*$.

Das Gupta proved

Theorem 4.2.4. Let $f_j(\mathbf{x},\mathbf{y})$ $(j=1,\ldots,n)$ be $\mathcal{G}^*$-invariant and unimodal functions on $\mathcal{R}^k \times \mathcal{R}^m$ for $\mathbf{x} \in \mathcal{R}^k$, $\mathbf{y} \in \mathcal{R}^m$; and for each $\mathbf{y}_1,\ldots,\mathbf{y}_n$ in $\mathcal{R}^m$ assume that

$$h(\mathbf{y}_1,\ldots,\mathbf{y}_n) = \int_{\mathcal{R}^k} \prod_{j=1}^{n} f_j(\mathbf{x},\mathbf{y}_j)\, d\mathbf{x} < \infty \qquad \text{(in the Lebesgue sense).}$$

Then

$$h(g_2(\mathbf{y}_1),\ldots,g_2(\mathbf{y}_n)) = h(\mathbf{y}_1,\ldots,\mathbf{y}_n) \tag{4.2.9}$$

holds for all $g_2 \in \mathcal{G}_2$, and

$$h(\mathbf{y}_1^*,\ldots,\mathbf{y}_n^*) \geqslant h(\mathbf{y}_1,\ldots,\mathbf{y}_n), \tag{4.2.10}$$

where for $j=1,\ldots,n$

$$\mathbf{y}_j^* = \sum_{i=1}^{N} \alpha_i g_{2i}(\mathbf{y}_j), \qquad \alpha_i \geqslant 0, \quad \sum_{i=1}^{N} \alpha_i = 1. \tag{4.2.11}$$

The proof of the theorem follows steps similar to those given in the proofs of Theorems 4.1.1 and 4.2.2. At various stages Fubini's theorem, the Brunn–Minkowski theorem, and the condition of convexity are used. The details are not given here.

Theorem 4.2.2 (hence Theorem 4.1.1) follows from this theorem as a special case by taking $k=m$, $n=2$, $\mathbf{y}_1=\mathbf{y}$, $f_1(\mathbf{x},\mathbf{y}_1)=f(\mathbf{x}+\mathbf{y})$, and $f_2(\mathbf{x},\mathbf{y}_2)=\chi_A(\mathbf{x})$, which is the indicator function of the set A. (Here A is a $\mathcal{G}^*$-invariant and convex set in $\mathcal{R}^k$.) Also, Das Gupta (1976a, b) showed that for the general case a marginal function of a unimodal function need not be unimodal even when the symmetry condition is satisfied.

4.3. INEQUALITIES FOR ELLIPTICALLY CONTOURED DISTRIBUTIONS

In their paper Das Gupta, Eaton, Olkin, Perlman, Savage, and Sobel (1972) gave a number of probability inequalities for a class of elliptically contoured distributions. Let $\boldsymbol{\Sigma}=(\sigma_{ij})$ denote a $k \times k$

positive definite matrix. They called a density function $f(\mathbf{x})$ elliptically contoured if $f(\mathbf{x})$ is of the form (with respect to Lebesgue measure)

$$f(\mathbf{x}) = |\mathbf{\Sigma}|^{-1/2} g(\mathbf{x}'\mathbf{\Sigma}^{-1}\mathbf{x}), \tag{4.3.1}$$

where g satisfies

$$\int_0^{\infty} r^{k-1} g(r^2)\, dr < \infty. \tag{4.3.2}$$

Among many other distributions, the multivariate normal and multivariate t distributions (with density given in (3.1.2)) are members of this family. Certain properties of this family of distributions were given by Kelker (1970). A basic property is that if $\mathbf{X} = (\mathbf{X}^{(1)}, \mathbf{X}^{(2)})'$ has an elliptically contoured distribution with parameter matrix $\mathbf{\Sigma}$, which is partitioned in the form

$$\mathbf{\Sigma} = \begin{pmatrix} \mathbf{\Sigma}_{11} & \mathbf{\Sigma}_{12} \\ \mathbf{\Sigma}_{21} & \mathbf{\Sigma}_{22} \end{pmatrix}, \tag{4.3.3}$$

then the marginal distributions of $\mathbf{X}^{(1)}$ and $\mathbf{X}^{(2)}$ are again elliptically contoured, and they depend on $\mathbf{\Sigma}$ only through $\mathbf{\Sigma}_{11}$ and $\mathbf{\Sigma}_{22}$, respectively. This follows from the identity

$$\begin{aligned} \mathbf{x}'\mathbf{\Sigma}^{-1}\mathbf{x} &= (\mathbf{x}^{(1)})'\mathbf{\Sigma}_{11}^{-1}\mathbf{x}^{(1)} \\ &\quad + \left(\mathbf{x}^{(2)} - \mathbf{\Sigma}_{21}\mathbf{\Sigma}_{11}^{-1}\mathbf{x}^{(1)}\right)'\mathbf{\Sigma}_{22\cdot1}^{-1}\left(\mathbf{x}^{(2)} - \mathbf{\Sigma}_{21}\mathbf{\Sigma}_{11}^{-1}\mathbf{x}^{(1)}\right), \end{aligned} \tag{4.3.4}$$

where

$$\mathbf{\Sigma}_{22\cdot1} = \mathbf{\Sigma}_{22} - \mathbf{\Sigma}_{21}\mathbf{\Sigma}_{11}^{-1}\mathbf{\Sigma}_{12}. \tag{4.3.5}$$

Similar things can be said for the conditional distributions.

The main theorem given by Das Gupta, Eaton, Olkin, Perlman, Savage, and Sobel (1972) is a generalization of a theorem of Šidák (see Theorem 2.2.5). The generalization was done along two different directions: The underlying distribution is elliptically contoured but not necessarily normal, and the set A in $\mathfrak{R}^{k-1}$ is not necessarily rectangular. For comparison and notational consistency we shall partition the elements of $\mathbf{X}$ and $\mathbf{\Sigma}$ into the form (which is identical to that given in Theorem 2.2.5)

$$\mathbf{X} = \begin{pmatrix} X_1 \\ \mathbf{X}^{(2)} \end{pmatrix}, \qquad \mathbf{\Sigma} = \begin{pmatrix} \sigma_{11} & \mathbf{\Sigma}_{12} \\ \mathbf{\Sigma}_{21} & \mathbf{\Sigma}_{22} \end{pmatrix}, \tag{4.3.6}$$

where $\mathbf{X}^{(2)}$ is $(k-1)\times 1$ and $\mathbf{\Sigma}_{22}$ is $(k-1)\times(k-1)$.

Theorem 4.3.1. Let $\mathbf{X}$ (given in (4.3.6)) have a density function $f(\mathbf{x})$ which is elliptically contoured, and for $\lambda \in [0,1]$ define

$$\boldsymbol{\Sigma}_\lambda = \begin{pmatrix} \sigma_{11} & \lambda\boldsymbol{\Sigma}_{12} \\ \lambda\boldsymbol{\Sigma}_{21} & \boldsymbol{\Sigma}_{22} \end{pmatrix}. \tag{4.3.7}$$

If $A \subset \mathscr{R}^{k-1}$ is symmetric about the origin and convex, then the probability

$$P_{\boldsymbol{\Sigma}_\lambda}\left[|X_1| \leqslant a, \mathbf{X}^{(2)} \in A\right]$$

is nondecreasing in λ.

Outline of the Proof. The proof of this theorem is, of course, entirely different from that given in Theorem 2.2.5 because the condition of normality and its properties (see Lemma 2.1.1) are no longer at our disposal. The proof given by Das Gupta, Eaton, Olkin, Perlman, Savage and Sobel depends basically on the following steps:

(1) Since A can be expressed as the intersection of a decreasing sequence of symmetric convex polyhedrons and since the inequality is preserved when passing to the limit, it suffices to consider the case in which A is such a polyhedron. Also, since $\boldsymbol{\Sigma}_{22}$ is positive definite, there exists a nonsingular matrix $\mathbf{M}$ such that $\boldsymbol{\Sigma}_{22} = \mathbf{M}\mathbf{M}'$. With this $\mathbf{M}$, if we make the transformation $X_1 \to X_1/(\sigma_{11})^{1/2}$ and $\mathbf{X}^{(2)} \to \mathbf{M}^{-1}\mathbf{X}^{(2)}$, then this transformation leaves the hypothesis of the theorem unchanged. Hence for convenience it may be assumed that $\boldsymbol{\Sigma}_\lambda$ is of the form

$$\boldsymbol{\Sigma}_\lambda = \begin{pmatrix} 1 & \boldsymbol{\Lambda}' \\ \boldsymbol{\Lambda} & \mathbf{I} \end{pmatrix}, \qquad \boldsymbol{\Lambda} = (\lambda, 0, \ldots, 0)'.$$

(2) Clearly this $\boldsymbol{\Sigma}_\lambda$ can be written as $\boldsymbol{\Sigma}_\lambda = \mathbf{T}_\lambda \mathbf{T}_\lambda'$, where

$$\mathbf{T}_\lambda = \begin{pmatrix} (1-\lambda^2)^{1/2} & \boldsymbol{\Lambda}' \\ \mathbf{0} & \mathbf{I} \end{pmatrix}.$$

Let $\mathbf{Y}$ be a linear transformation of $\mathbf{X}$ given by

$$\mathbf{Y} = (Y_1, Y_2, \mathbf{Y}^*)' = \mathbf{T}_\lambda^{-1}\mathbf{X}. \tag{4.3.8}$$

Then the probability content under consideration can be written as the expectation of a conditional probability function $\xi(\lambda)$

$$\begin{aligned} &P_{\Sigma_\lambda}\left[|X_1| \leqslant a, \mathbf{X}^{(2)} \in A\right] \\ &\quad = P\left[|(1-\lambda^2)^{1/2} Y_1 + \lambda Y_2| \leqslant a, (Y_2, \mathbf{Y}^*)' \in A\right] \\ &\quad = E\left\{ P\left[|(1-\lambda^2)^{1/2} Y_1 + \lambda Y_2| \leqslant a, (Y_2, \mathbf{Y}^*) \in A \,\Big|\, \left(\sum_{i=1}^{k} Y_i^2 \right)^{1/2} = r \right] \right\} \\ &\quad \equiv E\xi(\lambda). \end{aligned} \tag{4.3.9}$$

Let S denote the surface of the sphere with radius $r = (\sum_{i=1}^k Y_i^2)^{1/2}$. From the fact that the density function of $\mathbf{Y}$ is $g(\sum_{i=1}^k y_i^2)$ the conditional density of $\mathbf{Y}$, given $(\sum_{i=1}^k Y_i^2)^{1/2} = r$, is uniform on S. Let η denote this uniform surface measure on S; then the conditional probability function in (4.3.9) is of the form

$$\xi(\lambda) = \int_S \chi_{[-a,a]}\left[(1-\lambda^2)^{1/2} y_1 + \lambda y_2\right] \chi_A(y_2, \mathbf{y}^*)\, d\eta(\mathbf{y}), \tag{4.3.10}$$

where χ denotes the indicator function.

(3) Now comes the crucial part of the proof: Using the facts concerning approximate identity under convolutions, one constructs (for fixed r) a sequence of probability functions $\{\xi_\varepsilon(\lambda) : \varepsilon > 0\}$ such that $\xi_\varepsilon(\lambda) \to \xi(\lambda)$ as $\varepsilon \to 0$, except on the set of boundary points, which has probability measure zero. Then by showing that

$$\frac{d}{d\lambda} \xi_\varepsilon(\lambda) \geqslant 0 \qquad \text{for every} \quad \varepsilon > 0,$$

one concludes that $(d/d\lambda)\xi(\lambda) \geqslant 0$, and the proof of the theorem follows. For $i = 1,2$ let f_i: $\mathcal{R}^n \to [0,\infty)$ satisfy $\int f_i(\mathbf{x})\, d\mathbf{x} < \infty$, and let $f_1 * f_2$ denote the convolution as defined in (4.2.1). For $n = 1$ and $n = k-1$ consider the approximate identity given by

$$h_\varepsilon^{(n)}(\mathbf{x}) = (2\pi\varepsilon)^{-n/2} \exp\left(-\sum_{i=1}^{n} \frac{x_i^2}{2\varepsilon} \right), \tag{4.3.11}$$

and the functions

$$\zeta_\varepsilon = \chi_{[-a,a]} * h_\varepsilon^{(1)}, \qquad \psi_\varepsilon = \chi_A * h_\varepsilon^{k-1}. \tag{4.3.12}$$

Then define

$$\xi_\varepsilon(\lambda) = \int_S \zeta_\varepsilon\left[(1-\lambda^2)^{1/2}y_1 + \lambda y_2\right]\psi_\varepsilon(y_2, \mathbf{y}^*)\, d\eta(\mathbf{y}). \qquad (4.3.13)$$

It follows that (a) both ζ_ε and ψ_ε are unimodal functions, and (b) as $\varepsilon \to 0$, $\zeta_\varepsilon(\mathbf{y}) \to \chi_{[-a,a]}(\mathbf{y})$ and $\psi_\varepsilon(\mathbf{y}) \to \chi_A(\mathbf{y})$ for all $\mathbf{y}$ that are not boundary points; hence the dominated convergence theorem implies that $\xi_\varepsilon(\lambda) \to \xi(\lambda)$ as $\varepsilon \to 0$.

(4) It now suffices to show that

$$\frac{d}{d\lambda}\xi_\varepsilon(\lambda) \geqslant 0.$$

This involves another transformation of the variables, differentiation under the integral sign (which is permissible), and an application of the Anderson–Sherman theorem (Theorem 4.2.1). The details are omitted. ■

The following corollary is a generalization of Corollary 1 of Theorem 2.2.5. Its proof is immediate.

Corollary. Let $\mathbf{X}$ have a density $f(\mathbf{x})$, which is elliptically contoured as defined in (4.3.1), and for $\lambda_1, \ldots, \lambda_k$ in $[0,1]$ and $\boldsymbol{\lambda} = (\lambda_1, \ldots, \lambda_k)'$ define $\boldsymbol{\Sigma}_{\boldsymbol{\lambda}}$ the covariance matrix such that the (i,j)th element in $\boldsymbol{\Sigma}$ is replaced by $\lambda_i\lambda_j\sigma_{ij}$, $i \neq j$. Then the probability

$$P_{\boldsymbol{\Sigma}_{\boldsymbol{\lambda}}}\left[\bigcap_{i=1}^{k} \{|X_i| \leqslant a_i\}\right]$$

is nondecreasing in λ_i for each i.

As noted by Das Gupta, Eaton, Olkin, Perlman, Savage, and Sobel, Theorem 4.3.1 is rich in implications. In their paper they applied this theorem to obtain certain new inequalities. Some of them are described below.

Theorem 4.3.2. Under the same notation and assumptions as those stated in Theorem 4.3.1, the probability

$$P_{\boldsymbol{\Sigma}_\lambda}\left[|X_1| > a, \mathbf{X}^{(2)} \notin A\right]$$

is nondecreasing in λ.

Proof. This follows from Theorem 4.3.1, the identity

$$P(A \cap B) - P(A)P(B) = P(A^c \cap B^c) - P(A^c)P(B^c) \qquad (4.3.14)$$

for all events A, B (A^c indicates the complement of A), and the fact that the marginal distributions of X_1 and $\mathbf{X}^{(2)}$ depend on $\boldsymbol{\Sigma}$ only through σ_{11} and $\boldsymbol{\Sigma}_{22}$, respectively. ■

The following inequality offers an upper bound for rectangular probability contents of elliptically contoured distributions. Its proof depends on a particular transformation of the variables and an application of Theorem 4.1.1. Note that in this theorem the condition specified in (4.3.2) is not essential.

Theorem 4.3.3. Let $\mathbf{X}$ have a density $f(\mathbf{x})$ of the form in (4.3.1), where $\boldsymbol{\Sigma}=(\rho_{ij})$ is a positive definite correlation matrix. If g is monotonically nonincreasing, then

$$P_{\boldsymbol{\Sigma}}\left[\bigcap_{i=1}^{k}\{|X_i|\leqslant a_i\}\right]\leqslant P_{\mathbf{I}}\left[\bigcap_{i=1}^{k}\{|X_i|\leqslant a_i/\gamma_i\}\right],\qquad(4.3.15)$$

where $\mathbf{I}$ is the identity matrix and $\gamma_1=1$, $\gamma_m^2=|\boldsymbol{\Sigma}(m)|/|\boldsymbol{\Sigma}(m-1)|$ with $\boldsymbol{\Sigma}(m)=(\rho_{ij})$, $1\leqslant i,j\leqslant m$, $2\leqslant m\leqslant k$.

Proof. Let $\mathbf{T}=(\tau_{ij})$ denote an upper triangular matrix satisfying $\mathbf{T}\mathbf{T}'=\boldsymbol{\Sigma}$, and let $\mathbf{Y}=\mathbf{T}^{-1}\mathbf{X}$. Then the density of $\mathbf{Y}$ is $g(\Sigma_{i=1}^{k}y_i^2)$, and the probability content can be written as

$$\begin{aligned}P_{\boldsymbol{\Sigma}}\left[\bigcap_{i=1}^{k}\{|X_i|\leqslant a_i\}\right]&=P_{\mathbf{I}}\left[\bigcap_{i=1}^{k}\left\{\left|\sum_{j=i}^{k}\tau_{ij}Y_j\right|\leqslant a_i\right\}\right]\\&=\int\cdots\int_B\left[\int_H g(y_1^2+\cdots+y_k^2)\,dy_1\right]\prod_{i=2}^{k}dy_i;\end{aligned}\qquad(4.3.16)$$

where

$$B=\left\{(y_2,\ldots,y_k)'\left|\,\left|\sum_{j=i}^{k}\tau_{ij}y_j\right|\leqslant a_i,\ i=2,\ldots,k\right.\right\}$$

and for fixed $y_2,\ldots,y_k$

$$H=\left\{y_1\left|\,\left|\sum_{j=1}^{k}\tau_{1j}y_j\right|\leqslant a_1\right.\right\}.$$

Note that H is a shift in location of the symmetric interval

$H_0=\{y_1||\tau_{11}y_1|\leqslant a_1\}$. If g is monotonically nonincreasing, then the following inequality holds

$$\int_H g(y_1^2+\cdots+y_k^2)\,dy_1 \leqslant \int_{H_0} g(y_1^2+\cdots+y_k^2)\,dy_1.$$

By a repetitive argument when integrating out one variable at a time this yields

$$P_{\mathbf{I}}\left[\bigcap_{i=1}^{k}\left\{\left|\sum_{j=i}^{k}\tau_{ij}Y_j\right|\leqslant a_i\right\}\right] \leqslant P_{\mathbf{I}}\left[\bigcap_{i=1}^{k}\{|\tau_{ii}Y_i|\leqslant a_i\}\right]. \quad (4.3.17)$$

Now let $\mathbf{T}(m)$ denote the matrix (τ_{ij}) for $1\leqslant i, j\leqslant m$. Since $\mathbf{T}$ is an upper triangular matrix, we have for every m

$$|\mathbf{\Sigma}(m)| = |\mathbf{T}(m)||\mathbf{T}'(m)| = \prod_{i=1}^{m}\tau_{ii}^2.$$

Therefore $\tau_{11}=1$ and $\tau_{mm}^2=|\mathbf{\Sigma}(m)|/|\mathbf{\Sigma}(m-1)=\gamma_m^2$ for $m\geqslant 2$. This, together with (4.3.16) and (4.3.17), completes the proof. ■

Remark. For the special case in which $\mathbf{X}$ is a multivariate normal variable with mean vector $\mathbf{0}$ and covariance matrix $\mathbf{\Sigma}$, where the variances are one and the correlations are $\rho\geqslant 0$, it is known that (see Problem 10, Chapter 2)

$$P\left[\bigcap_{i=1}^{k}\{|X_i|\leqslant a_i\}\right] \leqslant \prod_{i=1}^{k}P\left[|X_i|\leqslant a_i/(1-\rho)^{1/2}\right]. \quad (4.3.18)$$

On the other hand, the bound in (4.3.15) is given with the γ values $\gamma_1=1$ and

$$\gamma_i=\left[\{1+(i-1)\rho\}/\{1+(i-2)\rho\}\right]^{1/2}(1-\rho)^{1/2}, \qquad i=2,\ldots,k. \quad (4.3.19)$$

This is so because the determinant of the matrix $\mathbf{\Sigma}(m)$ with equal correlations ρ is $(1-\rho)^{m-1}[1+(m-1)\rho]$ for $m\geqslant 2$. Since $(1/\gamma_i)<1/(1-\rho)^{1/2}$ holds for all $i\geqslant 2$ when $\rho>0$, the bound given in (4.3.15) is sharper for all $a_1,\ldots,a_k$.

The next theorem is a generalization of the corollary to Theorem 4.1.3 given in Section 4.1.

Theorem 4.3.4. Let $\mathbf{X}$ have a density $f(\mathbf{x})$ that is elliptically contoured, and assume that $A\subset\mathfrak{R}^k$ is symmetric about the origin and convex. If $\mathbf{\Sigma}_1$ and $\mathbf{\Sigma}_2$ are positive definite matrices such that $\mathbf{\Sigma}_2-\mathbf{\Sigma}_1$ is positive semidefinite, then

$$P_{\mathbf{\Sigma}=\mathbf{\Sigma}_1}[\mathbf{X}\in A] \geqslant P_{\mathbf{\Sigma}=\mathbf{\Sigma}_2}[\mathbf{X}\in A]. \tag{4.3.20}$$

Proof. Let g (as defined in (4.3.1)), $\mathbf{\Sigma}_1$, and $\mathbf{\Sigma}_2$ be given such that (4.3.2) is satisfied. Then there exists a smooth function h with a compact support such that the integral

$$\int_0^\infty r^{k-1}|g(r^2)-h(r^2)|\,dr \tag{4.3.21}$$

can be made arbitrarily small, so that

$$\int |\mathbf{\Sigma}|^{-1/2}|g(\mathbf{x}'\mathbf{\Sigma}^{-1}\mathbf{x})-h(\mathbf{x}'\mathbf{\Sigma}^{-1}\mathbf{x})|\,d\mathbf{x}$$

can be made arbitrarily small simultaneously for all $\mathbf{\Sigma}$. Therefore it suffices to assume that g is a smooth function with a compact support. The proof of the theorem depends on an inequality for the conditional probability of a new $(k+1)$-dimensional variable $(U,\mathbf{X})'$, which follows from Theorem 4.3.1.

Since $\mathbf{\Sigma}_1-\mathbf{\Sigma}_2$ is positive semidefinite, it may be written as the sum of positive semidefinite matrices with rank one. Consequently, we may simply assume that $\mathbf{\Sigma}_2=\mathbf{\Sigma}_1+\boldsymbol{\alpha}\boldsymbol{\alpha}'$, where $\boldsymbol{\alpha}=(\alpha_1,\ldots,\alpha_k)'$, and then proceed by induction. Let us define two $(k+1)\times(k+1)$ matrices

$$\mathbf{\Lambda}_1=\begin{pmatrix}1 & \boldsymbol{\alpha}'\\ \boldsymbol{\alpha} & \mathbf{\Sigma}_2\end{pmatrix}, \qquad \mathbf{\Lambda}_2=\begin{pmatrix}1 & \mathbf{0}\\ \mathbf{0} & \mathbf{\Sigma}_2\end{pmatrix}. \tag{4.3.22}$$

Choose $c>0$ so that

$$c|\mathbf{\Lambda}_1|^{-1/2}g\left((u,\mathbf{x}')\mathbf{\Lambda}_1^{-1}\binom{u}{\mathbf{x}}\right) \tag{4.3.23}$$

is a density function, and let $(U,\mathbf{X})'$ be the $(k+1)$-dimensional random variable which has this density. Then from (4.3.21) and (4.3.23) this density function is elliptically contoured, and Theorem 4.3.1 implies that

$$P_{\mathbf{\Lambda}_1}[|U|\leqslant a,\mathbf{X}\in A] \geqslant P_{\mathbf{\Lambda}_2}[|U|\leqslant a,\mathbf{X}\in A] \tag{4.3.24}$$

for all $a\geqslant 0$. Dividing both sides by $P[|U|\leqslant a]$ for $a>0$ (which does not depend on $\mathbf{\Lambda}_i$), (4.3.24) reduces to

$$P_{\mathbf{\Lambda}_1}[\mathbf{X}\in A\,||U|\leqslant a] \geqslant P_{\mathbf{\Lambda}_2}[\mathbf{X}\in A\,||U|\leqslant a]. \tag{4.3.25}$$

Then by letting $a\to 0$, we obtain

$$P_{\mathbf{\Lambda}_1}[\mathbf{X}\in A|U=0] \geqslant P_{\mathbf{\Lambda}_2}[\mathbf{X}\in A|U=0]. \tag{4.3.26}$$

But the conditional density of $\mathbf{X}$, given $U=0$ under Λ_i, is $|\boldsymbol{\Sigma}_i|^{-1/2}g(\mathbf{x}'\boldsymbol{\Sigma}_i^{-1}\mathbf{x})$ for $i=1,2$. Hence (4.3.20) follows. ■

An immediate consequence of this theorem is the following corollary.

Corollary. Let $\mathbf{X}$ have an elliptically contoured density that depends on a parameter matrix $\boldsymbol{\Sigma}$. Let $\mathbf{S}$ be a random matrix that is independent of $\mathbf{X}$, is positive definite with probability one, and has a distribution that does not depend on $\boldsymbol{\Sigma}$. If $\boldsymbol{\Sigma}_1$ and $\boldsymbol{\Sigma}_2$ are positive definite and $\boldsymbol{\Sigma}_2-\boldsymbol{\Sigma}_1$ is positive semidefinite, then

$$P_{\boldsymbol{\Sigma}=\boldsymbol{\Sigma}_1}\left[\mathbf{X}'\mathbf{S}^{-1}\mathbf{X}\leqslant a\right] \geqslant P_{\boldsymbol{\Sigma}=\boldsymbol{\Sigma}_2}\left[\mathbf{X}'\mathbf{S}^{-1}\mathbf{X}\leqslant a\right] \tag{4.3.27}$$

holds for all $a>0$.

The proof of this corollary follows by first applying Theorem 4.3.4 to establish the desired inequality (with $A=\mathbf{X}'\mathbf{S}_0^{-1}\mathbf{X}$) for given $\mathbf{S}=\mathbf{S}_0$, then taking expectations on both sides. It is important here that the distribution of $\mathbf{S}$ does not depend on $\boldsymbol{\Sigma}$. Otherwise the same proof does not go through.

We now consider a sequence of independent k-dimensional random variables with covariance structures similar to those described in (3.1.6). For $m=1,\ldots,n$ let $\boldsymbol{\Sigma}(m)=(\sigma_{ij}^{(m)})$ denote a $k\times k$ positive definite matrix and $\boldsymbol{\lambda}(m)=(\lambda_1(m),\ldots,\lambda_k(m))'$ a real vector such that $\lambda_i(m)\in[0,1]$; define $\boldsymbol{\Sigma}_{\boldsymbol{\lambda}(m)}(m)=(\sigma_{ij}^{(m)}(\boldsymbol{\lambda}(m)))$ to be the matrix such that

$$\sigma_{ii}^{(m)}(\boldsymbol{\lambda}(m)) = \sigma_{ii}^{(m)}, \qquad \sigma_{ij}^{(m)}(\boldsymbol{\lambda}(m)) = \lambda_i(m)\lambda_j(m)\sigma_{ij}^{(m)}, \qquad i\neq j. \tag{4.3.28}$$

Furthermore, let $(\mathbf{V}_1,\ldots,\mathbf{V}_n)'$ denote a random vector such that $\mathbf{V}_m=(V_{1m},\ldots,V_{km})'$ for $m=1,\ldots,n$.

Theorem 4.3.5. If $\mathbf{V}_1,\ldots,\mathbf{V}_n$ are independent random variables with elliptically contoured densities

$$|\boldsymbol{\Sigma}_{\boldsymbol{\lambda}(m)}(m)|^{-1/2}g\left(\mathbf{v}'(\boldsymbol{\Sigma}_{\boldsymbol{\lambda}(m)}(m))^{-1}\mathbf{v}\right), \qquad m=1,\ldots,n,$$

then for $t>0$ the probability

$$P\left[\bigcap_{i=1}^{k}\left\{\sum_{m=1}^{n}|V_{im}|^t\leqslant a_i\right\}\right] \tag{4.3.29}$$

is nondecreasing in each $\lambda_i(m)\in[0,1]$.

Proof. The proof is quite similar to that given in Theorem 3.1.2 except that it is in a slightly more general form. Conditionally on $B=[\mathbf{V}_m=\mathbf{v}_m,\ m=2,\ldots,n]$, the probability function in (4.3.29) can be written as

$$E_{\lambda(2),\ldots,\lambda(n)}\left\{P_{\lambda(1)}\left[\bigcap_{i=1}^{k}\left\{|V_{i1}|^t\leqslant a_i-\sum_{m=2}^{n}|V_{im}|^t\right\}|B\right]\right\}.$$

By the corollary to Theorem 4.3.1 this is nondecreasing in each $\lambda_i(1)$, $i\leqslant k$. The proof is then completed by repeating this argument. ■

An important application of this theorem can be made to the multivariate chi-square distribution when the random vector $\mathbf{X}$ has a normal distribution. Let the $k\times k$ random matrix $\mathbf{S}$ have a Wishart distribution $W_k(n,\Sigma)$ so that $E(\mathbf{S}/n)=\Sigma$. Then the joint distribution of the diagonal elements S_{ii} $(i=1,\ldots,k)$, after proper scaling, is a multivariate chi-square.

Corollaries. (1) If $\mathbf{S}$ has a $W_k(n,\Sigma)$ distribution, then

$$P_{\Sigma}\left[\bigcap_{i=1}^{k}\{S_{ii}\leqslant a_i\}\right]\geqslant\prod_{i=1}^{k}P[S_{ii}\leqslant a_i]. \tag{4.3.30}$$

(2) If $\mathbf{S}$ has a $W_k(n,\Sigma)$ distribution and if† $\Sigma=(\sigma_{ij})=\mathbf{D}+\boldsymbol{\alpha}\boldsymbol{\alpha}'$, where $\mathbf{D}$ is a diagonal matrix with positive diagonal elements and $\boldsymbol{\alpha}$ is of the form $\boldsymbol{\alpha}=(\alpha_1,\ldots,\alpha_k)'$, then

$$P_{\Sigma}\left[\bigcap_{i=1}^{k}\{S_{ii}>a_i\}\right]\geqslant\prod_{i=1}^{k}P[S_{ii}>a_i]. \tag{4.3.31}$$

(3) If $\mathbf{S}$ has a $W_k(n,\Sigma)$ distribution, Σ_1 and Σ_2 are positive definite, and $\Sigma_2-\Sigma_1$ is positive semidefinite, then

$$P_{\Sigma=\Sigma_1}\left[\bigcap_{i=1}^{k}\{S_{ii}\leqslant a_i\}\right]\geqslant P_{\Sigma=\Sigma_2}\left[\bigcap_{i=1}^{k}\{S_{ii}\leqslant a_i\}\right]. \tag{4.3.32}$$

The proofs follow from a similar conditional argument and Theorems 4.3.5, 2.3.2, and 4.3.4, respectively. Note that the inequalities in (4.3.30) and (4.3.31) were given in Chapter 3 under a different set of conditions (see (3.2.5) and (3.2.6)).

†This condition was originally imposed by Das Gupta, Eaton, Olkin, Perlman, Savage, and Sobel (1972). It is equivalent to saying that Σ has the structure l.

Finally, we state a generalization of Slepian's inequality. Its proof, given by Das Gupta, Eaton, Olkin, Perlman, Savage, and Sobel (1972), depends on a geometrical argument and is not given here.

Theorem 4.3.6. Let $\mathbf{X}$ have a density function $f(\mathbf{x})$ which is of the form given in (4.3.1). Let $\mathbf{R}=(\rho_{ij})$, $\mathbf{T}=(\tau_{ij})$ be two positive definite matrices such that $\rho_{ii}=\tau_{ii}=1$. If $\rho_{ij}\geqslant\tau_{ij}$ holds for all i,j, then

$$P_{\Sigma=\mathbf{R}}\left[\bigcap_{i=1}^{k}\{X_i\leqslant a_i\}\right]\geqslant P_{\Sigma=\mathbf{T}}\left[\bigcap_{i=1}^{k}\{X_i\leqslant a_i\}\right] \tag{4.3.33}$$

holds for every $\mathbf{a}=(a_1,\ldots,a_k)'$. Furthermore, the inequality is strict if $\rho_{ij}>\tau_{ij}$ holds for some i,j and if the support of f is unbounded.

PROBLEMS

1. Let $\mathbf{T}$ be a $k\times k$ positive semidefinite matrix. Show that the set $A=\{\mathbf{x}|\mathbf{x}'\mathbf{T}\mathbf{x}\leqslant a\}$ is convex for all $a\geqslant 0$.
2. Verify the statement in (4.1.7).
3. For fixed θ let $g_\theta(x)$ be of the form
$$g_\theta(x) = P(x)Q(\theta)\exp(T(x)R(\theta)).$$
If the second derivative exists, find a necessary and sufficient condition such that $g_\theta(x)$ is log-concave.
4. Let $F_1(x)$ and $F_2(x)$ be two continuous distribution functions of nonnegative random variables. Show that if they are log-concave, then their convolution is log-concave (Barlow and Proschan, 1975, p. 104).
5. Let $f(x)$ be the density function of a continuous random variable X given by
$$f(x) = \begin{cases} 0 & \text{for } x<-\frac{1}{30} \text{ or } x>\frac{5}{6}, \\ 5 & \text{for } -\frac{1}{30}\leqslant x\leqslant 0, \\ 1 & \text{for } 0<x\leqslant\frac{5}{6}, \end{cases}$$
which is unimodal. Show that $f*f(x)$ has two relative maxima at $-\frac{1}{30}$ and $\frac{4}{5}$ and a minimum at 0, and hence that the convolution is not unimodal (Chung, 1953).
6. Let $f(\mathbf{x})$ be an elliptically contoured density function. Show that if the function g given in (4.3.1) is monotonically nonincreasing, then $f(\mathbf{x})$ is unimodal.

7. Show that the multivariate t density function given in (3.1.2) is elliptically contoured for all $n>0$ by verifying that the condition in (4.3.2) is satisfied.
8. Let $f(\mathbf{x})$: $\mathfrak{R}^n \to [0,\infty)$ be a density function, and denote the convolution by $\psi_\varepsilon = f * h_\varepsilon^{(n)}$, where $h_\varepsilon^{(n)}$ is the approximate identity defined in (4.3.11). Show that if f is continuous, then $\psi_\varepsilon(\mathbf{x})$ converges to $f(\mathbf{x})$ pointwise in $\mathbf{x}$ as $\varepsilon \to 0$.
9. Let $\boldsymbol{\Sigma}_1 = (\sigma_{ij})$ and $\boldsymbol{\Sigma}_2 = (\tau_{ij})$ be two $k \times k$ positive definite matrices, where

$$\sigma_{ij} = \begin{cases} \sigma_1^2 & \text{for} \quad i=j, \\ \rho_1\sigma_1^2 & \text{for} \quad i \neq j, \end{cases} \qquad \text{and} \qquad \tau_{ij} = \begin{cases} \sigma_2^2 & \text{for} \quad i=j, \\ \rho_2\sigma_2^2 & \text{for} \quad i \neq j. \end{cases}$$

Show that $\boldsymbol{\Sigma}_2 - \boldsymbol{\Sigma}_1$ is positive semidefinite if and only if

$$\frac{\sigma_2^2}{\sigma_1^2} \geqslant \max\left\{ \frac{1+(k-1)\rho_1}{1+(k-1)\rho_2}, \frac{1-\rho_1}{1-\rho_2} \right\}.$$

10. The difference of two correlation matrices is never positive semidefinite except for the trivial case in which they are identical (Das Gupta, Eaton, Olkin, Perlman, Savage, and Sobel, 1972).
11. Verify the identity in (4.3.14) and explain why an extension to the case of three events fails.
12. Show that a covariance matrix $\boldsymbol{\Sigma}$ has the structure l if and only if there exist a diagonal matrix $\mathbf{D}$ with positive diagonal elements and a real vector $\boldsymbol{\alpha} = (\alpha_1, \ldots, \alpha_k)'$ such that $\boldsymbol{\Sigma} = \mathbf{D} + \boldsymbol{\alpha}\boldsymbol{\alpha}'$.

REFERENCES

Anderson, T. W. (1955). The integral of a symmetric unimodal function over a symmetric convex set and some probability inequalities. *Proc. Amer. Math. Soc.* **6**, 170–176.

Barlow, R. E., and Proschan, F. (1975). *Statistical Theory of Reliability and Life Testing*. Holt, New York.

Borell, C. (1975). Convex set functions in d-space. *Period. Math. Hungar.* **6**, 111–136.

Chung, K. L. (1953). Sur les lois de probabilité unimodales. *C. R. Séances Acad. Sci.* **236**, 583–584.

Das Gupta, S. (1976a). A generalization of Anderson's theorem on unimodal distributions. *Proc. Amer. Math. Soc.* **60**, 85–91.

Das Gupta, S. (1976b). S-Unimodal function: Related inequalities and statistical applications. *Sankhyā Ser. B* **38**, 301–314.

Das Gupta, S., Eaton, M. L., Olkin, I., Perlman, M. D., Savage, L. J., and Sobel, M. (1972). Inequalities on the probability content of convex regions for elliptically contoured distributions. *Proc. Sixth Berkeley Symp. Math. Statist. Probab.* **2** (L. M. LeCam, J. Neyman, and E. L. Scott, eds.), 241–265. Univ. of California Press, Berkeley, California.

Gnedenko, B. V., and Kolmogorov, A. N. (1968). *Limit Distributions for Sums of Independent Random Variables* (K. L. Chung, translator), rev. ed. Addison-Wesley, Reading, Massachusetts.

Kelker, D. (1970). Distribution theory of spherical distributions and a location-scale parameter generalization. *Sankhyā Ser. A* **32**, 419–430.

Mudholkar, G. S. (1966). The integral of an invariant unimodal function over an invariant convex set—an inequality and applications. *Proc. Amer. Math. Soc.* **17**, 1327–1333.

Mudholkar, G. S. (1969). A generalized monotone character of d.f.'s and moments of statistics from some well-known populations. *Ann. Inst. Statist. Math.* **21**, 277–285.

Prékopa, A. (1971). Logarithmic concave measures with applications. *Acta Sci. Math.* **32**, 301–316.

Sherman, S. (1955). A theorem on convex sets with applications. *Ann. Math. Statist.* **26**, 763–766.

CHAPTER

5

Inequalities via Dependence, Association, and Mixture

In this chapter we study bivariate and multivariate probability inequalities from a different angle. To provide some motivation let us first consider the simplest case of a bivariate normal variable $(X_1,X_2)'$. If X_1 and X_2 are stochastically dependent, then their joint distribution function is bounded either below or above by the product of the marginal distributions, depending on whether the correlation (covariance) is positive or negative (see Section 2.1). In the general case it is, of course, too much to expect that such inequalities depend on covariance alone (see, e.g., Kruskal (1958)). But they may be obtained when X_1,X_2 are dependent "in a certain fashion." In a paper Lehmann (1966) studied various concepts of dependence for the bivariate case and gave sufficient conditions for obtaining bounds for the joint distribution functions. These conditions and their implications will be studied in Section 5.1.

For the multivariate case Esary, Proschan, and Walkup (1967) introduced the concept of association of random variables. When $k=2$ (the bivariate case), the condition for association is weaker than that for most of the dependence properties considered by Lehmann. Therefore this concept is another useful tool for deriving probability inequalities. Additional results were later obtained by Esary and Proschan (1972) and Jogdeo (1977). Their results will be discussed in Section 5.2.

In Section 5.3 inequalities through the mixture of distributions will be studied, and the relationship with results given in the previous chapters will be discussed.

5.1. BIVARIATE DEPENDENCE

Let $(X_1,X_2)'$ be a bivariate random variable that has density $f(x_1,x_2)$ and distribution function $F(x_1,x_2)$. Let $f_1(x_1)$, $f_2(x_2)$, $F_1(x_1)$, and $F_2(x_2)$ denote the marginal densities and marginal distributions. We consider some sufficient conditions such that $F(x_1,x_2)$ is bounded below (or above) by $\prod_{i=1}^{2} F_i(x_i)$, which is the distribution function if X_1 and X_2 are actually independent.

Definition 5.1.1. The random variable $(X_1,X_2)'$ (or the distribution F) is said to be positively quadrant dependent (negatively quadrant dependent), denoted by PQD (NQD) if

$$F(x_1,x_2) \geqslant (\leqslant) \prod_{i=1}^{2} F_i(x_i) \tag{5.1.1}$$

holds for all $(x_1,x_2)'$. The dependence is strict if strict inequalities hold for some $(x_1,x_2)'$.

Note that by this definition $(X_1,X_2)'$ is PQD even if X_1 and X_2 are independent (i.e., if the inequality is an equality for all $(x_1,x_2)'$). Certain necessary and sufficient conditions for PQD random variables to be independent were studied by Jogdeo (1968).

We now consider stronger conditions which imply PQD or NQD. For notational convenience only the case of PQD need be discussed because if $(X_1,X_2)'$ is NQD, we can always change the sign of one variable; then $(X_1,-X_2)'$ is PQD. From the inequality in (5.1.1) it is clear that if $(X_1,X_2)'$ is PQD, then X_1,X_2 tend to "hang together." This is so because the condition for PQD is equivalent to

$$P[X_2 \leqslant x_2 | X_1 \leqslant x_1] \geqslant P[X_2 \leqslant x_2], \tag{5.1.2}$$

$$P[X_1 \leqslant x_1 | X_2 \leqslant x_2] \geqslant P[X_1 \leqslant x_1] \tag{5.1.3}$$

for all $(x_1, x_2)'$, provided that the conditional probabilities are defined. It is therefore relevant to consider the following properties that depend on conditional probabilities or regression.

Definition 5.1.2. (1) X_2 is said to be left-tail decreasing in X_1 (denoted by LTD($X_2|X_1$)) if $P[X_2 \leqslant x_2 | X_1 \leqslant x_1]$ is nonincreasing in x_1 for all x_2.

(2) X_2 is said to be right-tail increasing in X_1 (RTI($X_2|X_1$)) if $P[X_2 > x_2 | X_1 > x_1]$ is nondecreasing in x_1 for all x_2.

(3) X_2 is said to be positively regression dependent on X_1 (PRD($X_2|X_1$)) if $P[X_2 \leqslant x_2 | X_1 = x_1]$ is nonincreasing in x_1 for all x_2.

These conditions are defined through the conditional probabilities. Note that unlike the PQD property they are not symmetric in X_1 and X_2, and the PRD property simply says that the family of conditional distributions $\{P[X_2 \leqslant x_2 | X_1 = x_1] : x_1 \in \mathfrak{X}_1\}$ is stochastically increasing.† In view of the well-known relationship between a stochastically increasing family of distribution functions and a family of densities with the monotone likelihood ratio (MLR) property (see Lehmann, 1959, p. 74), this then suggests a stronger condition on the joint density function.

Definition 5.1.3. The random variable $(X_1, X_2)'$ (or its distribution) is said to be positively likelihood ratio dependent (PLRD) if

$$f(x_1, x_2^*) f(x_1^*, x_2) \leqslant f(x_1, x_2) f(x_1^*, x_2^*) \tag{5.1.4}$$

holds for all $x_1 > x_1^*$ and $x_2 > x_2^*$.

This condition says that the likelihood is larger when both coordinates take larger values together and smaller values together at the same time. Clearly the inequality in (5.1.4) is satisfied if and only if

$$\begin{vmatrix} f(x_1, x_2) & f(x_1, x_2^*) \\ f(x_1^*, x_2) & f(x_1^*, x_2^*) \end{vmatrix} \geqslant 0 \tag{5.1.5}$$

†A family of univariate distribution functions $\{F(x|\theta) = F_\theta(x) : \theta \in \Lambda\}$ is said to be stochastically increasing if for θ_1, θ_2 in Λ, $\theta_1 < \theta_2$ implies $F_{\theta_1}(x) \geqslant F_{\theta_2}(x)$ for all x, which says that the random variable is stochastically larger if the parameter value is larger. It is known that this property is equivalent to saying that $E_\theta \phi(X)$ is nondecreasing in θ for every nondecreasing function ϕ, for which the expectation exists. For details see Lehmann (1959, Chap. 3).

holds for all $x_1 > x_1^*$ and $x_2 > x_2^*$. A function f (not necessarily a density) with such a property is said to be totally positive of order two, denoted by TP_2 (Karlin, 1968, p. 11). Therefore $(X_1, X_2)'$ is PLRD if and only if its density is TP_2. On the other hand, this TP_2 property is equivalent to saying that the two families of conditional density functions

$$\{f_2(x_2|x_1) : x_1 \in \mathscr{X}_1\}, \qquad \{f_1(x_1|x_2) : x_2 \in \mathscr{X}_2\} \tag{5.1.6}$$

have the MLR property, where $\mathscr{X}_i$ is the domain of x_i such that the conditional densities are properly defined $(i = 1, 2)$. This can be checked by dividing both sides in (5.1.4) by $f_1(x_1)f_1(x_1^*)$ or $f_2(x_2)f_2(x_2^*)$, respectively.

In the following theorem, due to Lehmann (1955, 1966) and Esary and Proschan (1972), we show the implications of these concepts of dependence. Implication h is defined under the assumption that the covariance exists.

Theorem 5.1.1. The following implications are true:

$$\text{PLRD} \begin{array}{c} \overset{a}{\Rightarrow} \\ \underset{b}{\Rightarrow} \end{array} \begin{array}{l} \text{PRD}(X_2|X_1) \\ \text{PRD}(X_1|X_2), \end{array} \tag{5.1.7}$$

$$\text{PRD}(X_2|X_1) \begin{array}{c} \overset{c}{\Rightarrow} \\ \underset{d}{\Rightarrow} \end{array} \begin{array}{l} \text{LTD}(X_2|X_1) \\ \text{RTI}\,(X_2|X_1) \end{array} \begin{array}{c} \overset{e}{\Rightarrow} \\ \underset{f}{\Rightarrow} \end{array} \text{PQD} \overset{h}{\Rightarrow} \text{cov}(X_1, X_2) \geqslant 0. \tag{5.1.8}$$

Proof. (1) $\overset{a}{\Rightarrow}$ and $\overset{b}{\Rightarrow}$. This amounts to proving the statement that if the family of conditional densities has the MLR property, then the family of conditional distributions is stochastically increasing. This result is well known and its proof can be found in Lehmann (1959, Chap. 3).

(2) $\overset{c}{\Rightarrow}$ and $\overset{d}{\Rightarrow}$. The condition for $\text{LTD}(X_2|X_1)$ is equivalent to $P[X_2 > x_2 | X_1 \leqslant x_1]$ is nondecreasing in x_1 for all x_2, which in turn is equivalent to

$$P[X_2 > x_2 | X_1 \leqslant x_1^*] \leqslant P[X_2 > x_2 | x_1^* < X_1 \leqslant x_1] \tag{5.1.9}$$

for all x_2 and $x_1 > x_1^*$. Moreover, the condition for $\text{RTI}(X_2|X_1)$ is equivalent to

$$P[X_2 > x_2 | x_1^* < X_1 \leqslant x_1] \leqslant P[X_2 > x_2 | X_1 > x_1] \tag{5.1.10}$$

for all x_2 and $x_1 > x_1^*$. Now the proof follows from (5.1.9), (5.1.10), and the fact that for any interval I we have

$$P[X_2 > x_2 | X_1 \in I] = \int_I P[X_2 > x_2 | X_1 = x_1]\, dP[X_1 \leqslant x_1] / P[X_1 \in I].$$

(3) $\overset{e}{\Rightarrow}$ and $\overset{f}{\Rightarrow}$. This follows from

$$P[X_2 \leqslant x_2 | X_1 \leqslant x_1] \geqslant P[X_2 \leqslant x_2],$$

which follows immediately by letting the value of x_1 tend to ∞ on the right-hand side.

(4) $\overset{h}{\Rightarrow}$. The proof of this implication follows from an identity due to Hoeffding, which says that if the covariance of the random variables X_1 and X_2 exists, then

$$E(X_1X_2) - (EX_1)(EX_2) = \int_{-\infty}^{\infty}\int_{-\infty}^{\infty} [F(x_1,x_2) - F_1(x_1)F_2(x_2)]\, dx_1\, dx_2. \blacksquare \tag{5.1.11}$$

A natural question one may wish to ask is whether or not these implications are strict. The answer is given in the following theorem. Its proof was also due to Lehmann (1955, 1966) and Esary and Proschan (1972).

Theorem 5.1.2. All the implications stated in Theorem 5.1.1 are strict. Moreover,

$$\mathrm{LTD}(X_2|X_1) \nRightarrow \mathrm{RTI}(X_2|X_1), \qquad \mathrm{RTI}(X_2|X_1) \nRightarrow \mathrm{LTD}(X_2|X_1). \tag{5.1.12}$$

Proof. It suffices to give a counterexample for each case. A counterexample for $\overset{h}{\nLeftarrow}$ is easy and is omitted. Now let $(X_1, X_2)'$ be a discrete random variable with a density as given in Table 5.1.1.

TABLE 5.1.1

A DENSITY FUNCTION $f(x_1, x_2)$

	x_2		
x_1	1	2	3
1	q_1	0	r
2	0	q_2	0
3	s	0	q_3

It is easy to check that

(1) If $q_1 = q_2 = q_3 = \frac{1}{4}$ and $r = s = \frac{1}{8}$, then $(X_1, X_2)'$ is PQD, but we have neither LTD$(X_2|X_1)$ nor RTI$(X_2|X_1)$; this shows $\overset{e}{\not\Leftarrow}$ and $\overset{f}{\not\Leftarrow}$.

(2) If $r = 0$ and $s = q_1 = q_2 = q_3 = \frac{1}{4}$ ($s = 0$ and $r = q_1 = q_2 = q_3 = \frac{1}{4}$), then we have LTD$(X_2|X_1)$ but not RTI$(X_2|X_1)$, (RTI$(X_2|X_1)$) but not (LTD$(X_2|X_1)$); hence (5.1.12) is proved.

(3) We have LTD$(X_2|X_1)$ (RTI$(X_2|X_1)$) but not PRD$(X_2|X_1)$ if $r = 0$, $s = \frac{1}{8}$, and $q_1 = q_2 = q_3 = \frac{7}{24}$ (if $s = 0$, $r = \frac{1}{8}$, and $q_1 = q_2 = q_3 = \frac{7}{24}$). Hence $\overset{c}{\not\Leftarrow}$ and $\overset{d}{\not\Leftarrow}$.

To show $\overset{a}{\not\Leftarrow}$ and $\overset{b}{\not\Leftarrow}$ we simply give an example along the line discussed in Lehmann (1959, p. 75). Let $(X_1, X_2)'$ have a density

$$f(x_1, x_2) = \begin{cases} \left[\pi\{1 + (x_2 - x_1)^2\}\right]^{-1}, & x_2 \in \mathcal{R}^1, \quad x_1 \in [0,1], \\ 0 & \text{otherwise.} \end{cases} \tag{5.1.13}$$

Then the conditional density of X_2, given $X_1 = x_1 \in [0, 1]$, is Cauchy. It follows that the condition for PRD$(X_2|X_1)$ is satisfied, but the family of conditional densities does not have the MLR property; hence $\overset{a}{\not\Leftarrow}$. Interchanging the indices in (5.1.13) yields $\overset{b}{\not\Leftarrow}$. ■

From Theorem 5.1.1 the PLRD property becomes a useful tool for establishing inequalities of the form (5.1.1). Since this property is directly related to the MLR property, it is usually the easiest to check. Also, this PLRD property may be obtained through a mixture of density functions, as shown by Lehmann (1966).

Theorem 5.1.3. Let $\{g_\theta(x_1) : \theta \in \Lambda\}$ and $\{h_\theta(x_2) : \theta \in \Lambda\}$ be two families of univariate densities that have the MLR property. Let $\tau(\theta)$ be a probability distribution on Λ. Then

$$f(x_1, x_2) = \int_\Lambda g_\theta(x_1) h_\theta(x_2)\, d\tau(\theta) \tag{5.1.14}$$

has the PLRD property.

Proof. Let $x_1 > x_1^*$, $x_2 > x_2^*$ be arbitrary but fixed, and consider a product probability measure on $\Lambda \times \Lambda$ given by $\tau(\theta)\tau(\xi)$ for $(\theta, \xi) \in \Lambda \times \Lambda$. Then the proof of the theorem will be completed if we show

$$I=\int_{\Lambda\times\Lambda}\int \phi(x_1,x_2,x_1^*,x_2^*,\theta,\xi)\,d\tau(\theta)\,d\tau(\xi)\geqslant 0,$$

where

$$\phi(x_1,x_2,x_1^*,x_2^*,\theta,\xi)=\big[\,g_\theta(x_1^*)h_\theta(x_2^*)g_\xi(x_1)h_\xi(x_2) - g_\theta(x_1^*)h_\theta(x_2)g_\xi(x_1)h_\xi(x_2^*)\big].$$

Let us denote

$$B_1=\{(\theta,\xi)|(\theta,\xi)\in\Lambda\times\Lambda,\theta<\xi\},$$

$$B_2=\{(\theta,\xi)|(\theta,\xi)\in\Lambda\times\Lambda,\theta>\xi\}.$$

Then the integral is of the form

$$I=\int_{B_1}\int \phi(x_1,x_2,x_1^*,x_2^*,\theta,\xi)\,d\tau(\theta)\,d\tau(\xi) + \int_{B_2}\int \phi(x_1,x_2,x_1^*,x_2^*,\theta,\xi)\,d\tau(\theta)\,d\tau(\xi).$$

After interchanging the variables θ and ξ in the second integral and combining, we have

$$I=\int_{B_1}\int\big[\,g_\theta(x_1^*)g_\xi(x_1)-g_\xi(x_1^*)g_\theta(x_1)\big] \cdot\big[\,h_\theta(x_2^*)h_\xi(x_2)-h_\xi(x_2^*)h_\theta(x_2)\big]\,d\tau(\theta)\,d\tau(\xi),$$

which is nonnegative because by the MLR property of g and h the integrand is nonnegative for $(\theta,\xi)\in B_1$. ■

A theorem given by Lehmann (1966) says that the PQD property is invariant under certain monotone transformations of random variables. Therefore once the inequality in (5.1.1) is obtained, it may be preserved. In general we may consider two functions $g_1,g_2\colon \mathcal{R}^n\to\mathcal{R}^1$ of the forms

$$g_1(x_{11},\ldots,x_{1n}),\qquad g_2(x_{21},\ldots,x_{2n}).$$

Definition 5.1.4. g_1 and g_2 are said to be concordant (discordant) for the jth coordinate if with all other coordinates held fixed they are either nondecreasing or nonincreasing in the same (in the opposite) direction.

Theorem 5.1.4. For $j=1,\ldots,n$ let $\mathbf{X}_j=(X_{1j},X_{2j})'$ be independent random variables, and assume that $(X_{1j},X_{2j})'$ is PQD for each j. Then

(a) For all g_1 and g_2 such that g_1 and g_2 are concordant for each coordinate and such that the expectations exist, we have

$$E\prod_{i=1}^{2} g_i(X_{i1},\ldots,X_{in}) \geqslant \prod_{i=1}^{2} Eg_i(X_{i1},\ldots,X_{in}). \tag{5.1.15}$$

(b) As a more general case for all g_1 and g_2 such that g_1 and g_2 are concordant for each coordinate

$$(g_1(X_{11},\ldots,X_{1n}),g_2(X_{21},\ldots,X_{2n}))'$$

is PQD.

Proof. First note that for each j $(X_{1j},X_{2j})'$ is PQD if and only if $(-X_{1j},-X_{2j})'$ is PQD. Therefore without loss of generality we may assume that g_1,g_2 are nondecreasing in each coordinate. For $n=1$, it is clear that $(g_1(X_{11}),g_2(X_{21}))'$ is PQD if $(X_{11},X_{21})'$ is. From Theorem 5.1.1 this implies that if $(X_{11},X_{21})'$ is PQD, then

$$E\prod_{i=1}^{2} g_i(X_{i1}) \geqslant \prod_{i=1}^{2} Eg_i(X_{i1})$$

holds for all nondecreasing functions g_1 and g_2, provided that the expectations exist. This establishes (5.1.15) for $n=1$. Now for general n we proceed by induction through a conditional argument. For every fixed $(X_{i2},\ldots,X_{in})'=(x_{i2},\ldots,x_{in})'$ let us denote

$$g_i^*(x_{i2},\ldots,x_{in}) = Eg_i(X_{i1},x_{i2},\ldots,x_{in}), \qquad i = 1,2.$$

It follows that for every fixed $(x_{i2},\ldots,x_{in})'$

$$E\prod_{i=1}^{2} g_i(X_{i1},x_{i2},\ldots,x_{in}) \geqslant \prod_{i=1}^{2} g_i^*(x_{i2},\ldots,x_{in}).$$

We then consider the expectation of $\prod_{i=1}^{2} g_i^*(X_{i2},\ldots,X_{in})$. Since g_i^* has the same monotonicity properties in $x_{i2},\ldots,x_{in}$ $(i=1,2)$, we may proceed similarly by considering one argument at a time. This proves (a). The proof of (b) follows similarly by replacing $g_i(X_{i1},\ldots,X_{in})$ with $[1-\chi_{(-\infty,x_i]}(g_i(X_{i1},\ldots,X_{in}))]$, where χ_A is the indicator function of the set A. ∎

An immediate application of this theorem gives the following special results; these results have been found useful in certain applications.

Example 5.1.1. (1) If g is nondecreasing, then $(X,g(X))'$ is PQD. In particular, $(X,X)'$ is PQD.

(2) If U_1, U_2, and U are independent, then for $a_1a_2>0$ $(U_1+a_1U, U_2+a_2U)'$ is PQD. A special case of this result occurs when U_1 or (and) U_2 is a constant with probability one.

Remark. A direct comparison between Theorem 5.1.4 and Lemma 3.2.1 indicates that this theorem implies Lemma 3.2.1 as a special case. This is so because for independent random variables $U_1,\ldots,U_n$ we can always define $\mathbf{X}_j=(U_j,U_j)'$ in Theorem 5.1.4 for $j=1,\ldots,n$, and the implication follows from the fact that $(U_j,U_j)'$ is PQD for each j. Lemma 3.2.1 also follows from the concept of association of random variables (to be introduced in the next section), and the proof of (d) in Theorem 5.2.2 may be regarded as an alternative proof for Lemma 3.2.1, using the concept of association.

We have thus far considered inequalities for PQD random variables only. If the inequality given in (5.1.1) is in the direction of "$\leqslant$", i.e., if the distribution is bounded above by the product of the marginal distributions, then the random variable $(X_1,X_2)'$ is NQD, and this NQD property can also be preserved through appropriate transformations. In fact Theorem 5.1.4 can be modified to read that under the same conditions on the sequence of random variables $\{\mathbf{X}_j\}_{j=1}^n$ if g_1 and g_2 are discordant in each coordinate, then the inequality in (5.1.15) is reversed. This also implies that

$$(g_1(X_{11},\ldots,X_{1n}),g_2(X_{21},\ldots,X_{2n}))'$$

is NQD. This follows from the fact that g_1 and $-g_2$ are concordant.

5.2. ASSOCIATION OF RANDOM VARIABLES

It is clear that from letting $n=1$ in Theorem 5.1.4, $(X_1,X_2)'$ is PQD if and only if $\operatorname{cov}(g_1(X_1),g_2(X_2))\geqslant 0$ holds for all nondecreasing real-valued functions g_1,g_2, provided the expectations exist. Since $(X,X)'$ is PQD, this obviously implies a special case of the inequality of Chebyshev (see Lemma 2.2.1), which says that $\operatorname{cov}(g_1(X),g_2(X))\geqslant 0$. A concept of association of random variables (denoted by "A") that is stronger than the PQD property was introduced and studied by Esary, Proschan, and Walkup (1967) and Esary and Proschan (1972). In general, for any $k\geqslant 2$ let $\mathbf{X}=(X_1,\ldots,X_k)'$ denote a k-dimensional random variable.

Definition 5.2.1. $X_1,\ldots,X_k$ are said to be associated if

$$\operatorname{cov}(g_1(\mathbf{X}),g_2(\mathbf{X})) \geqslant 0 \tag{5.2.1}$$

for all g_1 and g_2 monotonically nondecreasing in each argument, such that the expectations exist.

In the bivariate case ($k=2$) this is directly related to the other concepts of dependence already discussed in Section 5.1. Esary, Proschan, and Walkup (1967) showed that if X_1 and X_2 are binary variables, i.e., if they each take only the values zero or one, then the conditions for PLRD, association, and PQD are all equivalent to $\operatorname{cov}(X_1,X_2)\geqslant 0$. Moreover, they showed that in the general case the implications are strict. The following theorem, together with Theorems 5.1.1.and 5.1.2, summarizes all the implications.

Theorem 5.2.1. The following implications are true:

$$\begin{matrix} \mathrm{LTD}(X_2|X_1) & \overset{e}{\Rightarrow} & & \\ & & \mathrm{A} & \overset{g}{\Rightarrow} \mathrm{PQD}. \\ \mathrm{RTI}(X_2|X_1) & \underset{f}{\Rightarrow} & & \end{matrix} \tag{5.2.2}$$

Moreover, all implications are strict.

Outline of the Proof. The proof of $\overset{g}{\Rightarrow}$ is obvious; the proof of $\overset{f}{\Rightarrow}$ is quite involved; therefore the major steps will be given below without details. The proof of $\overset{e}{\Rightarrow}$ follows from $\overset{f}{\Rightarrow}$ and the following facts:

(a) $\mathrm{LTD}(X_2|X_1)$ is equivalent to $\mathrm{RTI}(-X_2|-X_1)$, and
(b) X_1,X_2 are associated if and only if $-X_1, -X_2$ are.

To prove $\overset{f}{\Rightarrow}$ the first step is to define a new random variable $(X_1^*,X_2^*)'$ from $(X_1,X_2)'$ such that X_1^* and X_2^* are truncations of X_1 and X_2 and take values only on $\{0,\ldots,m\}$ and $\{0,\ldots,n\}$. Then it can be shown that the association of X_1,X_2 is equivalent to that of X_1^*,X_2^* for all choices of m, n, and truncation points and that $\mathrm{RTI}(X_2|X_1)$ implies $\mathrm{RTI}(X_2^*|X_1^*)$. Therefore the problem reduces to proving that $\mathrm{RTI}(X_2^*|X_1^*)$ implies the association of X_1^* and X_2^*. This can be done by defining two binary random variables $V_1(X_1^*,X_2^*)$, $V_2(X_1^*,X_2^*)$ and studying the distribution property of (V_1,V_2).

To show that all implications are strict we again consider $(X_1,X_2)'$ with a joint density as given in Table 5.1.1. For $q_1=q_3=\frac{15}{64}$, $q_2=\frac{18}{64}$, and $r=s=\frac{8}{64}$, X_1,X_2 are not associated but $(X_1,X_2)'$ is PQD. Finally, let $q_1 = q_2 = q_3 = \frac{1}{4}$ and $r = s = \frac{1}{8}$; then X_1, X_2 are associated, but we have neither LTD$(X_2|X_1)$ nor RTI$(X_2|X_1)$. ■

An obvious advantage to the concept of association of random variables is that now we no longer need to restrict ourselves to the bivariate case only, and multivariate probability inequalities can be obtained by simply showing that the random variables are associated. Esary, Proschan, and Walkup (1967) gave several conditions equivalent to association. In particular, they showed that if (5.2.1) holds for all g_1, g_2 that are bounded, continuous, and nondecreasing (in each argument), then $X_1, \ldots, X_k$ are associated. Therefore we may conclude that $X_1, \ldots, X_k$ are associated if and only if $-X_1, \ldots, -X_k$ are. They also showed that the property of association can be created and preserved through suitable combinations and transformations of random variables.

Theorem 5.2.2. The following statements are true:

(a) Any subset of associated random variables is a set of associated random variables.

(b) The set consisting of a single random variable is associated.

(c) If two sets of associated random variables are independent, then their union is a set of associated random variables.

(d) Independent random variables are associated.

(e) Nondecreasing functions of associated random variables are associated random variables.

Proof. (a) follows from the definition of association by choosing nondecreasing functions g_1, g_2 that depend only on the variables in the subset. (b) follows immediately from Lemma 2.2.1 or Theorem 5.1.4. To prove (c) let the components of $\mathbf{X} = (X_1, \ldots, X_k)'$ be associated and the components of $\mathbf{Y} = (Y_1, \ldots, Y_m)'$ be associated, and let $\mathbf{X}$, $\mathbf{Y}$ be independent. Let us define $V_1 = g_1(\mathbf{X}, \mathbf{Y})$, $V_2 = g_2(\mathbf{X}, \mathbf{Y})$, where g_1, g_2 are nondecreasing in each argument. Then it is easy to check that (by taking expectations conditionally on $\mathbf{X}$ first)

$$\operatorname{cov}(V_1, V_2) = E\{\operatorname{cov}[V_1, V_2]|\mathbf{X}\} + \operatorname{cov}[E\{V_1|\mathbf{X}\}, E\{V_2|\mathbf{X}\}]. \tag{5.2.3}$$

It is clear that for every fixed $\mathbf{X} = \mathbf{x}$ the conditional covariance of V_1, V_2 is nonnegative and the conditional expectations of V_1, V_2 are nondecreasing in $\mathbf{x}$ (in each argument). Therefore the right-hand side of (5.2.3) is nonnegative, and the statement in (c) follows. The statement in (d) is an immediate consequence of (b) and (c). The proof of (e) will be completed by showing that the covariance of two nondecreasing functions of nondecreasing functions of associated random variables is nonnegative. The details are postponed, and will be given in the proof of Theorem 5.2.3. ■

The next theorem, due to Jogdeo (1977), says that the property of association may be preserved under concordant transformations. The following definition given by Jogdeo (1977) may be regarded as a multivariate generalization of the concept of concordant functions given by Lehmann (see Definition 5.1.4).

Definition 5.2.2. A set of m functions $g_1,\dots,g_m$, each defined on $\mathfrak{R}^{kn}\to\mathfrak{R}^1$, is said to be k-concordant if the functions are monotone in each of the kn arguments and the direction of monotonicity is the same for each block of k arguments, $jk+1,\dots,jk+k$, where $0\leqslant j\leqslant(n-1)$.

Now let us consider n random variables $\mathbf{U}_1,\dots,\mathbf{U}_n$, where each is k-dimensional, and consider the random variables

$$V_i = g_i(\mathbf{U}_1,\dots,\mathbf{U}_n), \qquad i = 1,2.$$

Jogdeo (1977) obtained the following lemma. The proof given here is different from his original proof, and it is an immediate consequence of Theorem 5.2.2.

Lemma 5.2.1. Assume that $\mathbf{U}_1,\dots,\mathbf{U}_n$ are independent and that the components of $\mathbf{U}_j$ are associated for each j. If g_1,g_2 are k-concordant, then $\operatorname{cov}(V_1,V_2)\geqslant 0$, provided that it exists.

Proof. If for some j, g_1 and g_2 are nonincreasing in the arguments of the jth block, we may replace $\mathbf{U}_j$ by $-\mathbf{U}_j$ and the hypothesis of the problem remains unchanged. (This is so because the components of $\mathbf{U}_j$ are associated if and only if the components of $-\mathbf{U}_j$ are.) Therefore without loss of generality we may assume that g_1, g_2 are nondecreasing in each argument. By Theorem 5.2.2(c), the components of $(\mathbf{U}_1,\dots,\mathbf{U}_n)'$ are associated. Hence by Definition 5.2.1 $\operatorname{cov}(V_1,V_2)\geqslant 0$. ■

Applying Lemma 5.2.1, Jogdeo (1977) proved the following theorem.

Theorem 5.2.3. Assume that $g_1,\ldots,g_m$ are k-concordant functions defined for nk-tuples, and that $\mathbf{U}_1,\ldots,\mathbf{U}_n$ are independent k-dimensional random variables such that the components of $\mathbf{U}_i$ are associated for each i. Then the random variables

$$X_i = g_i(\mathbf{U}_1,\ldots,\mathbf{U}_n), \qquad i = 1,\ldots,m, \tag{5.2.4}$$

are associated.

Proof. Consider any two nondecreasing functions $h_1,h_2\colon \mathcal{R}^m \to \mathcal{R}^1$ such that the expectations exist. Obviously we can write

$$h_i(X_1,\ldots,X_m) = h_i^*(\mathbf{U}_1,\ldots,\mathbf{U}_n), \qquad i = 1,2,$$

and it is easy to check that h_1^*,h_2^* are k-concordant. Therefore by Lemma 5.2.1 we have

$$\operatorname{cov}\left[h_1^*(\mathbf{U}_1,\ldots,\mathbf{U}_n),h_2^*(\mathbf{U}_1,\ldots,\mathbf{U}_n)\right] \geqslant 0,$$

which implies

$$\operatorname{cov}\left[h_1(X_1,\ldots,X_m),h_2(X_1,\ldots,X_m)\right] \geqslant 0. \qquad \blacksquare$$

Remark. The above proof was given in Jogdeo (1977). Since without loss of generality we can assume that $g_1,\ldots,g_m$ are nondecreasing in each argument and since the components of $(\mathbf{U}_1,\ldots,\mathbf{U}_n)'$ are associated (by Theorem 5.2.2(c)), this proof is essentially for the statement that "nondecreasing functions of associated random variables are associated random variables" (Theorem 5.2.2(e)). The proof for this statement was given previously by Esary, Proschan, and Walkup (1967).

Our main reason for studying association here is, of course, to obtain probability inequalities. The next theorem is specifically designed for this purpose.

Theorem 5.2.4. If $X_1,\ldots,X_k$ are associated random variables, then

$$P\left[\bigcap_{i=1}^{k}\{X_i \leqslant a_i\}\right] \geqslant P\left[\bigcap_{i\in C}\{X_i \leqslant a_i\}\right]P\left[\bigcap_{i\notin C}\{X_i \leqslant a_i\}\right]$$

$$\geqslant \prod_{i=1}^{k} P\left[X_i \leqslant a_i\right] \tag{5.2.5}$$

holds for all $\mathbf{a}$ and all subsets C of $\{1,\ldots,k\}$.

Proof. Let us define

$$g_1(\mathbf{x}) = 1 - \chi_{A_1}(\mathbf{x}), \qquad g_2(\mathbf{x}) = 1 - \chi_{A_2}(\mathbf{x}),$$

where $\chi_A(\mathbf{x})$ is the indicator function of the set A and

$$A_1 = \{\mathbf{x} | x_i \leqslant a_i, i \in C\}, \qquad A_2 = \{\mathbf{x} | x_i \leqslant a_i, i \notin C\}.$$

Then g_1, g_2 are nondecreasing in each argument. Therefore we must have from the definition of association

$$Eg_1(\mathbf{X})g_2(\mathbf{X}) \geqslant [Eg_1(\mathbf{X})][Eg_2(\mathbf{X})],$$

which gives the first inequality in (5.2.5). The second inequality in (5.2.5) follows immediately from Theorem 5.2.2(a) by induction. ■

Remark. We mention here several more general results developed recently. Ahmed, Langberg, León, and Proschan (1978, 1979) extended the concepts of RTI and PQD to the multivariate case, obtained certain basic properties, and considered applications. Ahmed, León, and Proschan (1978) generalized the concept of association of random variables, given in Definition 5.2.1, to a partially ordered space, and obtained some new results which are useful for the mixture of distributions. Another more general result is known as the FKG inequality. Let $f: \mathfrak{R}^k \to [0, \infty)$ be a density function, and for $\mathbf{y} = (y_1, \ldots, y_k)'$, $\mathbf{y}^* = (y_1^*, \ldots, y_k^*)'$ in $\mathfrak{R}^k$ define $\mathbf{x} = (x_1, \ldots, x_k)'$ and $\mathbf{x}^* = (x_1^*, \ldots, x_k^*)'$ such that

$$x_i = \max(y_i, y_i^*), \qquad x_i^* = \min(y_i, y_i^*), \qquad i = 1, \ldots, k.$$

f is said to satisfy the FKG condition if

$$f(\mathbf{y})f(\mathbf{y}^*) \leqslant f(\mathbf{x})f(\mathbf{x}^*) \tag{5.2.6}$$

holds for all $\mathbf{y}, \mathbf{y}^*$. This condition is a generalization of the PLRD condition stated in (5.1.4). The FKG inequality, obtained by Fortuin, Kastelyn, and Ginibre (1971), says that if the density function of $\mathbf{X}$ satisfies the FKG condition, then the components of $\mathbf{X}$ are associated random variables. In view of Theorem 5.2.4 this inequality is then a multivariate generalization of Lehmann's result that PLRD implies PQD (see Theorem 5.1.1). For details of the discussion of the application of FKG inequality in statistics see Kemperman (1977) and Perlman and Olkin (1980).

Theorem 5.2.4 may be used to derive inequalities for the probability of rectangles after creating the association of the absolute values of the random variables, and this approach was adopted by

Jogdeo (1977). He first obtained a result analogous to the Anderson–Sherman theorem (Theorems 4.1.1 and 4.2.1), and then proceeded to apply that result to produce the association property.

Definition 5.2.3. A function $f: \mathcal{R}^k \to \mathcal{R}^1$ is said to be decreasing (increasing) in absolute value if it is sign invariant (see Definitions 4.1.3 and 4.1.4) and if $|\mathbf{x}| > |\mathbf{x}^*|$ implies $f(\mathbf{x}) \leqslant f(\mathbf{x}^*)$ $(f(\mathbf{x}) \geqslant f(\mathbf{x}^*))$. A set $A \subset \mathcal{R}^k$ is said to be decreasing (increasing) in absolute value if its indicator function is decreasing (increasing) in absolute value.

We recall that Sherman's (1955) result says that the convolution (for the definition, see (4.2.1)) of two symmetric unimodal functions is unimodal. Jogdeo (1977) obtained the following theorem, which says that a similar assertion is true for functions decreasing in absolute value.

Theorem 5.2.5. Let $f_1, f_2: \mathcal{R}^k \to [0, \infty)$ be two density functions. If they are decreasing in absolute value, then their convolution $f_1 * f_2$ is decreasing in absolute value.

Proof. The fact that $f_1 * f_2$ is sign invariant can be verified easily (see the proof of Theorem 4.1.4). To show the decreasing property let us first consider the case $f_1 = c_1\chi_{A_1}(\mathbf{x})$ and $f_2 = c_2\chi_{A_2}(\mathbf{x})$, where χ_{A_1}, χ_{A_2} are indicator functions of two sets A_1, A_2 that are decreasing in absolute value and c_1, c_2 are positive constants such that f_1, f_2 are densities. We can write

$$\chi_{A_1} * \chi_{A_2}(\mathbf{y}) = \int_{A_2+\mathbf{y}} \chi_{A_1}(\mathbf{x})\, d\mathbf{x} \equiv h(\mathbf{y}).$$

Now let $\mathbf{y} = (y_1, y_2, \ldots, y_k)'$, $\mathbf{y}^* = (y_1^*, y_2, \ldots, y_k)'$ be such that $|y_1| > |y_1^*|$ and $y_2, \ldots, y_k$ are arbitrary but fixed. Since $h(\mathbf{y})$ can be viewed as the volume of the intersection of the sets A_1 and $A_2 + \mathbf{y}$ from the decreasing property of A_1, A_2 we must have $h(\mathbf{y}^*) \geqslant h(\mathbf{y})$. This process can be continued by considering one argument at a time; hence we may conclude that the assertion of the theorem holds when f_1, f_2 are indicator functions of sets that are decreasing in absolute value.

Now let f be any density function that is decreasing in absolute value. Then the set $D_u = \{\mathbf{x} | f(\mathbf{x}) \geqslant u\}$ is. By a characterization of Khintchine (1938), the density f can be considered as a mixture of uniform distributions on sets that are decreasing in absolute value. It follows that for any such two densities f_1 and f_2, $f_1 * f_2$ can be regarded as a mixture of convolutions of uniform distributions on sets that are decreasing in absolute value. Hence the proof extends to the general case. ■

We note that in spite of their similarities this theorem is different from Sherman's result. This is so because a function symmetric about the origin need not be sign invariant. On the other hand, a function decreasing in absolute value may not be unimodal, as illustrated in the example given below:

Example 5.2.1. Let

$$f(x_1, x_2) = \begin{cases} \frac{1}{5} & \text{for} \quad (x_1, x_2)' \in S, \\ 0 & \text{otherwise}, \end{cases}$$

where

$$S = \{(x_1, x_2)' \,|\, |x_1| \leqslant 1, |x_2| \leqslant 2 \text{ or } |x_1| \leqslant 2, |x_2| \leqslant 1\}.$$

Then f is decreasing in absolute value but is not unimodal.

Applying Theorem 5.2.5, Jogdeo (1977) obtained the next two theorems (Theorems 5.2.6 and 5.2.7).

Theorem 5.2.6. Let f be the density function of $\mathbf{X}$, and let $\mathbf{y}$, $\mathbf{y}^*$ be two real vectors. If f is decreasing in absolute value and $|\mathbf{y}| > |\mathbf{y}^*|$, then $(\mathbf{X}+\mathbf{y})$ is stochastically larger than $(\mathbf{X}+\mathbf{y}^*)$ in absolute value, i.e.,

$$Eg(\mathbf{X}+\mathbf{y}) \geqslant Eg(\mathbf{X}+\mathbf{y}^*) \tag{5.2.7}$$

holds for all g that are increasing in absolute value.

Proof. For every $g = c\chi_A$, where $c > 0$ and A is a set increasing in absolute value, the assertion follows immediately from Theorem 5.2.5 with $f_1 = f$ and $f_2 = 1 - g$. In the general case f_1, f_2 can again be regarded as a mixture of uniform distributions on sets that are decreasing in absolute value; therefore the proof extends to the general case. ■

An immediate consequence of this result is the following corollary. (A similar result was given in Theorem 4.1.4.)

Corollary. Let $A \subset \mathfrak{R}^k$ be decreasing in absolute value, and let f, the density of $\mathbf{X}$, satisfy the conditions stated in Theorem 5.2.6. Then

$$P[(\mathbf{X}+\mathbf{y}^*)\in A] \geqslant P[(\mathbf{X}+\mathbf{y})\in A] \tag{5.2.8}$$

holds for all $\mathbf{y}$, $\mathbf{y}^*$ such that $|\mathbf{y}|>|\mathbf{y}^*|$.

We note that Theorems 5.2.5 and 5.2.6 actually follow from a more general result of Eaton and Perlman (1977) (see Theorem 6.2.7) as special cases. For details, see Section 6.2.

Theorem 5.2.7. Let $\mathbf{X}=(X_1,\ldots,X_k)'$ have independent components, each having a symmetric unimodal density function, and denote $\mathbf{Z}=\mathbf{X}+\mathbf{U}$ where $\mathbf{U}$ is independent of $\mathbf{X}$. If the components of $|\mathbf{U}|$ are associated, then the components of $|\mathbf{Z}|$ are associated.

Proof. Let g_1, g_2: $\mathfrak{R}^k \to \mathfrak{R}^1$ be nondecreasing. Then from an argument similar to that used in obtaining (5.2.3) we can write

$$\begin{aligned}\operatorname{cov}[g_1(|\mathbf{Z}|),g_2(|\mathbf{Z}|)] &= E\{\operatorname{cov}[g_1(|\mathbf{Z}|),g_2(|\mathbf{Z}|)]|\mathbf{U}\}\\ &\quad +\operatorname{cov}[E\{g_1(|\mathbf{Z}|)|\mathbf{U}\},E\{g_2(|\mathbf{Z}|)|\mathbf{U}\}].\end{aligned} \tag{5.2.9}$$

For given $\mathbf{U}=\mathbf{u}$, $|Z_1|,\ldots,|Z_k|$ are independent; hence they are associated, and the first term on the right-hand side of (5.2.9) is nonnegative. For the second term we realize Theorem 5.2.6 implies that $E\{g_i(|\mathbf{Z}|)|\mathbf{U}=\mathbf{u}\}$ is nondecreasing in $|\mathbf{u}|$ (in each argument) for $i=1,2$; since the components of $|\mathbf{U}|$ are associated, by Theorem 5.2.2(e) it is also nonnegative. Therefore the right-hand side of (5.2.9) is nonnegative. ∎

We note that by combining Theorem 5.2.7 with Theorem 5.2.4, Theorem 5.2.7 can be applied to derive inequalities for the probabilities of rectangles through association. In particular, if $\mathbf{Z}$ is a multivariate normal or multivariate t variable and if the covariance matrix has the structure l (see Definition 2.2.1), then $\mathbf{Z}$ can be written in the form $\mathbf{Z}=\mathbf{X}+\mathbf{U}$ such that the components of $\mathbf{X}$ are independent normal variables and the components of $|\mathbf{U}|$ are associated. Therefore certain inequalities already given in Chapters 2 and 3 now follow from this general result as special cases.

5.3. POSITIVE DEPENDENCE BY MIXTURE OF DISTRIBUTIONS

The conditions for positive dependence and association studied in the last two sections basically reflect the property that the random variables are more likely to hang together. Other concepts of positive (and negative) dependence were studied by several other authors. Harris (1970), Brindley and Thompson (1972), and Marshall (1975) defined and studied a multivariate generalization of the property of monotone failure rate; their approaches were motivated by applications in reliability theory. Yanagimoto (1972) and Shaked (1977a) unified some of the notions of positive dependence by introducing a family of concepts of dependence. Shaked's definition depends on the TP_2 property of the joint density function or a particular integral of the density. His results also have an application in reliability theory.

An important positive dependence property can be defined through the mixture of distribution functions. Let $f(\mathbf{x})$ denote the distribution function of $\mathbf{X}=(X_1,\ldots,X_k)'$. F is called a mixture of distributions if there exist distribution functions $G_u^{(i)}(x_i)$ (that depend on u and i) and $H(u)$ such that F is of the form

$$F(\mathbf{x}) = \int \prod_{i=1}^{k} G_u^{(i)}(x_i)\,dH(u). \tag{5.3.1}$$

We have already seen a number of special cases of this type of distribution. For example, as in Section 3.3, if $\mathbf{X}$ is a normal variable (whose covariance matrix has a certain structure) or a t, chi-square, or F variable, then its distribution function is a mixture. In those cases we were able to obtain probability inequalities through a conditional argument because the components of $\mathbf{X}$ are conditionally independent for fixed $U=u$. In statistics mixture of distributions arises in a variety of circumstances. In the area of statistical decision theory the variable u plays the role of a parameter, and $H(u)$ the prior distribution. For most problems in multiple comparisons, which will be discussed in Chapter 8, the probability of a correct decision can usually be given in a form defined by a mixture of distributions. This is so because of the following fact: For independent random variables $Y_1,\ldots,Y_k$, U, and Borel-measurable functions $\psi_1,\ldots,\psi_k$, if we define

$$X_i = \psi_i(Y_i, U), \qquad i = 1,\ldots,k,$$

then the distribution of $\mathbf{X}$ is a mixture.

Let us now consider k families of distribution functions

$$\mathcal{G}_i = \left\{ G_u^{(i)}(x_i) : u \in \mathcal{U} \right\}, \qquad i = 1,\ldots,k. \tag{5.3.2}$$

In applications $G_u^{(i)}(x_i)$ is the distribution function of X_i, given $U=u$. In most common applications it can be assumed that the $\mathcal{G}_i$s are stochastically increasing families. This is equivalent to saying that according to Definition 5.1.2 X_i is positively regression dependent on U for each i. If this condition is met, then by applying Lemma 2.2.1, we can prove

Theorem 5.3.1. Let $\mathbf{X}$ have a distribution F, which is of the form given in (5.3.1). If $\mathcal{G}_i$, defined in (5.3.2), is stochastically increasing for each i, then the inequalities

$$P\left[\bigcap_{i=1}^{k} \{X_i \leqslant a_i\}\right] \geqslant P\left[\bigcap_{i\in C} \{X_i \leqslant a_i\}\right] P\left[\bigcap_{i\notin C} \{X_i \leqslant a_i\}\right]$$

$$\geqslant \prod_{i=1}^{k} P[X_i \leqslant a_i], \tag{5.3.3}$$

$$P\left[\bigcap_{i=1}^{k} \{X_i > a_i\}\right] \geqslant P\left[\bigcap_{i\in C} \{X_i > a_i\}\right] P\left[\bigcap_{i\notin C} \{X_i > a_i\}\right]$$

$$\geqslant \prod_{i=1}^{k} P[X_i > a_i] \tag{5.3.4}$$

hold for every subset C of $\{1,\ldots,k\}$.

This result provides lower bounds for the joint distribution when the conditional distributions are monotone in u in the same direction. On the other hand, if in a given problem $G_u^{(i)}(x_i)$ is nondecreasing in u for some i and nonincreasing in u for some other i, then again by Lemma 2.2.1 an upper bound for the joint probability can be obtained. A special case of this result was already observed for the multivariate t distribution in Chapter 3 (see Problems 1 and 2 of Chapter 3).

The remainder of this section concerns the special but important case in which for given $U=u$, $X_1,\ldots,X_k$ are conditionally independent and identically distributed. We shall also call their distribution a mixture with a common marginal distribution. Let us consider the more general case in which $\mathbf{X}$ is an sk-dimensional vector, i.e., $\mathbf{X}=(\mathbf{X}^{(1)},\ldots,\mathbf{X}^{(k)})'$, where each of the $\mathbf{X}^{(i)}$s is an s-dimensional random variable. Then the joint distribution function of $\mathbf{X}$ is of the form

$$F(\mathbf{x}) = \int \prod_{i=1}^{k} G_{\mathbf{u}}(\mathbf{x}^{(i)})\, dH(\mathbf{u}), \qquad \mathbf{x} = (\mathbf{x}^{(1)},\ldots,\mathbf{x}^{(k)})'; \quad (5.3.5)$$

here $G_{\mathbf{u}}$ is a member of the family of distributions $\mathcal{G} = \{G_{\mathbf{u}} : \mathbf{u} \in \mathcal{U}\}$ for each given $\mathbf{u}$, and $H(\mathbf{u})$ is a probability distribution on $\mathcal{U}$, which is a subset in $\mathcal{R}^q$. Let $A \subset \mathcal{R}^s$ be Borel measurable; we then consider the probability of the joint event $\cap_{i=1}^{k}\{\mathbf{X}^{(i)} \in A\}$.

This type of random variable (or event) has been called positively dependent (Jensen, 1971), events which are almost independent (Dykstra, Hewett, and Thompson (1973)), conditionally independent and identically distributed (Tong, 1977), positively dependent by mixture (Shaked, 1977b), and possibly other names. We have already seen some special cases for normal variables and t variables (Theorem 2.3.4) and a more general result due to Šidák (1973) (Theorem 3.3.1). There we simply say that the random variables are "exchangeable in a certain fashion" without elaboration. In the following we shall investigate this concept further.

In probability and statistics it is customary to say that a finite number of k random variables $\mathbf{X}^{(1)},\ldots,\mathbf{X}^{(k)}$ are exchangeable if their joint distribution function is permutation invariant. As pointed out by Johnson and Kotz (1972, p. 3), "symmetric" (instead of "exchangeable") seems to be a better word. The definition of exchangeability for an infinite sequence of random variables, as given below, appears to be more appropriate.

Definition 5.3.1 (see, e.g., Loève (1963, p. 364)). Let $\mathbf{X}^{(1)},\mathbf{X}^{(2)},\ldots$ be an infinite sequence of random variables. We say that they are exchangeable if the joint distribution of $(\mathbf{X}^{(j_1)},\ldots,\mathbf{X}^{(j_k)})'$ is identical to that of $(\mathbf{X}^{(1)},\ldots,\mathbf{X}^{(k)})'$ for every subset $(j_1,\ldots,j_k)$ of $\{1,2,\ldots\}$ and every finite k.

It is clear that (by letting $k=1$) this condition implies that the marginal distributions are identical, but the converse is obviously false. A theorem of de Finetti stated below ties together this concept of exchangeability and the mixture of distributions with a common marginal distribution. A more precise proof of the theorem can be found in Loève (1963, p. 365), and some complements and applications of the theorem were discussed by Kingman (1978).

Theorem 5.3.2. For an infinite sequence of random variables the concept of exchangeability is equivalent to that of conditional independence with a common marginal distribution.

This theorem says that the joint distribution F of $(\mathbf{X}^{(1)},\ldots,\mathbf{X}^{(k)})'$ is of the form given in (5.3.5) if and only if $\{\mathbf{X}^{(1)},\ldots,\mathbf{X}^{(k)}\}$ is a subset of an infinite sequence of exchangeable random variables. It is important to note that if a finite sequence of random variables $\mathbf{X}^{(1)},\ldots,\mathbf{X}^{(n)}$ is exchangeable (i.e., its joint distribution is permutation invariant), then the distribution of a subset of this finite sequence may not be a mixture with a common marginal distribution. A simple example is that for which the n-dimensional random variable $(X_1,\ldots,X_n)'$ has a multivariate normal distribution with equal means, equal variances, and equal correlations ρ. In this case its joint distribution is a mixture with a common marginal distribution if and only if $\rho \geqslant 0$, but it is permutation invariant for all $\rho \in (-1/(n-1), 1)$.

The following theorem gives an inequality for random variables whose joint distribution is a mixture with a common marginal distribution.

Theorem 5.3.3. Let $\mathbf{X}=(\mathbf{X}^{(1)},\ldots,\mathbf{X}^{(k)})'$ have a joint distribution F which is a mixture with a common marginal distribution as given in (5.3.5). Let $A \subset \mathfrak{R}^s$ be a Borel-measurable set, and define

$$\beta(r) = P\left[\bigcap_{i=1}^{r} \{\mathbf{X}^{(i)} \in A\}\right] \tag{5.3.6}$$

for $r=1,\ldots,k$. Then

$$\beta(r)\beta(k-r) \geqslant \beta(r-1)\beta(k-r+1) \tag{5.3.7}$$

holds for all $r \geqslant k/2$, and the inequalities

$$\beta(k) \geqslant [\beta(r)]^{k/r} \geqslant \cdots \geqslant [\beta(2)]^{k/2} \geqslant [\beta(1)]^{k} \tag{5.3.8}$$

hold for all $k > r \geqslant 2$.

The proof of (5.3.7) is left to the reader. The proof of (5.3.8) follows immediately from Theorem 3.3.1 and the lemma given below, the proof of which is omitted.

Lemma 5.3.1. *F* is a mixture with a common marginal distribution as given in (5.3.5) if and only if there exist p-dimensional independent and identically distributed random variables $\mathbf{Y}_1,\ldots,\mathbf{Y}_k$, a q-dimensional random variable $\mathbf{U}$ that is independent of the $\mathbf{Y}_i$s, and a Borel-measurable function $\psi: \mathfrak{R}^{p+q} \to \mathfrak{R}^s$ such that $\psi(\mathbf{Y}_1,\mathbf{U}),\ldots,\psi(\mathbf{Y}_k,\mathbf{U})$ have a joint distribution F.

We note that here $\mathbf{X}^{(1)},\ldots,\mathbf{X}^{(k)}$ play the role played by $\psi(\mathbf{Y}_1,\mathbf{U}),\ldots,\psi(\mathbf{Y}_k,\mathbf{U})$ in Theorem 3.3.1. Hence the inequality in (5.3.8) is essentially another version of Theorem 3.3.1, except that it is now given under conditions on the joint distribution function. A sharper and more general result involves the concept of majorization, and that result will be given in the next chapter (Theorem 6.2.11).

PROBLEMS

1. Show that if $(X_1,X_2)'$ is a bivariate normal variable with correlation ρ, then all the conditions for the various concepts of dependence given in Theorems 5.1.1 and 5.2.1 are satisfied if and only if $\rho \geqslant 0$; and that X_1,X_2 are independent if and only if $\rho = 0$.
2. Show that if X_1,X_2 are binary random variables, then they are associated if and only if their covariance is nonnegative, that in fact $(X_1,X_2)'$ is PLRD if and only if the covariance is nonnegative, and that X_1,X_2 are independent if and only if the covariance is zero (Esary, Proschan, and Walkup (1967)).
3. Verify all the statements in the proof of Theorem 5.1.2.
4. Verify the details in the identity given in (5.2.3).
5. Let g be a monotonically nonincreasing function. Show that if $X_1,\ldots,X_k$ are associated, then $g(X_1),\ldots,g(X_k)$ are associated; in particular, $-X_1,\ldots,-X_k$ are associated.

Problems 6–8 were given in Lehmann (1966) and Esary, Proschan, and Walkup (1967) as examples.

6. Let $X_1,\ldots,X_k$ be independent random variables, and let $X_{(1)},\ldots,X_{(k)}$ denote the order statistics. Show that $X_{(1)},\ldots,X_{(k)}$

are associated. Therefore

$$P\left[\bigcap_{i=1}^{k}\{X_{(i)}\leqslant a_i\}\right]\geqslant\prod_{i=1}^{k}P[X_{(i)}\leqslant a_i],$$

and $\mathrm{cov}(X_{(i)},X_{(j)})\geqslant 0$ holds for all $i\neq j$ (Bickel, 1967).

7. Let $X_1,\ldots,X_k$ be independent random variables, and let $S_j=\sum_{i=1}^{j}X_i$ $(j=1,\ldots,k)$ denote the cumulative sums. Show that $S_1,\ldots,S_k$ are associated; as a consequence, the inequality

$$P\left[\bigcap_{j=1}^{k}\{S_j\leqslant a_j\}\right]\geqslant\prod_{j=1}^{k}P[S_j\leqslant a_j]$$

holds (Robbins, 1954).

8. Show that the F-variables $F_1,\ldots,F_k$ defined in (3.2.8) are associated.

9. Let $(X_1,\ldots,X_k)'$ be a multivariate normal variable. Show that if the covariance matrix $\mathbf{\Sigma}$ has the structure l, then $|X_1|,\ldots,|X_k|$ are associated; if $\mathbf{\Sigma}$ has the structure l with $\lambda_i\geqslant 0$, then $X_1,\ldots,X_k$ are associated.

10. For $i=1,\ldots,k$ let $\mathbf{X}^{(i)}$ denote the random variable $\psi(\mathbf{Y}_i,\mathbf{U})$ that was defined in Theorem 3.3.1. Let C be a subset of $\{1,\ldots,k\}$, and for a Borel-measurable set A define

$$Z_1=\prod_{i\in C}\chi_A(\mathbf{X}^{(i)}),\qquad Z_2=\prod_{i\notin C}\chi_A(\mathbf{X}^{(i)}),$$

where

$$\chi_A(\mathbf{x}^{(i)})=\begin{cases}1 & \text{if } \mathbf{x}^{(i)}\in A,\\ 0 & \text{otherwise.}\end{cases}$$

Show that Z_1 and Z_2 are associated.

11. Let $\mathbf{X}=(X_1,\ldots,X_k)'$ be a multivariate normal variable with equal means μ, equal variances σ^2, and equal correlations $\rho\geqslant 0$. Express the joint distribution F of $\mathbf{X}$ in the form of (5.3.5) (with $s=1$) by specifying the functional forms of $G_{\mathbf{u}}$ and $H(\mathbf{u})$. Also, specify the transformation needed when defining the X_is by $X_i=\psi(U_i,U_0)$, where $U_0,U_1,\ldots,U_k$ are independent $N(0,1)$ variables.

12. Let $\mathbf{t}=(t_1,\ldots,t_k)'$ be the multivariate t variable defined in (3.1.1), where $\mathbf{R}=(\rho_{ij})$ is of the form

$$\rho_{ij}=\begin{cases}1 & \text{for } i=j,\\ \rho\geqslant 0 & \text{for } i\neq j.\end{cases}$$

Express the joint distribution function of $\mathbf{t}$ in the form of (5.3.5) (with $s=1$) by specifying $G_{\mathbf{u}}$ and $H(\mathbf{u})$.

13. Let $\{X_i\}_{i=1}^{\infty}$ be an infinite sequence of random variables such that the correlation between X_i and X_j is ρ for all $i \neq j$. Show that if $X_1, X_2, \ldots$ are exchangeable, then $\rho \geqslant 0$.

14. Give a proof for the inequality in (5.3.7). (Hint: See Problem 14, Chapter 2.)

REFERENCES

Ahmed, A. H. N., León, R. V., and Proschan, F. (1978). Generalization of associated random variables, with applications. Tech. Rep. No. M468. Department of Statistics, Florida State Univ., Tallahassee, Florida.

Ahmed, A. H. N., Langberg, N. A., León, R. V., and Proschan, F. (1978). Two concepts of positive dependence, with applications in multivariate analysis. Tech. Rep. No. M486, Department of Statistics, Florida State Univ., Tallahassee, Florida.

Ahmed, A. H. N., Langberg, N. A., León, R. V., and Proschan, F. (1979). Partial ordering of positive quadrant dependence, with applications. Tech. Rep. No. M482, Department of Statistics, Florida State Univ., Tallahassee, Florida.

Bickel, P. J. (1967). Some contributions to the theory of order statistics. *Proc. Fifth Berkeley Symp. Math. Statist. Probab.* **1** (L. M. LeCam, and J. Neyman, eds.), 575–591. Univ. of California Press, Berkeley, California.

Brindley, E. C., Jr., and Thompson, W. A., Jr. (1972). Dependence and aging aspects of multivariate survival. *J. Amer. Statist. Assoc.* **67**, 822–830.

Dykstra, R. L., Hewett, J. E., and Thompson, W. A., Jr. (1973). Events which are almost independent. *Ann. Statist.* **1**, 674–681.

Eaton, M. L., and Perlman, M. D. (1977). Reflection groups, generalized Schur functions and the geometry of majorization. *Ann. Probab.* **5**, 829–860.

Esary, J. D., and Proschan, F. (1972). Relationships among some concepts of bivariate dependence. *Ann. Math. Statist.* **43**, 651–655.

Esary, J. D., Proschan, F., and Walkup, D. W. (1967). Association of random variables, with applications. *Ann. Math. Statist.* **38**, 1466–1474.

Fortuin, C. M., Kastelyn, P. W., and Ginibre, J. (1971). Correlation inequalities on some partially ordered sets. *Comm. Math. Phys.* **22**, 89–103.

Harris, R. (1970). A multivariate definition for increasing hazard rate distribution functions. *Ann. Math. Statist.* **41**, 713–717.

Jensen, D. R. (1971). A note of positive dependence and the structure of bivariate distributions. *SIAM J. Appl. Math.* **20**, 749–753.

Jogdeo, K. (1968). Characterizations of independence in certain families of bivariate and multivariate distributions. *Ann. Math. Statist.* **39**, 433–441.

Jogdeo, K. (1977). Association and probability inequalities. *Ann. Statist.* **5**, 495–504.

Johnson, N. L., and Kotz, S. (1972). *Distributions in Statistics: Continuous Multivariate Distributions*. Wiley, New York.

Karlin, S. (1968). *Total Positivity*, Vol. 1. Stanford Univ. Press, Stanford, California.

Kemperman, J. H. B. (1977). On the FKG-inequality for measures on a partially ordered space. *Indag. Math.* **39**, 313–331.

Khintchine, A. Y. (1938). On unimodal distributions. *Izv. Nauchno. Issled Inst. Mat. Mech. Tomsk.* Gos. Univ. **2**, 1–7.

Kingman, J. F. C. (1978). Uses of exchangeability. *Ann. Probab.* **6**, 183–197.

Kruskal, W. H. (1958). Ordinal measures of association. *J. Amer. Statist. Assoc.* **53**, 814–864.

Lehmann, E. L. (1955). Ordered families of distributions. *Ann. Math. Statist.* **26**, 399–419.

Lehmann, E. L. (1959). *Testing Statistical Hypotheses.* Wiley, New York.

Lehmann, E. L. (1966). Some concepts of dependence. *Ann. Math. Statist.* **37**, 1137–1153.

Loève, M. (1963). *Probability Theory*, 3rd ed. Van Nostrand-Reinhold, New York.

Marshall, A. W. (1975). Multivariate distributions with monotone hazard rate. In *Reliability and Fault Tree Analysis* (R. E. Barlow, J. B. Fussell, and N. D. Singpurwalla, eds.), pp. 259–284. SIAM, Philadelphia, Pennsylvania.

Perlman, M. D., and Olkin, I. (1980). Unbiasedness of invariant tests for MANOVA and other multivariate problems. *Ann. Statist.* (to appear).

Robbins, H. (1954). A remark on the joint distribution of cumulative sums. *Ann. Math. Statist.* **25**, 614–616.

Shaked, M. (1977a). A family of concepts of dependence for bivariate distributions. *J. Amer. Statist. Assoc.* **72**, 642–654.

Shaked, M. (1977b). A concept of positive dependence for exchangeable random variables. *Ann. Statist.* **5**, 505–515.

Sherman, S. (1955). A theorem on convex sets with applications. *Ann. Math. Statist.* **26**, 763–766.

Šidák, Z. (1973). A chain of inequalities for some types of multivariate distributions, with nine special cases. *Apl. Mat.* **18**, 110–118.

Tong, Y. L. (1977). An ordering theorem for conditionally independent and identically distributed random variables. *Ann. Statist.* **5**, 274–277.

Yanagimoto, T. (1972). Families of positively dependent random variables. *Ann. Inst. Statist. Math.* **24**, 559–573.

CHAPTER

6

Inequalities Via Majorization and Weak Majorization

6.1. INTRODUCTION

The concept of majorization basically concerns the comparison of degrees of diversity among the components between two vectors. For fixed k let us consider two real vectors

$$\mathbf{a} = (a_1, \ldots, a_k)', \qquad \mathbf{b} = (b_1, \ldots, b_k)'; \tag{6.1.1}$$

and let $a_{[1]} \geqslant \cdots \geqslant a_{[k]}, b_{[1]} \geqslant \cdots \geqslant b_{[k]}$ denote the ordered values of the components of $\mathbf{a}$ and $\mathbf{b}$.

Definition 6.1.1. $\mathbf{a}$ is said to majorize $\mathbf{b}$, or $\mathbf{b}$ is said to be majorized by $\mathbf{a}$, in symbols $\mathbf{a} \succ \mathbf{b}$, if

$$\sum_{i=1}^{k} a_i = \sum_{i=1}^{k} b_i \quad \text{and}$$

$$\sum_{i=1}^{r} a_{[i]} \geqslant \sum_{i=1}^{r} b_{[i]} \qquad \text{for} \quad r = 1, \ldots, k-1. \tag{6.1.2}$$

The condition that the sums of the components be equal usually makes the comparison more meaningful. Subject to a common total, the components of **a** are more diverse if $\mathbf{a} \succ \mathbf{b}$. This can be seen partially from the facts that

$$\mathbf{a} \succ (\bar{a},\ldots,\bar{a}) \tag{6.1.3}$$

for every **a**, where

$$\bar{a} = \frac{1}{n} \sum_{i=1}^{n} a_i,$$

and

$$\left(\sum_{i=1}^{n} a_i, 0 \ldots, 0 \right) \succ \mathbf{a} \tag{6.1.4}$$

when the components of **a** are nonnegative. It is known that (see Marshall and Olkin (1979, p. 64)) if $\mathbf{a} \succ \mathbf{b}$, then

$$\sum_{i=1}^{n} |a_i - \bar{a}|^t \geqslant \sum_{i=1}^{n} |b_i - \bar{b}|^t \tag{6.1.5}$$

holds for all $t \geqslant 1$, where $\bar{a} = \bar{b}$. Therefore by taking $t=2$, majorization is stronger than the variance concept when used as a measurement of diversity. This also yields $\Sigma_{i=1}^{k} a_i^2 \geqslant \Sigma_{i=1}^{k} b_i^2$. Hence if $\mathbf{a} \succ \mathbf{b}$, then **b** is closer to the origin.

The condition for weak majorization is more general. We define

Definition 6.1.2. **a** is said to weakly majorize **b**, or **b** is said to be weakly majorized by **a**, in symbols $\mathbf{a} \succ\succ \mathbf{b}$, if

$$\sum_{i=1}^{r} a_{[i]} \geqslant \sum_{i=1}^{r} b_{[i]} \qquad \text{for} \quad r = 1,\ldots,k. \tag{6.1.6}$$

Here the condition that the components of the vectors have a common total is dropped. Therefore if two real vectors are not comparable through majorization, they might still be comparable through weak majorization. In particular, $\mathbf{a} \succ\succ \mathbf{b}$ holds if $\mathbf{a} \succ \mathbf{b}$ or (and) $\mathbf{a} \geqslant \mathbf{b}$ (i.e., $a_i \geqslant b_i$ for each i).

Our main concern in this chapter is to study probability inequalities in multivariate distributions, using majorization and weak majorization as a tool. If two parameter vectors or two sets can be partially ordered through majorization or weak majorization, then under suitable conditions the corresponding probability contents can be ordered. It is in this capacity that majorization and weak majorization play an important role in deriving probability inequalities. For a complete discussion of the theory of majorization and weak majorization and their applications, the reader is referred to the authoritative new book by Marshall and Olkin (1979). In this chapter we shall not make any attempt to cover all such major inequalities in this area (this is neither possible nor necessary). Instead we shall restrict our attention to those inequalities that appear to be more relevant in statistical applications.

In the next section we first study several basic preservation theorems under integral transformations. Their implications in location and scale parameter families will be discussed. In Section 6.3 concepts of stochastic ordering of random variables will be discussed; probability inequalities can be obtained by comparing random variables through the ordering of their parameter vectors. Section 6.4 concerns some probability inequalities for order statistics from heterogeneous populations; this in turn depends on inequalities for the number of successes in independent but not necessarily identical Bernoulli trials.

Before proceeding, we summarize in the next theorem several basic known properties of majorization and weak majorization. We recall the definition that a $k\times k$ matrix $\mathbf{P}=(p_{ij})$ is doubly stochastic if $p_{ij}\geqslant 0$ for all i,j and that all the row sums and column sums are one.

Theorem 6.1.1. (a) $\mathbf{a}\succ\mathbf{b}$ if and only if there exists a doubly stochastic matrix $\mathbf{P}$ such that $\mathbf{b}=\mathbf{Pa}$.

(b) $\mathbf{a}\succ\mathbf{b}$ if and only if there exists a finite number of real vectors $\mathbf{c}_1,\ldots,\mathbf{c}_n$ such that

$$\mathbf{a}=\mathbf{c}_1\succ\mathbf{c}_2\succ\cdots\succ\mathbf{c}_{n-1}\succ\mathbf{c}_n=\mathbf{b}, \tag{6.1.7}$$

and such that for all i, $\mathbf{c}_i$ and $\mathbf{c}_{i+1}$ differ in two coordinates only.

(c) Let $\mathbf{a}$, $\mathbf{b}$ be $k_1\times 1$ real vectors and $\mathbf{c},\mathbf{d}$ be $k_2\times 1$ real vectors, and let $(\mathbf{a},\mathbf{c})'$ and $(\mathbf{b},\mathbf{d})'$ be two $(k_1+k_2)\times 1$ real vectors. If $\mathbf{a}\succ\mathbf{b}$ and $\mathbf{c}\succ\mathbf{d}$, then (after the components are properly reordered) $(\mathbf{a},\mathbf{c})'\succ(\mathbf{b},\mathbf{d})'$ holds; in particular, since $\mathbf{c}\succ\mathbf{c}$, $(\mathbf{a},\mathbf{c})'\succ(\mathbf{b},\mathbf{c})'$ holds for all $\mathbf{c}$.

(d) $\mathbf{a}\succ\succ\mathbf{b}$ if and only if there exists a $\mathbf{c}$ such that $\mathbf{a}\succ\mathbf{c}$ and $\mathbf{c}\geqslant\mathbf{b}$ (i.e., $c_i\geqslant b_i$ for each i).

The proofs of (a) and (b) can be found in Hardy, Littlewood, and Pólya (1959, pp. 49 and 47, respectively). The proof of (c) follows from (a), and the proof of (d) can be found in Marshall and Olkin (1979, p. 123) or Nevius, Proschan, and Sethuraman (1977b).

The results in the above theorem will be referred to and used at various places in this chapter. Because of (b), we may assume that only two coordinates are different when proving inequalities, and this simplifies the work considerably. This result actually illustrates the concept of majorization: If $\mathbf{a} \succ \mathbf{b}$, we may conclude that there is more diversity among the components of $\mathbf{a}$. Now if we move downhill toward equality by leveling off two components at a time, we can change $\mathbf{a}$ into $\mathbf{b}$ in a finite number of steps. It is in this sense we say that if $\mathbf{a} \succ \mathbf{b}$, then $\mathbf{b}$ is an "average" of $\mathbf{a}$ (Hardy, Littlewood, and Pólya (1959, p. 49)).

6.2. SOME PRESERVATION THEOREMS UNDER INTEGRAL TRANSFORMS

The first theorem we study here is a special case of Mudholkar's generalization of Anderson's theorem (see Theorem 4.2.2). We recall that the generalization extends the condition of symmetry about the origin to an invariance condition under a group of linear transformations. Let us then consider the group of permutations, which is related to majorization, and apply Theorem 4.2.2 to obtain a probability inequality via majorization. This leads to the following theorem due to Mudholkar (1966).

Theorem 6.2.1. Let $f(\mathbf{x})\colon \mathcal{R}^k \to [0, \infty)$ be permutation invariant and unimodal. Let $A \subset \mathcal{R}^k$ be permutation invariant and convex. If $\int_A f(\mathbf{x})\, d\mathbf{x} < \infty$ (in the Lebesgue sense) and if $\boldsymbol{\xi} \succ \boldsymbol{\theta}$, then

$$\int_A f(\mathbf{x}+\boldsymbol{\theta})\, dx \geqslant \int_A f(\mathbf{x}+\boldsymbol{\xi})\, d\mathbf{x}, \tag{6.2.1}$$

or equivalently,

$$\int_{A+\boldsymbol{\theta}} f(\mathbf{x})\, d\mathbf{x} \geqslant \int_{A+\boldsymbol{\xi}} f(\mathbf{x})\, d\mathbf{x}. \tag{6.2.2}$$

Proof. The key argument of the proof depends on Birkhoff's theorem (see Marshall and Olkin (1979, p. 19)), which says that the set of $k \times k$ doubly stochastic matrices is a convex polyhedron with $N = k!$ permutation matrices $\mathbf{P}_1, \ldots, \mathbf{P}_N$ as the vertices. Thus every doubly stochastic matrix $\mathbf{P}$ is of the form

$$\mathbf{P} = \sum_{i=1}^{N} \alpha_i \mathbf{P}_i, \qquad \alpha_i \geqslant 0, \quad \sum_{i=1}^{N} \alpha_i = 1. \tag{6.2.3}$$

It then follows from Theorem 6.1.1 that if in Theorem 4.2.2 $\mathcal{G}$ is the group of permutations and if $\boldsymbol{\xi} \succ \boldsymbol{\theta}$, then we can write $\boldsymbol{\theta} = \boldsymbol{\alpha}(\boldsymbol{\xi})$ as given in (4.2.4). Now the proof follows immediately from Theorem 4.2.2 because both f and A are $\mathcal{G}$ (permutation) invariant. ■

A function ψ of k arguments is said to be Schur-convex (Schur-concave) if $\mathbf{a} \succ \mathbf{b}$ implies $\psi(\mathbf{a}) \geqslant \psi(\mathbf{b})$ $(\psi(\mathbf{a}) \leqslant \psi(\mathbf{b}))$. An important special case occurs if ψ is of the form

$$\psi(\mathbf{a}) = \sum_{i=1}^{k} h(a_i). \tag{6.2.4}$$

The following lemma was first noted by Schur (see Marshall and Olkin (1979, p. 64)).

Lemma 6.2.1. Let $I \subset \mathcal{R}^1$ be an interval. If $h: I \to \mathcal{R}^1$ is convex (concave), then $\psi(\mathbf{a})$ as defined in (6.2.4) is Schur-convex (Schur-concave) for $\mathbf{a} \in I \times \cdots \times I$.

The essential part of the proof is that for the case $k=2$. After that is done Theorem 6.1.1 can be applied to complete the argument. It should be noted that the converse is also true; i.e., if ψ is Schur-convex (Schur-concave) and h is continuous, then h is convex (concave) (see Problem 1). In the previous section we saw such a function; i.e., the function $\psi(\mathbf{a}) = \sum_{i=1}^{k} |a_i - \bar{a}|^t$ is Schur-convex for $t \geqslant 1$. It is known that all Schur (convex or concave) functions are permutation invariant. Also, it is known that

Lemma 6.2.2. A necessary and sufficient condition that a permutation-invariant and differentiable function ψ be Schur-convex (Schur-concave) is that

$$(a_i - a_j)\left(\frac{\partial}{\partial a_i}\psi(\mathbf{a}) - \frac{\partial}{\partial a_j}\psi(\mathbf{a})\right) \geqslant (\leqslant) 0 \tag{6.2.5}$$

for all $i \neq j$.

This was due to Schur and Ostrowski (see Marshall and Olkin (1979, p. 57)), and usually this condition is easy to verify.

According to this definition, the conclusion in Theorem 6.2.1 is simply that "the function $\psi(\boldsymbol{\theta})=\int_A f(\mathbf{x}+\boldsymbol{\theta})\,d\mathbf{x}$ is Schur-concave." This theorem implies the following corollary which was also given by Mudholkar (1966).

Corollary. Let $\mathbf{X}$ have a density f, and assume that f and A satisfy the conditions stated in Theorem 6.2.1. Let $\mathbf{Y}$ be another random variable that is independent of $\mathbf{X}$, and let $\mathbf{P}$ denote a doubly stochastic matrix. Then

$$P[(\mathbf{X}+\mathbf{PY})\in A] \geqslant P[(\mathbf{X}+\mathbf{Y})\in A]. \tag{6.2.6}$$

The proof follows immediately from the fact that

$$P[(\mathbf{X}+\mathbf{PY})\in A|\mathbf{Y}=\mathbf{y}] \geqslant P[(\mathbf{X}+\mathbf{Y})\in A|\mathbf{Y}=\mathbf{y}]$$

for every $\mathbf{y}$. This corollary is an analogue to Theorem 4.1.3, and the condition here is that $\mathbf{Y}$ majorize $\mathbf{Y}^*=\mathbf{PY}$ with probability one. Some weaker conditions will be considered in Section 6.3 under the concept of stochastic majorization, and more inequalities of this type will be studied there.

When applying Theorem 6.2.1, we no longer require that the set A be symmetric about the origin. Therefore we are now free to obtain inequalities for the probability contents of events which may not even contain the origin (e.g., the cumulative probabilities). But the condition of convexity is still there. In a paper Marshall and Olkin (1974) proved another theorem by replacing this convexity condition with a weaker condition. Their theorem is now stated.

Theorem 6.2.2. Let $\mathbf{X}$ have a density function $f(\mathbf{x})$ that is Schur-concave. If $A\subset\Re^k$ is a Lebesgue-measurable set such that its indicator function is Schur-concave, i.e., if

$$\mathbf{x}\in A \quad \text{and} \quad \mathbf{x}\succ\mathbf{y} \qquad \text{implies} \quad \mathbf{y}\in A, \tag{6.2.7}$$

then

$$\int_{A+\boldsymbol{\theta}} f(\mathbf{x})\,d\mathbf{x} = P[(\mathbf{X}-\boldsymbol{\theta})\in A] \tag{6.2.8}$$

is a Schur-concave function of $\boldsymbol{\theta}$.

Remark. If A is permutation invariant and convex, then its indicator function is Schur-concave (see Problem 8); hence it satisfies (6.2.7). Moreover, a density function f is Schur-concave if and only if the set $\{\mathbf{x}|f(\mathbf{x})\geqslant u\}$ satisfies (6.2.7) for every $u>0$. This implies that if f is permutation invariant and unimodal, then it is Schur-concave. Therefore the conditions in Theorem 6.2.2 are weaker than those in Theorem 6.2.1.

As a special consequence of Theorem 6.2.2, we observe an important result for location parameter families. Its proof follows from the translation invariance property and the fact that $\boldsymbol{\xi}\succ\boldsymbol{\theta}$ if and only if $-\boldsymbol{\xi}\succ-\boldsymbol{\theta}$.

Corollary. Let $\mathbf{X}$ have a density function $f_{\boldsymbol{\theta}}(\mathbf{x})=f(\mathbf{x}-\boldsymbol{\theta})$. If $f(\mathbf{x})$ and A satisfy the conditions in Theorem 6.2.2, then $P_{\boldsymbol{\theta}}[\mathbf{X}\in A]$ is a Schur-concave function of $\boldsymbol{\theta}$.

A more general version of Theorem 6.2.2 was later proved by Marshall and Olkin (1979, p. 100). This is stated below.

Theorem 6.2.3. If ϕ and f are Schur-concave functions, defined on $\mathfrak{R}^k$, then the function ψ defined by

$$\psi(\boldsymbol{\theta}) = \int_{\mathfrak{R}^k}\phi(\boldsymbol{\theta}-\mathbf{x})f(\mathbf{x})\,d\mathbf{x} \tag{6.2.9}$$

is Schur-concave (whenever the integral exists).

Theorem 6.2.2 follows from Theorem 6.2.3 by letting

$$\phi(\boldsymbol{\theta}-\mathbf{x}) = \chi_A(\mathbf{x}-\boldsymbol{\theta}),$$

where χ_A is the indicator function of A. Note that if A satisfies the condition given in (6.2.7), then $\chi_A(\mathbf{x})$, hence $\chi_A(-\mathbf{x})$, is a Schur-concave function of $\mathbf{x}$. Let $\boldsymbol{\xi}\succ\boldsymbol{\theta}$; the first step in proving Theorem 6.2.3 is to apply Theorem 6.1.1 so that all except two components of $\boldsymbol{\theta}$ and $\boldsymbol{\xi}$ are assumed to be equal. Then after using the permutation invariance properties of ϕ and f and after changing variables, the difference $\psi(\boldsymbol{\theta})-\psi(\boldsymbol{\xi})$ can be expressed in the form of an integral over a region in which the integrand is nonnegative (by the Schur-concavity of ϕ and f), which implies $\psi(\boldsymbol{\theta})\geqslant\psi(\boldsymbol{\xi})$. The steps involved are similar to those used in proving Theorem 5.1.3; for details of the proof the reader is referred to Marshall and Olkin (1979). Theorem 6.2.3 was first given by Marshall and Olkin (1974) with a different proof under the additional condition that ϕ, f also be nonnegative.

Note that the function $\psi(\boldsymbol{\theta})$ in (6.2.9) is the convolution of ϕ and f.

If $f(\mathbf{x})$ is of the form

$$f(\mathbf{x}) = \prod_{i=1}^{k} g(x_i) \tag{6.2.10}$$

for some g, then $f(\mathbf{x})$ is Schur-concave if and only if g is log-concave (see Problem 2). This fact is both interesting and important because, in addition to providing an easy check for the Schur-concavity of f, it ties the concepts of the Schur-concavity and unimodality of f together (see Theorem 4.1.2). Therefore if the random variables are independent with a common density g which is log-concave, then both Theorems 6.2.1 and 6.2.2 can be applied. But the application of Theorem 6.2.1 requires the convexity of A, and the application of Theorem 6.2.2 does not. This makes it possible to derive probability inequalities for nonconvex regions, provided that the condition stated in (6.2.7) be satisfied. In the two-dimensional case the condition in (6.2.7) simply requires that for every fixed c the set of points in A on the straight line $x_1+x_2=c$ be an interval with midpoint at $(c/2, c/2)$. For general k a simple case in which Theorem 6.2.2 applies and Theorem 6.2.1 does not is that for

$$A = \{\mathbf{x} | x_i \leqslant b, i=1,\ldots,k\} \cup \{\mathbf{x} | x_i \geqslant a, i=1,\ldots,k\}. \tag{6.2.11}$$

When a and b are chosen to be zero, then $P(A)$ is the probability that all random variables are simultaneously nonpositive or nonnegative.

The next theorem was originally given by Mudholkar (1969). It concerns independent observations from a location parameter family with possibly different location parameters. Its proof is now immediate. Note that the theorem remains true in the general case in which the X_is are not necessarily independent but their joint density $f(\mathbf{x}-\boldsymbol{\theta})$ is such that $f(\mathbf{x})$ is Schur-concave.

Theorem 6.2.4. Let $\mathbf{X}=(X_1,\ldots,X_k)'$ have a density function $\prod_{i=1}^{k} g(x_i-\theta_i)$ for $x_i \in \mathfrak{R}^1, \theta_i \in \mathfrak{R}^1$ $(i=1,\ldots,k)$; and let $T(\mathbf{x})\colon \mathfrak{R}^k \to \mathfrak{R}^1$ be permutation invariant such that it is either a nondecreasing function of a real-valued convex function of $\mathbf{x}$ or a nonincreasing function of a real-valued concave function of $\mathbf{x}$. If $g(x)$ is log-concave, then for every a the probability function $P_{\boldsymbol{\theta}}[T(\mathbf{X}) \leqslant a]$ is Schur-concave in $\boldsymbol{\theta}$; hence $E_{\boldsymbol{\theta}} T(\mathbf{X})$ is Schur-convex in $\boldsymbol{\theta}$.

The following corollary is a special application of Theorem 6.2.4 to the distributions of linear combinations of order statistics when the coefficients are either in ascending or in descending order. This was given by Mudholkar (1969). Again the condition on the distribution of the X_is can be replaced by the weaker condition that the joint density be of the form $f(\mathbf{x}-\boldsymbol{\theta})$ and that $f(\mathbf{x})$ be Schur-concave.

Corollary. Let $X_1,\ldots,X_k$ be independent continuous random variables with densities $g(x_i-\theta_i)$ for $x_i \in \mathcal{R}^1$, $\theta_i \in \mathcal{R}^1$ $(i=1,\ldots,k)$, and let $X_{(1)} \leqslant \cdots \leqslant X_{(k)}$ denote the order statistics. For fixed real numbers $c_1,\ldots,c_k$ let us define

$$T(\mathbf{X}) = \sum_{i=1}^{k} c_i X_{(i)}. \tag{6.2.12}$$

If $g(x)$ is log-concave, then for every a the probability function $P_{\boldsymbol{\theta}}[T(\mathbf{X}) \leqslant a]$ is Schur-concave (Schur-convex) in $\boldsymbol{\theta}$ for

$$0 \leqslant c_1 \leqslant \cdots \leqslant c_k \qquad (c_1 \geqslant \cdots \geqslant c_k \geqslant 0).$$

This corollary follows because for $0 \leqslant c_1 \leqslant \cdots \leqslant c_k$ $(c_1 \geqslant \cdots \geqslant c_k \geqslant 0)$, $\sum_{i=1}^{k} c_i x_{(i)}$ is convex (concave) in $\mathbf{x}$. This type of linear combination of order statistics includes the maximum and the minimum, as a special case we conclude that when the conditions are met, $P_{\boldsymbol{\theta}}\left[\bigcap_{i=1}^{k}\{X_i \leqslant a\}\right]$ and $P_{\boldsymbol{\theta}}\left[\bigcap_{i=1}^{k}\{X_i > a\}\right]$ are Schur-concave in $\boldsymbol{\theta}$.

In the next theorem (due to Marshall and Olkin (1974)) Theorem 6.2.2 is applied to obtain an inequality for cumulative probabilities. As a special consequence the upper bound given depends only on a cumulative probability with equal coordinates.

Theorem 6.2.5. Let $\mathbf{X}=(X_1,\ldots,X_k)'$ have a density $f(\mathbf{x})$ that is Schur-concave. Then the probability functions

$$P\left[\bigcap_{i=1}^{k}\{X_i \leqslant a_i\}\right], \qquad P\left[\bigcap_{i=1}^{k}\{X_i > a_i\}\right] \tag{6.2.13}$$

are Schur-concave in $\mathbf{a}=(a_1,\ldots,a_k)'$. In particular, we have

$$P\left[\bigcap_{i=1}^{k}\{X_i \leqslant a_i\}\right] \leqslant P\left[\bigcap_{i=1}^{k}\{X_i \leqslant \bar{a}\}\right], \tag{6.2.14}$$

$$P\left[\bigcap_{i=1}^{k}\{X_i > a_i\}\right] \leqslant P\left[\bigcap_{i=1}^{k}\{X_i > \bar{a}\}\right], \tag{6.2.15}$$

where

$$\bar{a} = \frac{1}{k} \sum_{i=1}^{k} a_i.$$

Proof. Let A denote the set

$$A = \{\mathbf{x} | x_i \leqslant 0, i=1,\ldots,k\}.$$

Then for $\mathbf{a} \succ \mathbf{b}$, from Theorem 6.2.2 we have

$$P\left[\bigcap_{i=1}^{k} \{X_i - a_i \leqslant 0\}\right] \leqslant P\left[\bigcap_{i=1}^{k} \{X_i - b_i \leqslant 0\}\right],$$

which is equivalent to

$$P\left[\bigcap_{i=1}^{k} \{X_i \leqslant a_i\}\right] \leqslant P\left[\bigcap_{i=1}^{k} \{X_i \leqslant b_i\}\right]. \tag{6.2.16}$$

The proof for the other inequality is similar. ■

Note that here the condition on f is satisfied when $X_1,\ldots,X_k$ are continuous independent and identically distributed random variables with a log-concave density. Since the upper bound given in (6.2.14) (and (6.2.15)) depends on an equicoordinate probability under a distribution that is permutation invariant, it involves a joint probability of exchangeable events only.

Let us now consider the special case in which f is elliptically contoured (see Section 4.3). If the $k \times k$ matrix $\mathbf{\Sigma}$ is a covariance matrix with equal variances and equal correlations and if the function g in (4.3.1) is nonincreasing, then $f(\mathbf{x})$ is Schur-concave. This result, given by Marshall and Olkin (1974), follows from the fact that the set $\{\mathbf{x} | \mathbf{x}'\mathbf{\Sigma}^{-1}\mathbf{x} \leqslant a\}$ is permutation invariant and convex for all $a > 0$ (Problem 1 of Chapter 4 and Problem 8). As a special consequence of this result we observe

Corollary 1. Let $\mathbf{X}$ be a multivariate normal variable with means zero, variances σ^2, and correlations $\rho \in (-1/(k-1), 1)$, or a multivariate t variable with correlations ρ and any number of degrees of freedom. Then Theorem 6.2.5 applies.

In view of the translation invariance property of multivariate normal distribution, this corollary can be restated in a different form as given below. Note that it also follows directly from the corollary of Theorem 6.2.2.

Corollary 2. Let **X** be a multivariate normal variable with mean $\boldsymbol{\mu}$, variances σ^2, and correlations $\rho \in (-1/(k-1), 1)$. Then for every fixed a the probability functions

$$P_{\boldsymbol{\mu}}\left[\bigcap_{i=1}^{k}\{X_i \leqslant a\}\right], \qquad P_{\boldsymbol{\mu}}\left[\bigcap_{i=1}^{k}\{X_i > a\}\right] \tag{6.2.17}$$

are Schur-concave in $\boldsymbol{\mu}$. Hence for every given $\boldsymbol{\mu}$ the upper bounds on the probabilities may be obtained from replacing $\boldsymbol{\mu}$ by

$$\bar{\boldsymbol{\mu}} = \left(\frac{1}{k}\sum_{i=1}^{k}\mu_i, \ldots, \frac{1}{k}\sum_{i=1}^{k}\mu_i\right)'. \tag{6.2.18}$$

An important special case of an earlier result of Olkin (1972) also follows from Theorem 6.2.5 as a special case. That result concerns multivariate beta and F distributions. It is easy to check that (by taking partial derivatives with respect to x_i and x_j and applying Lemma 6.2.2) the following density functions are Schur-concave for $r \geqslant 1$:

$$f(\mathbf{x}) = c\prod_{i=1}^{k} x_i^{r-1}\left(1 - \sum_{j=1}^{k} x_j\right)^{s-1}, \qquad x_i \geqslant 0, \quad \sum_{i=1}^{k} x_i \leqslant 1, \tag{6.2.19}$$

$$f(\mathbf{x}) = c\prod_{i=1}^{k} x_i^{r-1} \Big/ \left(1 + \sum_{j=1}^{k} x_j\right)^{s-1}, \qquad x_i \geqslant 0. \tag{6.2.20}$$

Therefore we observe

Corollary 3. Let **X** have a density $f(\mathbf{x})$ that is of the form given in (6.2.19) or (6.2.20). Then for $r \geqslant 1$ the probability functions in (6.2.13) are Schur-concave in **a**. In particular, the inequalities in (6.2.14) and (6.2.15) hold for all real vectors **a** with nonnegative components.

An application of this corollary yields a new inequality for the bivariate F distribution. For $k=2$ let $(F_1, F_2)'$ be a bivariate F-variable such that F_1 and F_2 are defined in Theorem 3.2.2 with

$n_1 = n_2$. Then it can be verified easily that

$$P\left[\bigcap_{i=1}^{2}\{F_i > a^*\}\right] = P\left[\bigcap_{i=1}^{2}\{X_i > a\}\right].$$

where $(X_1, X_2)'$ has density of the form (6.2.20) with $r = n_1/2$ and a depends on a^*. This implies the following inequality due to Olkin (1972):

$$P\left[\bigcap_{i=1}^{2}\{F_i > a^*\}\right] \geqslant P[F_1 > 2a^*] \tag{6.2.21}$$

(the lower bound depends only on a univariate F probability). An application of this inequality to an analysis-of-variance problem will be discussed in Chapter 8, and a comparison with the bound obtained in Theorem 3.2.2 (through the concept of mixture of distributions) will be noted.

The next theorem, from Marshall and Olkin (1974), concerns the preservation property through a mixture of distributions. Let us follow the same notation used in Theorem 3.3.1 or Lemma 5.3.1 and assume that the univariate random variables $Y_1, \ldots, Y_k$, U are independent and that $Y_1, \ldots, Y_k$ are identically distributed with a common density g. For $\psi: \mathcal{R}^2 \to \mathcal{R}^1$ let us define $X_i = \psi(Y_i, U)$, $i = 1, \ldots, k$.

Theorem 6.2.6. Let $\psi(y, u)$ be linear and increasing in y for all fixed u. If $Y_1, \ldots, Y_k$ and U are independent and the common density g of $Y_1, \ldots, Y_k$ is log-concave, then the density $f(\mathbf{x})$ of $\mathbf{X} = (X_1, \ldots, X_k)'$ is Schur-concave.

The proof depends on a conditional argument for given $U = u$, and the details are left to the reader. This theorem yields important applications when the distribution is a mixture. In particular, it implies that most of the density functions described in Section 3.3 are Schur-concave.

Eaton and Perlman (1977) took a more general approach and obtained a theorem that implies Theorems 6.2.1–6.2.3 as special cases. They considered a closed subgroup $\mathcal{G}$ of the group of $k \times k$ orthogonal matrices acting on $\mathcal{R}^k$. For $\mathbf{x}, \mathbf{x}^* \in \mathcal{R}^k$ define "$\mathbf{x} \succcurlyeq \mathbf{x}^*$" if $\mathbf{x}^* \in C(\mathbf{x})$, where $C(\mathbf{x})$ is the convex hull of the $\mathcal{G}$-orbit of $\mathbf{x}$. Thus the relation "$\succcurlyeq$" defines a partial ordering, and geometrically "$\mathbf{x} \succcurlyeq \mathbf{x}^*$"

implies that $\mathbf{x}^*$ is in some sense closer to the origin than $\mathbf{x}$ is. An extended real-valued function (which may take values $\pm\infty$) $\psi: \mathcal{R}^k \to [-\infty, \infty]$ is called $\mathcal{G}$-monotone decreasing if $\mathbf{x} \succcurlyeq \mathbf{x}^*$ implies that $\psi(\mathbf{x}) \leqslant \psi(\mathbf{x}^*)$. Now let $\mathcal{C}_{\mathcal{G}}$ denote the class of $\mathcal{G}$-monotone decreasing functions. The problem they studied concerns the conditions on $\mathcal{G}$ such that $\mathcal{C}_{\mathcal{G}}$ is closed under convolution (integrating with respect to the Lebesgue measure on $\mathcal{R}^k$). Eaton and Perlman (1977) proved that

Theorem 6.2.7. If $\mathcal{G}$ is a reflection group, then $\mathcal{C}_{\mathcal{G}}$ is closed under convolution.

As a consequence of this general theorem they showed

Corollary. Let $\mathcal{G}$ be a reflection group. If $f \geqslant 0$ and ϕ are $\mathcal{G}$-monotone decreasing functions on $\mathcal{R}^k$ such that their convolution $\psi(\mathbf{x}) = (\phi * f)(\mathbf{x})$ exists (possibly $\pm\infty$, but well defined) for every $\mathbf{x} \in \mathcal{R}^k$, then ψ is $\mathcal{G}$-monotone decreasing.

Roughly speaking, a reflection is an orthogonal transformation whose purpose is to map every point $\mathbf{x}$ to its mirror image with respect to a line in $\mathcal{R}^2$ or a plane in $\mathcal{R}^k$ containing the origin. It is known that in the two dimensional case ($k=2$) every orthogonal transformation is either a rotation or a reflection. It is also known that the permutation group, the group of all sign changes of coordinates, and the group generated by all permutations and sign changes of coordinates in $\mathcal{R}^k$ are reflection groups. Thus Theorem 6.2.7 implies the theorems of Mudholkar (Theorem 6.2.1) and Marshall and Olkin (Theorems 6.2.2 and 6.2.3) as special cases. This is so because if $\mathcal{G}$ is the permutation group, then "$\mathbf{x} \succcurlyeq \mathbf{x}^*$" reduces to "$\mathbf{x} \succ \mathbf{x}^*$" and "$\psi$ is $\mathcal{G}$-monotone decreasing" reduces to "ψ is Schur-concave." Theorem 6.2.7 also implies the theorem of Jogdeo (1977) (Theorem 5.2.5) as a special case because that theorem concerns sign-change invariance, and the condition for $\mathcal{G}$-monotone decreasing reduces there to the decreasing-in-absolute-value condition. Therefore by considering a reflection group of transformations, the theorem of Eaton and Perlman covers several basic inequalities under one roof.

The proof of Theorem 6.2.7 involves two path lemmas and a careful study of the structure of $C(\mathbf{x})$, the convex hull of $\mathcal{G}$-orbit of $\mathbf{x}$.

The path lemmas say basically that if $\mathbf{x} \succcurlyeq \mathbf{x}^*$, then one can transform $\mathbf{x}$ into $\mathbf{x}^*$ through a finite number of steps, which is an analogue to statement (b) in Theorem 6.1.1 concerning majorization. The details of the proof are quite involved mathematically and are beyond the scope of this book.

The next theorem, by Marshall and Proschan (1965), is useful for scale parameter families.

Theorem 6.2.8. Let $f(\mathbf{x})$, the density of $\mathbf{X}$, be permutation invariant, and let $\phi: \mathcal{R}^k \to \mathcal{R}^1$ be Borel-measurable, permutation invariant, and convex (concave). Then $E\phi(a_1 X_1, \ldots, a_k X_k)$ is Schur-convex (Schur-concave) in $\mathbf{a} = (a_1, \ldots, a_k)'$.

Proof. Let $\mathbf{a} \succ \mathbf{b}$, and assume that ϕ is convex (if ϕ is concave we replace ϕ by $-\phi$). By Theorem 6.1.1 we may assume that $\mathbf{a}$ and $\mathbf{b}$ differ in two coordinates only, i.e., $\mathbf{a}$ and $\mathbf{b}$ are of the form

$$\mathbf{a} = (a_1, a_2, a_3, \ldots, a_k)', \quad \mathbf{b} = (b_1, b_2, a_3, \ldots, a_k)',$$

and that $(a_1, a_2)' \succ (b_1, b_2)'$. Then there exists an $\alpha \in [0, 1]$ such that

$$b_1 = \alpha a_1 + (1 - \alpha) a_2, \quad b_2 = (1 - \alpha) a_1 + \alpha a_2.$$

Now for fixed $\mathbf{X}_3 = \mathbf{x}_3 \equiv (x_3, \ldots, x_k)'$ consider the conditional expectation. Since ϕ is convex, we have (by the permutation invariance property)

$$\begin{aligned}
E\big[&\phi(b_1 X_1, b_2 X_2, a_3 X_3, \ldots, a_k X_k) | \mathbf{X}_3 = \mathbf{x}_3\big] \\
&= E\big[\phi(\alpha(a_1 X_1, a_2 X_2) + (1 - \alpha)(a_2 X_1, a_1 X_2), \\
&\qquad a_3 X_3, \ldots, a_k X_k) | \mathbf{X}_3 = \mathbf{x}_3\big] \\
&\leq \alpha E\big[\phi(a_1 X_1, a_2 X_2, a_3 X_3, \ldots, a_k X_k) | \mathbf{X}_3 = \mathbf{x}_3\big] \\
&\quad + (1 - \alpha) E\big[\phi(a_2 X_1, a_1 X_2, a_3 X_3, \ldots, a_k X_k) | \mathbf{X}_3 = \mathbf{x}_3\big] \\
&= E\big[\phi(a_1 X_1, a_2 X_2, a_3 X_3, \ldots, a_k X_k) | \mathbf{X}_3 = \mathbf{x}_3\big].
\end{aligned}$$

The proof of the theorem is completed by taking expectations on both sides. ■

We immediately observe the following corollary. Its proof follows from replacing $a_i x_i$ by $(a_i x_i)^2$ in the ϕ function.

Corollary. Under the conditions stated in Theorem 6.2.8, the function $E\phi(a_1^2X_1^2,\ldots,a_k^2X_k^2)$ is either Schur-convex or Schur-concave in $\mathbf{a}^2=(a_1^2,\ldots,a_k^2)'$, depending on whether ϕ is convex or concave.

An application of Theorem 6.2.8 yields the following theorem, which appears to be useful for scale parameter families.

Theorem 6.2.9. Let $\mathbf{X}=(X_1,\ldots,X_k)'$ have a joint density

$$\prod_{i=1}^{k}\frac{1}{\theta_i}g\left(\frac{x_i}{\theta_i}\right)\qquad \text{for}\quad x_i\in\mathcal{X},\quad \theta_i\in\Lambda\subset(0,\infty)\quad (i=1,\ldots,k);$$

and let $\phi:\mathcal{R}^k\to\mathcal{R}^1$ be Borel-measurable, permutation invariant, and convex. Then $E_{\boldsymbol{\theta}}\phi(X_1,\ldots,X_k)$ is a Schur-concave function of $\boldsymbol{\theta}=(\theta_1,\ldots,\theta_k)'$.

Proof. Define $Y_i=X_i/\theta_i, i=1,\ldots,k$. Then the joint density of Y is $\prod_{i=1}^k g(y_i)$, which is permutation invariant. The proof follows from Theorem 6.2.8 and the identity

$$E_{\boldsymbol{\theta}}\phi(X_1,\ldots,X_k) = E\phi(\theta_1Y_1,\ldots,\theta_kY_k). \quad \blacksquare$$

Theorems 6.2.8, 6.2.9 are useful for deriving inequalities for the expectations of convex and concave functions, but unfortunately they do not yield useful probability inequalities. This is so simply because the indicator function of a proper subset of $\mathcal{R}^k$ is neither convex nor concave. Moreover, Theorem 6.2.8 fails when the condition on the convexity (concavity) of ϕ is replaced by Schur-convexity (Schur-concavity), as shown by Marshall and Proschan (1965). A natural problem is to find additional conditions of f and A such that $P_{\boldsymbol{\theta}}[\mathbf{X}\in A]$ is Schur-concave in $\boldsymbol{\theta}$. It appears that such a result will be useful and that the problem has not yet been fully studied.

In the following we see a different preservation theorem under integral transformations due to Proschan and Sethuraman (1977). This theorem applies to nonnegative random variables only, and it depends on the TP_2 property of the density function (for definition, see Section 5.1). Let $X_1,\ldots,X_k$ be independent random variables with densities $g_{\theta_1}(x_1),\ldots,g_{\theta_k}(x_k)$, respectively, where $g_\theta(x)$ belongs to a family of density functions for $x\in\mathcal{X}$, $\theta\in\Lambda$. Then the joint density of $\mathbf{X}=(X_1,\ldots,X_k)'$ is $\prod_{i=1}^k g_{\theta_i}(x_i)$.

Theorem 6.2.10. Suppose that either (a) $\mathcal{X}$ is the real line, $\Lambda \subset (0, \infty)$ is an interval, and η is the Lebesgue measure or (b) $\mathcal{X}$ is the set of integers, $\Lambda \subset (0, \infty)$ is an interval (or an interval of integers), and η is the counting measure. Furthermore, suppose that $g_\theta(x) = 0$ for $x < 0$ and $\theta \in \Lambda$, and that it satisfies the TP_2 property and the semigroup property

$$g_{\theta_1+\theta_2}(x) = \int g_{\theta_1}(x-y) g_{\theta_2}(y)\, d\eta(y). \tag{6.2.22}$$

Then, provided that the integral exists,

$$\psi(\boldsymbol{\theta}) = \int \phi(\mathbf{x}) \prod_{i=1}^{k} g_{\theta_i}(x_i) \prod_{i=1}^{k} d\eta(x_i) \tag{6.2.23}$$

is Schur-convex (Schur-concave) in $\boldsymbol{\theta}$ for $\boldsymbol{\theta} \in \Lambda \times \cdots \times \Lambda$ whenever ϕ is Schur-convex (Schur-concave).

Outline of the Proof. Let $\boldsymbol{\xi} \succ \boldsymbol{\theta}$. The first step of the proof of this theorem again depends on an application of Theorem 6.1.1 to assume that all except two of the components in $\boldsymbol{\xi}$ and $\boldsymbol{\theta}$ are equal. Then after using the permutation invariance properties, the TP_2 property, and the semigroup property, the difference $\psi(\boldsymbol{\theta}) - \psi(\boldsymbol{\xi})$ can be expressed in the form of an integral over a region in which the integrand is nonpositive when ϕ is Schur-convex and when $g_\theta(x) = 0$ for $x < 0$. (If ϕ is Schur-concave, then $-\phi$ is Schur-convex; hence we can consider the expectation of $-\phi$ and proceed similarly.) The steps involved are also similar to those in the proof of Theorem 5.1.3, and the reader is referred to either Proschan and Sethuraman (1977) or Marshall and Olkin (1979, p. 101) for details. ■

The corollary given below is immediate by letting $\phi(\mathbf{x}) = \chi_A(\mathbf{x})$, where χ_A is the indicator function of the set A.

Corollary. Let $\mathbf{X} = (X_1, \ldots, X_k)'$ have a density $f_{\boldsymbol{\theta}}(\mathbf{x}) = \prod_{i=1}^{k} g_{\theta_i}(x_i)$, and assume that g satisfies the conditions given in Theorem 6.2.10. Let $A \subset \mathcal{R}^k$ satisfy the condition in (6.2.7). Then $P_{\boldsymbol{\theta}}[\mathbf{X} \in A]$ is Schur-concave in $\boldsymbol{\theta}$. In particular, this corollary applies when A is one of the following subsets:

$$A = \{\mathbf{x} \mid b \leqslant x_i \leqslant a,\ i = 1, \ldots, k\},$$

$$A = \{\mathbf{x} \mid x_i \leqslant b,\ i = 1, \ldots, k\} \cup \{\mathbf{x} \mid x_i \geqslant a,\ i = 1, \ldots, k\}.$$

Theorem 6.2.10 has a number of applications to nonnegative random variables when their densities satisfy the TP_2 and semigroup properties. In the following example we state a few direct applications due to Nevius, Proschan, and Sethuraman (1977a).

Example 6.2.1. Let $\mathbf{X}_\theta$ be a random variable whose components are independent and have one of the following distributions:

(a) Poisson distribution:

$$g_\theta(x) = e^{-\theta}\frac{\theta^x}{x!}, \qquad x = 0, 1, \ldots, \quad \theta > 0. \tag{6.2.24}$$

(This was also considered by Rinott (1973).)

(b) Binomial distribution with a common fixed p:

$$g_\theta(x) = \binom{\theta}{x} p^x (1-p)^{\theta-x}, \qquad x = 0, \ldots, \theta, \quad \theta = \text{a positive integer}. \tag{6.2.25}$$

(c) Gamma distribution with a common fixed scale parameter α:

$$g_\theta(x) = \frac{1}{\Gamma(\theta)}\alpha^\theta x^{\theta-1} e^{-\alpha x}, \qquad x \geqslant 0, \quad \theta > 0. \tag{6.2.26}$$

Then the conditions in Theorem 6.2.10 are satisfied. Hence the probability inequalities in the above corollary apply. As an immediate consequence, for each $\boldsymbol{\theta} = (\theta_1, \ldots, \theta_k)'$ in (a) and (c) the probability $P_{\boldsymbol{\theta}}(A)$ is bounded above by $P_{\bar{\boldsymbol{\theta}}}(A)$, where

$$\bar{\boldsymbol{\theta}} = \left(\frac{1}{k}\sum_{i=1}^{k}\theta_i, \ldots, \frac{1}{k}\sum_{i=1}^{k}\theta_i\right)', \tag{6.2.27}$$

and in (b) it is bounded above when $|\theta_i - \theta_j| \leqslant 1$ for $i \neq j$, provided that $\sum_{i=1}^{k}\theta_i$ is kept fixed.

In Section 6.3 Theorem 6.2.10 will be applied to derive an inequality via stochastic majorization, and additional examples will be considered in Example 6.3.1.

Thus far we have studied inequalities by partial ordering via majorization of either parameter vectors or sets. The next theorem uses majorization in a different fashion. The theorem concerns the probability of exchangeable events, and we have already seen several partial results in previous chapters (Theorems 2.3.4, 3.3.1, and 5.3.3). Since their proofs depend on moment inequalities, we shall first prove a new moment inequality, then obtain a probability inequality as a consequence of this moment inequality.

Our basic tool for proving the moment inequality is Muirhead's theorem (see Hardy, Littlewood, and Pólya (1959, p. 44)), which is stated below.

Lemma 6.2.3. Let $w_1,\ldots,w_r$ be nonnegative real numbers. If $\mathbf{a}=(a_1,\ldots,a_r)'\succ\mathbf{b}=(b_1,\ldots,b_r)'$, then

$$\sum_L \left(\prod_{j=1}^{r} w_{l_j}^{a_j} \right) \geqslant \sum_L \left(\prod_{j=1}^{r} w_{l_j}^{b_j} \right) \tag{6.2.28}$$

holds, where the summations are taken over all $L=(l_1,\ldots,l_r)'$, the $r!$ permutations of the set of integers $\{1,\ldots,r\}$.

In the original form given by Muirhead it was stated that both a_j and b_j are nonnegative integers. In Hardy, Littlewood, and Pólya (1959, p. 44) a_j and b_j are not necessarily integers but are assumed to be nonnegative. This condition was later removed by Marshall and Proschan (1965). Hence in Lemma 6.2.3 there is no additional condition imposed on the values of a_j and b_j.

In obtaining a more complete solution for the inequality of Šidák (1973) (Theorem 3.3.1), Tong (1977) applied Muirhead's theorem to prove a moment inequality for nonnegative random variables. The same inequality was also mentioned in a technical report of Proschan and Sethuraman (1974).

Lemma 6.2.4. Let $\mathbf{W}=(W_1,\ldots,W_r)'$ have a joint density function that is permutation invariant. If $P[\bigcap_{j=1}^{r}\{W_j\geqslant 0\}]=1$, then

$$E\prod_{j=1}^{r} W_j^{a_j} \geqslant E\prod_{j=1}^{r} W_j^{b_j} \tag{6.2.29}$$

holds whenever $\mathbf{a}\succ\mathbf{b}$ (provided that the expectations exist).

The proof follows from the fact that for every ω in the sample space we have

$$\sum_L \left(\prod_{j=1}^{r} W_{l_j}^{a_j}(\omega) \right) \geqslant \sum_L \left(\prod_{j=1}^{r} W_{l_j}^{b_j}(\omega) \right),$$

and it is completed by taking expectations on both sides after applying the permutation invariance condition. This result covers the special case in which $W, W_1, \ldots, W_r$ are independent and identically distributed random variables. Hence if W is a nonnegative random variable with kth moment μ_k ($\mu_0 = 1$), then

$$\prod_{j=1}^{r} \mu_{a_j} \geqslant \prod_{j=1}^{r} \mu_{b_j} \tag{6.2.30}$$

holds whenever $\mathbf{a} \succ \mathbf{b}$, provided that the moments exist.

Now an application of this moment inequality yields the following theorem due to Tong (1977). It provides a comparison for the probabilities of positively dependent random variables by mixture, as discussed in Section 5.3, through partial ordering of the dimension vectors.

Theorem 6.2.11. Let $\mathbf{X} = (\mathbf{X}^{(1)}, \ldots, \mathbf{X}^{(k)})'$ be an $sk \times 1$ random vector such that each $\mathbf{X}^{(i)}$ is $s \times 1$ and the distribution of $\mathbf{X}$ is of the form

$$F(\mathbf{x}) = \int \prod_{i=1}^{k} G_{\mathbf{u}}(\mathbf{x}^{(i)})\, dH(\mathbf{u}), \qquad \mathbf{x} = (\mathbf{x}^{(1)}, \ldots, \mathbf{x}^{(k)})', \tag{6.2.31}$$

where H is a distribution function on $\mathfrak{U}$ and $G_{\mathbf{u}}$ is a distribution function for each $\mathbf{u} \in \mathfrak{U}$. Let $A \subset \mathfrak{R}^s$ be a Borel-measurable set. If $\mathbf{a} \succ \mathbf{b}$, then

$$\prod_{j=1}^{r} P\left[\bigcap_{i=1}^{a_j} \{\mathbf{X}^{(i)} \in A\}\right] \geqslant \prod_{j=1}^{r} P\left[\bigcap_{i=1}^{b_j} \{\mathbf{X}^{(i)} \in A\}\right]. \tag{6.2.32}$$

Proof. The proof follows immediately from Lemma 6.2.4 in general and from the inequality in (6.2.30) in particular, with $\mu_0 = 1$ and

$$\mu_m = EP\left[\bigcap_{i=1}^{m} \{\mathbf{X}^{(i)} \in A\} | \mathbf{U} = \mathbf{u}\right], \qquad m = 1, \ldots, k. \quad \blacksquare$$

From Lemma 5.3.1 we may conclude that if $\mathbf{X}^{(1)}, \ldots, \mathbf{X}^{(k)}$ are constructed as in Theorem 3.3.1 with $\mathbf{X}^{(i)} = \psi(\mathbf{Y}_i, \mathbf{U})$ ($i = 1, \ldots, k$), then Theorem 6.2.11 applies. This theorem is a refinement of Theorem 3.3.1; it bridges the gap between the bounds given in Theorem 3.3.1 and provides a partial ordering. A special application of this theorem to certain multiple decision problems will be made in Chapter 8.

We also note that for most mixtures of distributions discussed earlier, e.g., for the examples given in Section 3.3, Theorem 6.2.11 applies. Therefore a new partial ordering for the joint probabilities in multivariate distributions of normal, t, chi-square, F, beta, exponential, and Poisson, among others, can now be obtained for any Borel-measurable set A.

6.3. INEQUALITIES VIA STOCHASTIC ORDERING OF RANDOM VARIABLES

Our purpose in this section is to study probability inequalities through a stochastic ordering of random variables by a partial ordering of their parameter vectors via majorization and weak majorization. In the one-dimensional case a random variable X is said to be stochastically larger than Y if the distribution function of X is bounded above by the distribution function of Y. This is equivalent to saying that $P[X>x] \geqslant P[Y>x]$, and it is also equivalent to saying that "$E\phi(X) \geqslant E\phi(Y)$ holds for every nondecreasing real-valued function ϕ for which the expectation exists" (see Section 5.1). In symbols we write $X \overset{\text{st}}{\geqslant} Y$. A multivariate generalization of this definition appears to have been considered first by Lehmann (1955) in a paper on ordered families of distributions. It is stated below.

Definition 6.3.1. Let $\mathbf{X}$ and $\mathbf{Y}$ be two k-dimensional random variables. $\mathbf{X}$ is said to be stochastically larger than $\mathbf{Y}$, in symbols $\mathbf{X} \overset{\text{st}}{\geqslant} \mathbf{Y}$, if $\phi(X) \overset{\text{st}}{\geqslant} \phi(\mathbf{Y})$ for every $\phi: \mathfrak{R}^k \to \mathfrak{R}^1$ such that ϕ is nondecreasing in each argument.

It is clear that if $\mathbf{X}=(X_1,\ldots,X_k)' \overset{\text{st}}{\geqslant} \mathbf{Y}=(Y_1,\ldots,Y_k)'$, then $X_i \overset{\text{st}}{\geqslant} Y_i$ for each $i=1,\ldots,k$. But the converse is false; this is so because the marginal distributions do not determine the joint distribution uniquely unless, of course, the components are independent.

A concept of stochastic majorization was introduced by Nevius, Proschan, and Sethuraman (1977a). This concept serves as a useful tool for obtaining probability inequalities in certain multivariate distributions.

Definition 6.3.2. Let $\mathbf{X}$ and $\mathbf{Y}$ be two k-dimensional random variables. $\mathbf{X}$ is said to stochastically majorize $\mathbf{Y}$, in symbols $\mathbf{X} \overset{\text{st}}{\succ} \mathbf{Y}$, if

$\phi(\mathbf{X}) \overset{\text{st}}{\geqslant} \phi(\mathbf{Y})$ for every ϕ: $\mathcal{R}^k \to \mathcal{R}^1$ that is Borel-measurable and Schur-convex.

This definition is a stochastic analogue of a characterization of deterministic majorization, namely, $\mathbf{a} \succ \mathbf{b}$ if and only if $\phi(\mathbf{a}) \geqslant \phi(\mathbf{b})$ for every Schur-convex function ϕ. Roughly speaking, in the two-dimensional case if $\mathbf{X} \overset{\text{st}}{\succ} \mathbf{Y}$, then $\mathbf{Y}$ is more likely to take values in the neighborhood of the 45° straight line "$x_1 = x_2$" in $\mathcal{R}^2$. Several other versions of stochastic majorization are also possible, and a complete discussion of their relationships can be found in Marshall and Olkin (1979, Chap. 11).

In most applications $\mathbf{X}$ and $\mathbf{Y}$ are two random variables with distribution functions belonging to the same family of distributions with a certain parametric form. In this case we can usually order $\mathbf{X}$ and $\mathbf{Y}$ stochastically through a partial ordering of the parameter vectors. Once this is done, probability inequalities for Schur-convex and Schur-concave sets can be obtained (a set $A \subset \mathcal{R}^k$ is said to be Schur-convex (Schur-concave) if its indicator function is Schur-convex (Schur-concave)).

The next theorem is designed for this purpose. It was obtained by Nevius, Proschan, and Sethuraman (1977a).

Theorem 6.3.1. The following statements are equivalent:

(a) $\mathbf{X} \overset{\text{st}}{\succ} \mathbf{Y}$.

(b) $E\phi(\mathbf{X}) \geqslant E\phi(\mathbf{Y})$ for every Schur-convex function ϕ for which both of these expectations exist.

(c) $E\phi(\mathbf{X}) \geqslant E\phi(\mathbf{Y})$ for every bounded Schur-convex function ϕ.

(d) $P[\mathbf{X} \in A] \geqslant (\leqslant) P[\mathbf{Y} \in A]$ for every Borel-measurable Schur-convex (Schur-concave) set A.

Proof. The implications (a)⇒(b)⇒(c)⇒(d) are trivial. The implication (d)⇒(b) follows from the following fact: If $E\phi(\mathbf{X})$, $E\phi(\mathbf{Y})$ exist, $\phi(\mathbf{X})$ $(\phi(\mathbf{Y}))$ may be approximated in the L_1-norm under the corresponding probability measure by a positive linear combination of indicator functions of Schur-convex sets. This is a consequence of the fact that the set $\{\mathbf{x} | \phi(\mathbf{x}) > a\}$ is Schur-convex for all a. The inequality given in (d) for Schur-convex sets is then preserved. The implication (b)⇒(a) is a consequence of the fact that a nondecreasing function of a Schur-convex function is Schur-convex (see Problem 6). ■

Combining Theorems 6.2.10 and 6.3.1, Nevius, Proschan, and Sethuraman (1977a) gave the following preservation theorem.

Theorem 6.3.2. Let $X_1,\ldots,X_k$ be independent random variables with densities $g_{\theta_i}(x_i)$ for $x_i \in \mathscr{X} \subset [0,\infty)$ and $\theta_i \in \Lambda \subset (0,\infty)$ $(i=1,\ldots,k)$. Let $\mathbf{X}_{\boldsymbol{\theta}}=(X_1,\ldots,X_k)'$ be the k-dimensional random variable whose distribution depends on the parameter vector $\boldsymbol{\theta}$. If the density $g_{\boldsymbol{\theta}}(x)$ satisfies the conditions in Theorem 6.2.10, then $\boldsymbol{\xi} \succ \boldsymbol{\theta}$ implies $\mathbf{X}_{\boldsymbol{\xi}} \overset{\text{st}}{\succ} \mathbf{X}_{\boldsymbol{\theta}}$. Therefore if A is a Borel-measurable and Schur-convex (or Schur-concave) set, then $\boldsymbol{\xi} \succ \boldsymbol{\theta}$ implies

$$P_{\boldsymbol{\theta}}[\mathbf{X}\in A] \leqslant (\text{or} \geqslant) P_{\boldsymbol{\xi}}[\mathbf{X}\in A]. \tag{6.3.1}$$

The next theorem, which is in a slightly different form from that given by Nevius, Proschan, and Sethuraman (1977a), says that (a) stochastic majorization can be obtained and preserved through mixture of distributions and (b) as a partial converse, conditional stochastic majorization can be obtained from unconditional stochastic majorization.

Theorem 6.3.3. Let $\mathbf{X}$ be a k-dimensional random variable, and suppose that the distribution of $\mathbf{X}$ depends on a parameter vector that is either $\boldsymbol{\xi}$ or $\boldsymbol{\theta}$. Let $\boldsymbol{\xi} \succ \boldsymbol{\theta}$.

(a) Assume that $\mathbf{U}$ is another random variable (possibly multidimensional) whose distribution remains the same under $\boldsymbol{\xi}$ or $\boldsymbol{\theta}$. If the conditional distribution of $\mathbf{X}_{\boldsymbol{\xi}}$ stochastically majorizes that of $\mathbf{X}_{\boldsymbol{\theta}}$ for every given $\mathbf{U}=\mathbf{u}$, then $\mathbf{X}_{\boldsymbol{\xi}} \overset{\text{st}}{\succ} \mathbf{X}_{\boldsymbol{\theta}}$.

(b) If $\mathbf{X}_{\boldsymbol{\xi}} \overset{\text{st}}{\succ} \mathbf{X}_{\boldsymbol{\theta}}$ and if $U=h(\mathbf{X})=h(\Sigma_{i=1}^k X_i)$ is a discrete one-dimensional random variable (here U depends on $\mathbf{X}$ only through the sum of its components), then the conditional distribution of $\mathbf{X}_{\boldsymbol{\xi}}$, given $U=u$, stochastically majorizes that of $\mathbf{X}_{\boldsymbol{\theta}}$, given $U=u$, for every u such that $P[U=u]>0$ (this probability remains the same under $\boldsymbol{\xi}$ or $\boldsymbol{\theta}$).

Proof. (a) Let ϕ be a bounded Schur-convex function, and let $H(\mathbf{u})$ be the distribution of $\mathbf{U}$ when the parameter vector is either $\boldsymbol{\xi}$ or $\boldsymbol{\theta}$. Then by Theorem 6.3.1 we have

$$\begin{aligned} E_{\boldsymbol{\xi}}\phi(\mathbf{X}) &= \int \left[E_{\boldsymbol{\xi}}\{\phi(\mathbf{X})|\mathbf{U}=\mathbf{u}\} \right] dH(\mathbf{u}) \\ &\geqslant \int \left[E_{\boldsymbol{\theta}}\{\phi(\mathbf{X})|\mathbf{U}=\mathbf{u}\} \right] dH(\mathbf{u}) = E_{\boldsymbol{\theta}}\phi(\mathbf{X}). \end{aligned}$$

Again, by Theorem 6.3.1 we have $\mathbf{X}_{\boldsymbol{\xi}} \overset{\text{st}}{\succ} \mathbf{X}_{\boldsymbol{\theta}}$.

(b) The fact that $P[U=u]$ remains the same under $\boldsymbol{\xi}$ or $\boldsymbol{\theta}$ is easy to justify (see Problem 17). Let $\phi(\mathbf{x})$ be bounded and Schur-convex; let B denote any Borel-measurable subset of the real line such that $P[U \in B] > 0$. Define

$$\chi_B(\mathbf{x}) = \begin{cases} 1 & \text{if } h(\mathbf{x}) \in B, \\ 0 & \text{otherwise,} \end{cases}$$

which is the indicator function of the event $[U \in B]$. Since $\chi_B(\mathbf{x})$ depends on $\mathbf{x}$ only through the sum and is nonnegative, $\phi(\mathbf{x})\chi_B(\mathbf{x}) \geqslant \phi(\mathbf{y})\chi_B(\mathbf{y})$ holds for all $\mathbf{x}$, $\mathbf{y}$ such that $\mathbf{x} \succ \mathbf{y}$. Therefore as a function of $\mathbf{x}$, $\phi(\mathbf{x})\chi_B(\mathbf{x})$ is also Schur-convex. From the condition $\mathbf{X}_{\boldsymbol{\xi}} \overset{\text{st}}{\succ} \mathbf{X}_{\boldsymbol{\theta}}$ and Theorem 6.3.1 we have

$$E_{\boldsymbol{\xi}}\phi(\mathbf{X})\chi_B(\mathbf{X}) \geqslant E_{\boldsymbol{\theta}}\phi(\mathbf{X})\chi_B(\mathbf{X}).$$

Dividing both sides by $P[U \in B]$ the inequality reduces to

$$E_{\boldsymbol{\xi}}[\phi(\mathbf{X}) | U \in B] \geqslant E_{\boldsymbol{\theta}}[\phi(\mathbf{X}) | U \in B].$$

Therefore we can conclude that

$$E_{\boldsymbol{\xi}}[\phi(\mathbf{X}) | U = u] \geqslant E_{\boldsymbol{\theta}}[\phi(\mathbf{X}) | U = u]$$

holds for every u satisfying $P[U=u] > 0$. The proof is then completed by applying Theorem 6.3.1. ∎

For location parameter families we can restate the corollary of Theorem 6.2.2 in a version of stochastic majorization as given below.

Theorem 6.3.4. Let the random variable $\mathbf{X}_{\boldsymbol{\theta}}$ have a density $f(\mathbf{x}-\boldsymbol{\theta})$ for $\mathbf{x} \in \mathfrak{R}^k$, $\boldsymbol{\theta} \in \mathfrak{R}^k$ (a location parameter family). If $f(\mathbf{x})$ is Schur-concave in $\mathbf{x}$, then $\boldsymbol{\xi} \succ \boldsymbol{\theta}$ implies $\mathbf{X}_{\boldsymbol{\xi}} \overset{\text{st}}{\succ} \mathbf{X}_{\boldsymbol{\theta}}$.

From Theorems 6.3.2–6.3.4 we can perform stochastic comparisons of random variables, and hence can obtain probability inequalities for Schur-convex and Schur-concave sets, by a partial ordering of the parameter vectors via majorization. In addition to the cases already given in Example 6.2.1, we shall now consider a few other important applications to well-known distributions. ((a)–(c) were provided by Nevius, Proschan, and Sethuraman (1977a).)

Example 6.3.1. Let $\mathbf{X}_{\boldsymbol{\theta}}$ be a random variable with one of the following distributions. If $\boldsymbol{\xi} \succ \boldsymbol{\theta}$, then $\mathbf{X}_{\boldsymbol{\xi}} \overset{\text{st}}{\succ} \mathbf{X}_{\boldsymbol{\theta}}$.

(a) Multinomial distribution:

$$f_{\boldsymbol{\theta}}(\mathbf{x}) = \binom{n}{x_1,\ldots,x_k}\prod_{i=1}^{k}\theta_i^{x_i}, \qquad \begin{matrix} x_i \geqslant 0, & \sum_{i=1}^{k} x_i = n, \\ \theta_i \geqslant 0, & \sum_{i=1}^{k} \theta_i = 1. \end{matrix} \tag{6.3.2}$$

(This was also considered by Rinott (1973).)

(b) Multivariate hypergeometric distribution:

$$f_{\boldsymbol{\theta}}(\mathbf{x}) = \binom{N}{n}^{-1}\prod_{i=1}^{k}\binom{\theta_i}{x_i}, \qquad \begin{matrix} \theta_i \geqslant x_i \geqslant 0, & \\ \sum_{i=1}^{k} x_i = n, & \sum_{i=1}^{k}\theta_i = N. \end{matrix} \tag{6.3.3}$$

(c) Dirichlet distribution with a fixed $\lambda > 0$:

$$f_{\boldsymbol{\theta}}(\mathbf{x}) = \left[\frac{\Gamma\left(\lambda + \sum_{i=1}^{k}\theta_i\right)}{\left\{\Gamma(\lambda)\prod_{i=1}^{k}\Gamma(\theta_i)\right\}}\right]\prod_{i=1}^{k} x_i^{\theta_i - 1}\left(1 - \sum_{j=1}^{k} x_j\right)^{\lambda - 1},$$

$$x_i \geqslant 0, \quad \sum_{i=1}^{k} x_i \leqslant 1, \quad \theta_i > 0. \tag{6.3.4}$$

(d) A multivariate normal distribution with mean vector $\boldsymbol{\theta}$ and with equal variances and equal correlations.

(e) A noncentral multivariate t distribution, i.e., the distribution of $\mathbf{X}_{\boldsymbol{\theta}}$ with components

$$X_i = Y_i/S, \qquad i = 1,\ldots,k.$$

Here $\mathbf{Y} = (Y_1,\ldots,Y_k)'$ has the distribution stated in (d) with unit variances, and νS^2 is a chi-square variable with ν degrees of freedom and is independent of $\mathbf{Y}$.

Hence, $P_{\boldsymbol{\theta}}[\mathbf{X} \in A] \geqslant P_{\boldsymbol{\xi}}[\mathbf{X} \in A]$ whenever A is Schur-concave. In particular, this applies to the following sets:

$$A = \{\mathbf{x} \mid b \leqslant x_i \leqslant a, i = 1,\ldots,k\}, \tag{6.3.5}$$

$$A = \{\mathbf{x} \mid b \leqslant |x_i| \leqslant a, i = 1,\ldots,k\}, \tag{6.3.6}$$

$$A = \{\mathbf{x} \mid x_i \leqslant b, i = 1,\ldots,k\} \cup \{\mathbf{x} \mid x_i \geqslant a, i = 1,\ldots,k\}. \tag{6.3.7}$$

Proof. (a) follows from the fact that the multinomial distribution is the conditional distribution of independent Poisson variables, given their sum, Example 6.2.1, and Theorem 6.3.3. (Note that the distribution of the sum remains unchanged under $\boldsymbol{\theta}$ or $\boldsymbol{\xi}$.) (b) follows similarly and from the fact that the multivariate hypergeometric distribution is the conditional distribution of independent binomial variables, given their sum. (c) follows similarly and from the fact that the Dirichlet distribution is the distribution of $\mathbf{X}=(X_1,\ldots,X_k)'$, where

$$X_i = Y_i\Big/\left(\sum_{i=1}^{k} Y_i + U\right), \qquad i = 1,\ldots,k,$$

and $Y_1,\ldots,Y_k$ and U are independent gamma variables with shape parameters $\theta_1,\ldots,\theta_k$ and λ and a common scale parameter. (d) is an immediate consequence of Theorem 6.3.4. (e) follows from (d) and Theorem 6.3.3. ■

Remark. Let $\mathbf{Y}$ and S be defined as in (e) of Example 6.3.1. Let $\chi_{\boldsymbol{\theta}}^{*2}=\sum_{i=1}^{k} Y_i^2$ and $F_{\boldsymbol{\theta}}^{*}=\sum_{i=1}^{k} Y_i^2/kS^2$ denote the noncentral chi-square and noncentral F variables. Then $\boldsymbol{\xi}\succ\boldsymbol{\theta}$ implies $\chi_{\boldsymbol{\xi}}^{*2} \overset{\text{st}}{\geqslant} \chi_{\boldsymbol{\theta}}^{*2}$ and $F_{\boldsymbol{\xi}}^{*} \overset{\text{st}}{\geqslant} F_{\boldsymbol{\theta}}^{*}$. This result, from Marshall and Olkin (1974), follows from (d) of Example 6.3.1, the facts that (i) $\phi(\mathbf{y})=\sum_{i=1}^{k} y_i^2$ is a Schur-convex function, (ii) a nondecreasing function of a Schur-convex function is Schur-convex (see Problem 6), and Theorems 6.3.1 and 6.3.3. Note that the Y_is are not necessarily independent (i.e., the common correlation is not necessarily zero). For the special case in which they are independent, it is an immediate consequence of the well-known classical result that $\chi_{\boldsymbol{\theta}}^{*2}$ and $F_{\boldsymbol{\theta}}^{*}$ are stochastically larger if the noncentrality parameter is larger. This is so because if $\boldsymbol{\xi}\succ\boldsymbol{\theta}$, then $\sum_{i=1}^{k} \xi_i^2 \geqslant \sum_{i=1}^{k} \theta_i^2$ (see Section 6.1), hence the noncentrality parameter is larger under $\boldsymbol{\xi}$.

The concept of stochastic majorization was later generalized into stochastic weak majorization by Nevius, Proschan, and Sethuraman (1977b). Thus if two parameter vectors are not comparable via majorization because their sums are not equal, they might still be comparable via weak majorization. In this case again a deterministic property of the parameter vector can be transformed into a corresponding stochastic property (stochastic weak majorization) of the random variables.

Definition 6.3.3. **X** is said to stochastically weakly majorize **Y**, in symbols $\mathbf{X} \overset{\mathrm{st}}{\gg} \mathbf{Y}$, if $\phi(\mathbf{X}) \overset{\mathrm{st}}{\geqslant} \phi(\mathbf{Y})$ for every nondecreasing Borel-measurable, Schur-convex function ϕ.

Roughly speaking, if $\mathbf{X} \overset{\mathrm{st}}{\gg} \mathbf{Y}$, then **X** is more likely to take either larger values, or values with a larger degree of diversity among the components, or both.

The next two theorems are due to Nevius, Proschan, and Sethuraman (1977b). (An increasing Schur-convex (decreasing Schur-concave) set is a Schur-convex (Schur-concave) set whose indicator function is nondecreasing (nonincreasing) in each argument.)

Theorem 6.3.5. The following statements are equivalent:

(a) $\mathbf{X} \overset{\mathrm{st}}{\gg} \mathbf{Y}$.

(b) $E\phi(\mathbf{X}) \geqslant E\phi(\mathbf{Y})$ for every nondecreasing Schur-convex function ϕ for which both of these expectations exist.

(c) $E\phi(\mathbf{X}) \geqslant E\phi(\mathbf{Y})$ for every bounded nondecreasing Schur-convex function ϕ.

(d) $P[\mathbf{X} \in A] \geqslant (\leqslant) P[\mathbf{Y} \in A]$ for every Borel-measurable increasing Schur-convex (decreasing Schur-concave) set A.

The proof of this theorem is similar to that of Theorem 6.3.1.

Theorem 6.3.6. Let $\mathbf{X}_{\boldsymbol{\theta}} = (X_1, \ldots, X_k)'$ be the random variable defined in Theorem 6.3.2, and assume that the conditions given there are satisfied. Then $\boldsymbol{\xi} \gg \boldsymbol{\theta}$ implies $\mathbf{X}_{\boldsymbol{\xi}} \overset{\mathrm{st}}{\gg} \mathbf{X}_{\boldsymbol{\theta}}$. Therefore for every Borel-measurable increasing Schur-convex (decreasing Schur-concave) set A, we have

$$P_{\boldsymbol{\theta}}[\mathbf{X} \in A] \leqslant (\geqslant) P_{\boldsymbol{\xi}}[\mathbf{X} \in A]. \tag{6.3.8}$$

Proof. The proof of this theorem depends on the result that if $\boldsymbol{\xi}$, $\boldsymbol{\theta} \in \Lambda \times \cdots \times \Lambda$ and $\boldsymbol{\xi} \gg \boldsymbol{\theta}$, then there exists a $\boldsymbol{\tau} \in \Lambda \times \cdots \times \Lambda$ such that $\boldsymbol{\xi} \succ \boldsymbol{\tau}$ and $\boldsymbol{\tau} \geqslant \boldsymbol{\theta}$ (see Theorem 6.1.1). Therefore the proof can be done in two steps. For any bounded increasing Schur-convex function ϕ the first step is to assert that $E_{\boldsymbol{\xi}}\phi(\mathbf{X}) \geqslant E_{\boldsymbol{\tau}}\phi(\mathbf{X})$ from Theorems 6.3.1 and 6.3.2. The second step uses the condition of the monotonicity of ϕ. Since the density function $g_{\theta}(x)$ has the TP_2

property, the corresponding family of distributions is stochastically increasing. This implies

$$E_\tau\phi(\mathbf{X}) = \int \phi(\mathbf{x}) \prod_{i=1}^{k} g_{\tau_i}(x_i) \prod_{i=1}^{k} d\eta(x_i)$$

$$\geqslant \int \phi(\mathbf{x}) g_{\theta_1}(x_1) \prod_{i=2}^{k} g_{\tau_i}(x_i) \prod_{i=1}^{k} d\eta(x_i) \geqslant \cdots$$

$$\geqslant \int \phi(\mathbf{x}) \prod_{i=1}^{k} g_{\theta_i}(x_i) \prod_{i=1}^{k} d\eta(x_i) = E_{\boldsymbol\theta}\phi(\mathbf{X}).$$

After combining those two inequalities the proof follows from Theorem 6.3.5. ■

The next two theorems are modifications of Theorems 6.3.3 and 6.3.4, respectively, from stochastic majorization to stochastic weak majorization.

Theorem 6.3.7. Let $\mathbf{X}$ be a random variable whose distribution depends on a parameter vector $\boldsymbol\theta \in \Omega \subset \mathcal{R}^k$. Let U be a one-dimensional random variable with distribution $G_{\boldsymbol\theta}(u)$. Suppose that

(a) for $\boldsymbol\xi \succ\succ \boldsymbol\theta$ and every $U=u$ the conditional distribution of $\mathbf{X}$ under $\boldsymbol\xi$ stochastically weakly majorizes the conditional distribution of $\mathbf{X}$ under $\boldsymbol\theta$;

(b) for every given $\boldsymbol\theta$ the conditional expectation $E_{\boldsymbol\theta}\{\phi(\mathbf{X})|U=u\}$ is nondecreasing in u for every ϕ that is nondecreasing in each argument; and

(c) $\boldsymbol\xi \succ\succ \boldsymbol\theta$ implies $U_{\boldsymbol\xi} \overset{\text{st}}{\geqslant} U_{\boldsymbol\theta}$.

Then for $\boldsymbol\xi, \boldsymbol\theta \in \Omega$, $\boldsymbol\xi \succ\succ \boldsymbol\theta$ implies $\mathbf{X}_{\boldsymbol\xi} \overset{\text{st}}{\succ\succ} \mathbf{X}_{\boldsymbol\theta}$.

Proof. Let ϕ be bounded, nondecreasing, and Schur-convex. Then

$$E_{\boldsymbol\xi}\phi(\mathbf{X}) = \int \left[E_{\boldsymbol\xi}\{\phi(\mathbf{X})|U=u\} \right] dG_{\boldsymbol\xi}(u)$$

$$\geqslant \int \left[E_{\boldsymbol\theta}\{\phi(\mathbf{X})|U=u\} \right] dG_{\boldsymbol\xi}(u)$$

$$\geqslant \int \left[E_{\boldsymbol\theta}\{\phi(\mathbf{X})|U=u\} \right] dG_{\boldsymbol\theta}(u) = E_{\boldsymbol\theta}\phi(\mathbf{X});$$

here the second inequality follows from assumptions (b) and (c). The proof now follows from Theorem 6.3.5. ■

Theorem 6.3.8. Let the random variable $\mathbf{X}_{\boldsymbol{\theta}}$ have a density $f(\mathbf{x}-\boldsymbol{\theta})$ for $\mathbf{x}\in\mathcal{R}^k$, $\boldsymbol{\theta}\in\mathcal{R}^k$. If $f(\mathbf{x})$ is Schur-concave in $\mathbf{x}$, then $\boldsymbol{\xi}\succ\succ\boldsymbol{\theta}$ implies $\mathbf{X}_{\boldsymbol{\xi}} \overset{\text{st}}{\succ\succ} \mathbf{X}_{\boldsymbol{\theta}}$.

Proof. Let $\boldsymbol{\tau}$ satisfy $\boldsymbol{\xi}\succ\boldsymbol{\tau}$ and $\boldsymbol{\tau}\geqslant\boldsymbol{\theta}$, and let A be any Borel-measurable increasing Schur-convex set. Theorem 6.3.4 implies that $P_{\boldsymbol{\xi}}[\mathbf{X}\in A]\geqslant P_{\boldsymbol{\tau}}[\mathbf{X}\in A]$. On the other hand, from the translation invariance property we have $P_{\boldsymbol{\tau}}[\mathbf{X}\in A]\geqslant P_{\boldsymbol{\theta}}[\mathbf{X}\in A]$. Combining these two inequalities yields

$$P_{\boldsymbol{\xi}}[\mathbf{X}\in A] \geqslant P_{\boldsymbol{\theta}}[\mathbf{X}\in A], \tag{6.3.9}$$

and the proof of the theorem again follows from Theorem 6.3.5. ∎

Theorems 6.3.6–6.3.8 can be applied to establish stochastic weak majorization and thus to obtain probability inequalities over an increasing Schur-convex or a decreasing Schur-concave set when majorization and monotonicity go together in the same direction. To illustrate this point we consider the following example ((a) and (b) were given by Nevius, Proschan, and Sethuraman (1977b)).

Example 6.3.2. Let $\mathbf{X}_{\boldsymbol{\theta}}$ be a random variable with one of the following distributions.

(a) The Poisson distribution, binomial distribution, and gamma distribution with marginal densities as given in Example 6.2.1.

(b) The multinomial and multivariate hypergeometric distributions with densities as given in Example 6.3.1.

(c) The multivariate normal distribution with mean vector $\boldsymbol{\theta}$, a common variance, and a common correlation.

(d) The multivariate noncentral t distribution defined in (e) of Example 6.3.1.

If $\boldsymbol{\xi}\succ\succ\boldsymbol{\theta}$, then $\mathbf{X}_{\boldsymbol{\xi}} \overset{\text{st}}{\succ\succ} \mathbf{X}_{\boldsymbol{\theta}}$. Hence $P_{\boldsymbol{\theta}}[\mathbf{X}\in A]\leqslant(\geqslant)P_{\boldsymbol{\xi}}[\mathbf{X}\in A]$ whenever A is increasing and Schur-convex (decreasing and Schur-concave). In particular, the following sets are decreasing and Schur-concave:

(i) $A=\{\mathbf{x}|x_i\leqslant a, i=1,\ldots,k\}$;

(ii) $A=\{\mathbf{x}|\Sigma_{i=1}^k h(x_i)\leqslant a\}$, where h is a nondecreasing convex function;

(iii) $A=\{\mathbf{x}|\Sigma_{i=1}^k c_i x_{(i)}\leqslant a\}$, where $c_i\geqslant 0$ and c_i, $x_{(i)}$ satisfy $c_i\leqslant c_{i+1}$, $x_{(i)}\leqslant x_{(i+1)}$.

Proof. (a) follows immediately from Theorem 6.3.6; (b) follows from (a), Theorem 6.3.7, the arguments given in Example 6.3.1, and the facts that (i) the distribution of $U=\Sigma_{i=1}^{k}X_i$ depends on $\boldsymbol{\theta}$ only through $\Sigma_{i=1}^{k}\theta_i$ and is a stochastically increasing family, and (ii) condition (b) in Theorem 6.3.7 is satisfied (this was justified by Nevius, Proschan, and Sethuraman (1977b)); (c) is an immediate consequence of Theorem 6.3.8; (d) follows from (c) and the fact that weak majorization is preserved through mixture of distributions. ■

The concept of stochastic weak majorization can also be applied to obtain an inequality for a class of contaminated random variables. Let $\mathbf{X}$ and $\mathbf{U}$ be independent, and let $\mathbf{Z}=\mathbf{X}+\mathbf{U}$. In many statistical applications, as noted by Jogdeo (1977), $\mathbf{X}$ is contaminated because of certain experimental conditions, and what can actually be observed is the value of $\mathbf{Z}$. In this and the previous chapters we have already seen results concerning the probability $P[(\mathbf{X}+\mathbf{U})\in A]$, where $\mathbf{U}$ is a singular random variable (see Theorems 4.1.1, 6.2.1, 6.2.2, and 5.2.6). Those inequalities say that if the density of $\mathbf{X}$ and the set A possess a permutation invariance (or sign invariance) property and if $\boldsymbol{\xi}\succ\boldsymbol{\theta}$ (or $|\boldsymbol{\xi}|\geqslant|\boldsymbol{\theta}|$), then

$$P[(\mathbf{X}-\boldsymbol{\theta})\in A] \geqslant P[(\mathbf{X}-\boldsymbol{\xi})\in A]$$

holds. This can be rewritten in the form

$$P[(\mathbf{X}+\mathbf{U}_1)\in A] \geqslant P[(\mathbf{X}+\mathbf{U}_2)\in A],$$

where $\mathbf{U}_2=-\boldsymbol{\xi}$ a.s., $\mathbf{U}_1=-\boldsymbol{\theta}$ a.s., and $\boldsymbol{\xi}\succ\boldsymbol{\theta}$ (or $|\boldsymbol{\xi}|\geqslant|\boldsymbol{\theta}|$). The question is then what can be said when $\mathbf{U}_1$ and $\mathbf{U}_2$ are nonsingular random variables.

In the following we prove a theorem under the assumption that the density of $\mathbf{X}$ and the set A are both Schur-concave and decreasing in absolute value, which implies that they are both permutation invariant and sign invariant. Under these stronger conditions the inequality given depends on stochastic weak majorization of the absolute values of the components of the random variable $\mathbf{U}$.

Theorem 6.3.9. Let $f(\mathbf{x})$, the density function of $\mathbf{X}$, be Schur-concave and decreasing in absolute value. Let $A\subset\mathfrak{R}^k$ be Lebesgue-measurable, Schur-concave (i.e., it satisfies (6.2.7)), and decreasing in absolute value. If $\mathbf{X}$ is independent of $\mathbf{U}_1$, $\mathbf{U}_2$ and if $|\mathbf{U}_2|\overset{\text{st}}{\succ\succ}|\mathbf{U}_1|$, then

$$P[(\mathbf{X}+\mathbf{U}_1)\in A] \geqslant P[(\mathbf{X}+\mathbf{U}_2)\in A]. \qquad (6.3.10)$$

Proof. Let $\chi_A(\mathbf{x})$ be the indicator function of A. We need to show that

$$E\chi_A(\mathbf{X}+\mathbf{U}_1) \geqslant E\chi_A(\mathbf{X}+\mathbf{U}_2). \tag{6.3.11}$$

By independence, for every $\mathbf{U}$ we can write

$$E\chi_A(\mathbf{X}+\mathbf{U}) = E\left[E\chi_A(\mathbf{X}+\mathbf{U})|\mathbf{U}=\mathbf{u}\right] \equiv E\phi(\mathbf{U}).$$

From the conditions on A and the fact that the group generated by all permutations and sign changes of coordinates in $\mathcal{R}^k$ is a reflection group (see Section 6.2), Theorem 6.2.7 implies that $\phi(\mathbf{u})$ is Schur-concave and decreasing in absolute value. By Theorem 6.3.5 this then implies that

$$E\phi(\mathbf{U}_1) = E\phi(|\mathbf{U}_1|) \geqslant E\phi(|\mathbf{U}_2|) = E\phi(\mathbf{U}_2).$$

Hence (6.3.11) follows. ■

This theorem may be applied to derive probability inequalities over a region which is both permutation invariant and sign invariant. It should be noted that $\mathbf{U}_1$, $\mathbf{U}_2$ are not necessarily continuous random variables. As a matter of fact, an important special case is the singular case with $\mathbf{U}_1=\boldsymbol{\theta}$ a.s., $\mathbf{U}_2=\boldsymbol{\xi}$ a.s. In this special case the theorem applies when $|\boldsymbol{\xi}|\succ|\boldsymbol{\theta}|$, or $|\boldsymbol{\xi}|\geqslant|\boldsymbol{\theta}|$, or both. An individual application to the monotonicity of power functions in certain hypothesis-testing problems will be discussed in Chapter 8.

6.4. INEQUALITIES FOR HETEROGENEOUS DISTRIBUTIONS

In this section we employ majorization as a tool to obtain inequalities concerning order statistics (particularly the maximum and minimum) when the observations come from heterogeneous populations. As noted by Pledger and Proschan (1971), "a great body of theory and methods exists for order statistics and their spacings from a single underlying distribution"; but "a far smaller set of results is available for the case of order statistics from underlying heterogeneous distributions" (p. 90). In the latter case the exact distributions become complicated. Hence such inequalities become useful. In Section 6.2 we have already seen some inequalities of this type (Theorems 6.2.4, 6.2.5, 6.2.9, and their corollaries). Those inequalities were given through the partial ordering of the parameter vectors. The inequalities to be studied in this section depend mainly on the distribution functions themselves, which in turn depend on two inequalities for independent Bernoulli trials.

To begin, we first observe a theorem of Hoeffding (1956), concerning independent but not necessarily identical Bernoulli trials.

Theorem 6.4.1. Let $U_1,\ldots,U_k$ be independent Bernoulli variables with parameters $\theta_1,\ldots,\theta_k$, respectively ($\theta_i \in [0,1]$, $i=1,\ldots,k$). Let $S_k=\Sigma_{i=1}^k U_i$ be the total number of successes. Denoting $\bar{\theta}=(1/k)\Sigma_{i=1}^k \theta_i$, we have

$$P_{\boldsymbol{\theta}}[S_k \leqslant a] \leqslant \sum_{j=0}^{a}\binom{k}{j}\bar{\theta}^j(1-\bar{\theta})^{k-j} \qquad \text{for} \quad a \leqslant k\bar{\theta}-1, \tag{6.4.1}$$

$$P_{\boldsymbol{\theta}}[S_k \leqslant a] \geqslant \sum_{j=0}^{a}\binom{k}{j}\bar{\theta}^j(1-\bar{\theta})^{k-j} \qquad \text{for} \quad a \geqslant k\bar{\theta}, \tag{6.4.2}$$

$$P_{\boldsymbol{\theta}}[b \leqslant S_k \leqslant a] \geqslant \sum_{j=b}^{a}\binom{k}{j}\bar{\theta}^j(1-\bar{\theta})^{k-j} \qquad \text{for} \quad 0 \leqslant b \leqslant k\bar{\theta} \leqslant a \leqslant k, \tag{6.4.3}$$

where a and b are integers.

Note that (6.4.2) implies

$$P_{\boldsymbol{\theta}}[S_k > a] \leqslant \sum_{j=a+1}^{k}\binom{k}{j}\bar{\theta}^j(1-\bar{\theta})^{k-j} \qquad \text{for} \quad k\bar{\theta} \leqslant a < k. \tag{6.4.4}$$

Also, the inequality in (6.4.3) says that, subject to a fixed $\bar{\theta}$ (or $\Sigma_{i=1}^k \theta_i$), S_k is more likely to take one of the extreme values when the θ_is are equal; the bounds here depend on a binomial probability. This is consistent with the fact that the variance of S_k is maximized at $\theta_1=\cdots=\theta_k=\bar{\theta}$, which was the motivation given by Hoeffding (1956) for deriving these inequalities. The proof of the theorem involves a detailed study of the behavior of the expectation of $\phi(S_k)$ for an arbitrary or convex function ϕ, and it is not given here.

This theorem was later generalized by Gleser (1975). Gleser's result says that the probabilities in (6.4.1)–(6.4.4) can be obtained through the partial ordering of the parameter vector $\boldsymbol{\theta}=(\theta_1,\ldots,\theta_k)'$ via majorization. Since $\boldsymbol{\theta}\succ(\bar{\theta},\ldots,\bar{\theta})'$ for every $\boldsymbol{\theta}$, this result is then a refinement of Theorem 6.4.1.

Theorem 6.4.2. Let $U_1,\ldots,U_k$ be independent Bernoulli variables with parameters $\theta_1,\ldots,\theta_k$, respectively ($\theta_i\in[0,1]$, $i=1,\ldots,k$). Let $S_k=\Sigma_{i=1}^k U_i$. Then $P_{\boldsymbol{\theta}}[S_k\leqslant a]$ is Schur-concave in $\boldsymbol{\theta}$ for $a\leqslant[k\bar{\theta}-2]$ and Schur-convex in $\boldsymbol{\theta}$ for $a\geqslant[k\bar{\theta}+2]$, where $\bar{\theta}=(1/k)\Sigma_{i=1}^k\theta_i$, $a\leqslant k$ is a nonnegative integer, and $[x]$ is the largest integer less than or equal to x.

Proof. Let ϕ be an arbitrary real-valued function defined on the set of integers $\{0,1,\ldots,k\}$, and define $\psi(\boldsymbol{\theta})\equiv E_{\boldsymbol{\theta}}\phi(S_k)$. Let $\boldsymbol{\theta}^{ij}$ be the $(k-2)\times 1$ vector formed by deleting the ith and jth components of $\boldsymbol{\theta}$, and let $f(l|\boldsymbol{\theta}^{ij})$ be the probability of getting l successes ($l\leqslant k-2$) in the $k-2$ trials other than trials i and j. Then it is easy to see that for $l=0,1,\ldots,k$ we have

$$P[S_k=l] = (1-\theta_i)(1-\theta_j)f(l|\boldsymbol{\theta}^{ij}) + \theta_i\theta_j f(l-2|\boldsymbol{\theta}^{ij})$$
$$+ (\theta_i+\theta_j-2\theta_i\theta_j)f(l-1|\boldsymbol{\theta}^{ij}).$$

(Here $f(l|\boldsymbol{\theta}^{ij})=0$ if $l<0$ or $l>k-2$.)Using this identity when taking derivatives under the expectation sign with respect to θ_i and θ_j, we obtain

$$(\theta_i-\theta_j)\left(\frac{\partial}{\partial\theta_i}\psi(\boldsymbol{\theta})-\frac{\partial}{\partial\theta_j}\psi(\boldsymbol{\theta})\right)$$
$$=\sum_{l=0}^{k}\phi(l)\,\Delta f(l-2|\boldsymbol{\theta}^{ij})$$
$$=-(\theta_i-\theta_j)^2\sum_{l=0}^{k-2}f(l|\boldsymbol{\theta}^{ij})\,\Delta\phi(l), \tag{6.4.5}$$

where Δf is the second difference of f given by

$$\Delta f(l)=f(l+2)-2f(l+1)+f(l), \qquad l\leqslant k-2,$$

and a similar definition goes for $\Delta\phi$.

Now for a fixed nonnegative integer a let us define $\phi_a(S_k) = \chi_{\{0,\dots,a\}}(S_k)$, where χ is the indicator function. Then $P_{\boldsymbol\theta}[S_k \leqslant a] = E_{\boldsymbol\theta}\phi_a(S_k) = \psi(\boldsymbol\theta)$, and the second difference of ϕ_a is

$$\Delta\phi_a(l) = \begin{cases} 1 & \text{for } l = a, \\ -1 & \text{for } l = a-1, \\ 0 & \text{otherwise.} \end{cases}$$

With this particular function ϕ_a, (6.4.5) reduces to

$$(\theta_i - \theta_j)\left(\frac{\partial}{\partial\theta_i}\psi(\boldsymbol\theta) - \frac{\partial}{\partial\theta_j}\psi(\boldsymbol\theta)\right)$$
$$= -(\theta_i - \theta_j)^2\left[f(a|\boldsymbol\theta^{ij}) - f(a-1|\boldsymbol\theta^{ij})\right]. \tag{6.4.6}$$

A result of Samuels (1965) says that if $\sum_{l=1}^{k-2} lf(l|\boldsymbol\theta^{ij}) = c$, then $f(l|\boldsymbol\theta^{ij})$ is increasing (decreasing) in l for $l \leqslant [c]$ $(l \geqslant [c+1])$. Since here $c = n\bar\theta - \theta_i - \theta_j$ and $\theta_i + \theta_j \leqslant 2$, the right-hand side of (6.4.6) is $\leqslant 0 (\geqslant 0)$ for $a \leqslant [k\bar\theta - 2]$ $(a \geqslant [k\bar\theta + 2])$. The proof of the theorem now follows immediately from Lemma 6.2.2. ■

We note that when $a = [k\theta - 1]$ or $a = [k\theta + 1]$, Theorem 6.4.1 applies but the proof of Theorem 6.4.2 fails (because it depends on the fact that $\theta_i + \theta_j \leqslant 2$). As a matter of fact in such cases the assertion in Theorem 6.4.2 fails to hold. Two counterexamples were offered by Gleser (1975).

Let us now see an inequality of Sen (1970) concerning the bounds for distributions of order statistics from heterogeneous populations. Its proof depends on an application of Hoeffding's inequality (Theorem 6.4.1). We have already seen in section 6.2 (Theorem 6.2.4) that for location parameter families the distribution functions of the maximum and minimum may be bounded above or below when the parameters of the populations are equal with a fixed sum. Sen's result may be regarded as its nonparametric analogue.

Theorem 6.4.3. Let $X_1, \dots, X_k$ be independent random variables with continuous distributions $F_1, \dots, F_k$, respectively. Let $X_{(1)} \leqslant \cdots \leqslant X_{(k)}$ denote the order statistics. Define

$$\bar F(x) = \frac{1}{k}\sum_{i=1}^{k} F_i(x), \qquad -\infty < x < \infty, \tag{6.4.7}$$

and for every positive integer r denote

$$\xi_r = \text{the } (r/k)\text{th percentile of } \bar{F} \qquad \left(\text{i.e., } \bar{F}(\xi_r) = r/k\right),$$

$$P_r(c|\mathbf{F}) = P_{(F_1,\ldots,F_k)}\left[X_{(r)} \leqslant c\right], \tag{6.4.8}$$

$$P_r(c|\bar{\mathbf{F}}) = P_{(\bar{F},\ldots,\bar{F})}\left[X_{(r)} \leqslant c\right]. \tag{6.4.9}$$

Then

(a) for the minimum and maximum the inequalities

$$P_1(c|\mathbf{F}) \geqslant P_1(c|\bar{\mathbf{F}}), \qquad P_k(c|\mathbf{F}) \leqslant P_k(c|\bar{\mathbf{F}}) \tag{6.4.10}$$

hold for all c;

(b) for $2 \leqslant r \leqslant k-1$

$$P_r(c|\mathbf{F}) - P_r(d|\mathbf{F}) \geqslant P_r(c|\bar{\mathbf{F}}) - P_r(d|\bar{\mathbf{F}}) \tag{6.4.11}$$

holds for all d, c satisfying $d \leqslant \xi_{r-1} < \xi_r \leqslant c$.

Proof. The proof of (a) follows from the fact that, given k nonnegative real numbers, their geometric mean is bounded above by their arithematic mean. To prove (b), for $2 \leqslant r \leqslant k-1$ we can write (for every fixed x)

$$\begin{aligned} 1 - P_r(x|\mathbf{F}) &= P_{(F_1,\ldots,F_k)}\left[\text{at most } (r-1) \text{ of the } X_i\text{s are } \leqslant x\right] \\ &\equiv Q(x|\mathbf{F}). \end{aligned}$$

Similarly we write

$$\begin{aligned} 1 - P_r(x|\bar{\mathbf{F}}) &= P_{(\bar{F},\ldots,\bar{F})}\left[\text{at most } (r-1) \text{ of the } X_i\text{s are} \leqslant x\right] \\ &\equiv Q(x|\bar{\mathbf{F}}). \end{aligned}$$

By definition, for $d \leqslant \xi_{r-1} < \xi_r \leqslant c$ we have

$$k\bar{F}(d) = \sum_{i=1}^{k} F_i(d) \leqslant \sum_{i=1}^{k} F_i(\xi_{r-1}) = k\bar{F}(\xi_{r-1}) = (r-1),$$

$$k\bar{F}(c) \geqslant r.$$

The proof now follows from Theorem 6.4.1, which implies that

$$\begin{aligned} Q(x|\mathbf{F}) &\geqslant Q(x|\bar{\mathbf{F}}) \qquad \text{for} \quad k\bar{F}(x) \leqslant (r-1), \\ Q(x|\mathbf{F}) &\leqslant Q(x|\bar{\mathbf{F}}) \qquad \text{for} \quad k\bar{F}(x) \geqslant r. \end{aligned}$$

■

This theorem says that when the populations are heterogeneous, the minimum tends to be stochastically smaller, the maximum tends to be stochastically larger, and the other order statistics are more likely to take moderate values.

Since in the above theorem the bounds depend on the distributions of order statistics from homogeneous populations with an average distribution $\bar{F}$, the theorem does not provide a comparison for the distributions of order statistics under two different possible sets of heterogeneous populations. In this case if one set of populations is less heterogeneous or diverse than the other in a certain fashion, one would hope that similar results might be possible through majorization when applying Theorem 6.4.2. This would be a generalization of Theorem 6.4.3. But it is easy to see that we cannot use the same type of argument here that was used in proving (6.4.11). This is so because in Gleser's inequality nothing can be said about the probability $P_{\theta}[S_k \leqslant a]$ when $a=[a\bar{\theta}-1]$ or $a=[a\bar{\theta}+1]$. Therefore we need a different argument for the distribution of $X_{(r)}$; thus a complete generalization of Theorem 6.4.3 appears to be not yet available. However, for the maximum and the minimum the result given in the following theorem follows immediately from the fact that $\phi(\mathbf{a})=\prod_{i=1}^{k} a_i$ is Schur-concave for $a_i \geqslant 0$, $i=1,\ldots,k$.

Theorem 6.4.4. Let $F_1,\ldots,F_k$, $G_1,\ldots,G_k$ be continuous distribution functions, and assume that

$$(F_1(x),\ldots,F_k(x))' \succ (G_1(x),\ldots,G_k(x))' \tag{6.4.12}$$

holds at $x=c$. Then

$$P_{(F_1,\ldots,F_k)}[X_{(1)} \leqslant c] \geqslant P_{(G_1,\ldots,G_k)}[X_{(1)} \leqslant c], \tag{6.4.13}$$

$$P_{(F_1,\ldots,F_k)}[X_{(k)} \leqslant c] \leqslant P_{(G_1,\ldots,G_k)}[X_{(k)} \leqslant c]. \tag{6.4.14}$$

This theorem provides a chain of inequalities for the distributions of the extreme order statistics through majorization. A special application of this result may occur in the following situation: The k populations are first divided into several clusters, and the distributions within each cluster are then replaced by their average for that cluster. If we proceed similarly when subdividing these clusters into smaller clusters, then we obtain sharper bounds for the distributions of the maximum and the minimum. This may be regarded as a nonparametric analogue to a result of Olkin, Sobel, and Tong (1976) (in which the parameters are averaged instead of the distribution functions).

For the theorems so far discussed in this section the inequalities depend on the concept of arithmetic average; e.g., the bounds given depend on the arithmetic mean of the parameters or distribution functions. Pledger and Proschan (1971) obtained a theorem for comparing order statistics from heterogeneous populations via majorization of the log values of the parameters; hence their results depend on the concept of geometric average, and the comparisons are made when the product of the parameters, instead of their sum, is kept fixed. Their approach to the problem was motivated by a problem in reliability theory. We first observe their result concerning independent Bernoulli trials, which is related to the problem of an r-out-of-k system in reliability theory.

Theorem 6.4.5. Let $U_1,\ldots,U_k$ be independent Bernoulli variables with parameters $\theta_1,\ldots,\theta_k$, respectively. Let $\boldsymbol{\tau}=(\tau_1,\ldots,\tau_k)'$, where $\tau_i=-\ln\theta_i$ ($\theta_i\in(0,1]$, $i=1,\ldots,k$) and $S_k=\Sigma_{i=1}^k U_i$. Then $P_{\boldsymbol{\theta}}[S_k\geqslant r]$ is Schur-convex in $\boldsymbol{\tau}$ for all $1\leqslant r\leqslant k-1$; hence $P_{\boldsymbol{\theta}}[S_k\leqslant r]$ is Schur-concave in $\boldsymbol{\tau}$ for all $0\leqslant r\leqslant k-2$.

Proof. For $1\leqslant r\leqslant k-1$ let $P_{\boldsymbol{\theta}}[S_k\geqslant r]=h_r(\boldsymbol{\theta})\equiv h_r^*(\boldsymbol{\tau})$. Clearly we have

$$h_r(\boldsymbol{\theta}) = \left(\prod_{i=1}^k \theta_i\right)\left[1+\sum_{i=1}^k(\theta_i^{-1}-1)\right.$$
$$\left.+\cdots+\sum_{i_1<\cdots<i_{k-r}}\prod_{l=1}^{k-r}(\theta_{i_l}^{-1}-1)\right].$$

Hence from $e^{\tau_i}=\theta_i^{-1}$ ($i=1,\ldots,k$) we can write

$$-\ln h_r^*(\boldsymbol{\tau}) = \sum_{i=1}^k \tau_i - \ln c_r(\boldsymbol{\tau}),$$

where

$$c_r(\boldsymbol{\tau}) = 1+\sum_{i=1}^k(\exp(\tau_i)-1)$$
$$+\cdots+\sum_{i_1<\cdots<i_{k-r}}\prod_{l=1}^{k-r}(\exp(\tau_{i_l})-1)>0.$$

Straightforward calculation shows that for all $i \neq j$

$$(\tau_i - \tau_j)\left[\frac{\partial}{\partial \tau_i}\ln h_r^*(\tau) - \frac{\partial}{\partial \tau_j}\ln h_r^*(\tau)\right] \geqslant 0.$$

Now the proof of the theorem follows from Lemma 6.2.2 and the fact that the log function is nondecreasing. ■

This theorem is an analogue to Theorems 6.4.1 and 6.4.2. As a special consequence, let $\bar{\theta}^* = (\prod_{i=1}^k \theta_i)^{1/k}$ denote the geometric mean; then

$$P_{\boldsymbol{\theta}}[S_k \leqslant a] \leqslant \sum_{j=0}^{a} \binom{k}{j} \bar{\theta}^{*j}(1-\bar{\theta}^*)^{k-j}, \qquad 0 \leqslant a \leqslant k-2, \tag{6.4.15}$$

and again the upper bound is a binomial probability. Now for $0 \leqslant a \leqslant k\bar{\theta} - 1$ a direct comparison between (6.4.1) and (6.4.15) seems interesting. Since it is well known that for k positive real numbers $\theta_1, \ldots, \theta_k$ their geometric mean $\bar{\theta}^*$ is less than or equal to their arithmetic mean $\bar{\theta}$ and that the family of binomial distributions is stochastically increasing in the parameter θ, we conclude that the bound in (6.4.1) is always sharper. But (6.4.15) also applies when a is in the range $k\bar{\theta} \leqslant a \leqslant k-2$, and (6.4.1) does not apply to such values of a.

PROBLEMS

1. Prove that if $\psi(\mathbf{x}) = \sum_{i=1}^k h(x_i)$ is Schur-convex (Schur-concave) for $\mathbf{x} \in I \times \cdots \times I$ and if h is continuous on I, where I is an interval, then h is convex (concave) on I (Marshall and Olkin, 1979, p. 67).
2. Let $\mathbf{X}$ have a continuous density function $f(\mathbf{x}) = \prod_{i=1}^k g(x_i)$. Show that f is Schur-concave if and only if g is log-concave (Marshall and Olkin, 1979, p. 73).
3. Let $X_1, \ldots, X_k$ be independent random variables with continuous distributions $F(x - \theta_i)$, $i = 1, \ldots, k$, respectively. For fixed i let $\boldsymbol{\theta}^i$ denote the parameter vector $(\theta_1, \ldots, \theta_{i-1}, \theta_{i+1}, \ldots, \theta_k)'$. Show that (for every fixed θ_i) if $\boldsymbol{\xi} \succ \boldsymbol{\tau}$ and F is log-concave, then

$$P_{\boldsymbol{\theta}^i = \boldsymbol{\tau}}\left[X_i = \max_{1 \leqslant j \leqslant k} X_j\right] \geqslant P_{\boldsymbol{\theta}^i = \boldsymbol{\xi}}\left[X_i = \max_{1 \leqslant j \leqslant k} X_j\right]$$

(Olkin, Sobel, and Tong, 1976).

4. Show that in Problem 3 the inequality remains true for scale parameter families with distributions $F(x/\alpha_i)$, $i=1,\ldots,k$, $F(0)=0$, $\alpha>0$, and with $\theta_i=\ln\alpha_i$ $(i=1,\ldots,k)$.
5. Show that in Problems 3 and 4 the inequalities remain true if "$\boldsymbol{\xi}\succ\boldsymbol{\tau}$" is replaced by the weaker condition "$\boldsymbol{\xi}\succ\succ\boldsymbol{\tau}$".
6. Show that a nondecreasing function of a Schur-convex function is Schur-convex.
7. Show that the union and intersection of two Schur-concave sets are Schur-concave.
8. Show that if $A\subset\mathfrak{R}^k$ is permutation invariant and convex, then its indicator function is Schur-concave.
9. Show that if $A\subset\mathfrak{R}^k$ satisfies the condition in (6.2.7), then its indicator function is Schur-concave, but not concave unless $A=\mathfrak{R}^k$.
10. Let $X_1,\ldots,X_k$ be independent random variables with continuous densities $g(x-\theta_i)$ for $x\in\mathfrak{R}^1$, $\theta_i\in\mathfrak{R}^1$. Let $X_{(1)}\leqslant\cdots\leqslant X_{(k)}$ denote the order statistics, and $\hat{\delta}_i=X_{(k)}-X_{(i)}$ denote the spacings for $i=1,\ldots,k-1$. Show that if g is log-concave, then $P_{\boldsymbol{\theta}}[\sum_{i=1}^{k-1}\hat{\delta}_i\leqslant a]$ is Schur-concave in $\boldsymbol{\theta}$ for all a (Tong, 1978).
11. (Continuation) Find an example to illustrate that $P_{\boldsymbol{\theta}}[(X_{(k)}-X_{(k-1)})\leqslant a]$ is neither Schur-concave nor Schur-convex in $\boldsymbol{\theta}$ (Tong, 1978).
12. Let $\mathbf{X}=(X_1,\ldots,X_k)'$ and $\mathbf{Y}=(Y_1,\ldots,Y_k)'$ be two multivariate normal variables with means $\mathbf{0}$ and covariance matrices $\boldsymbol{\Sigma}_{\mathbf{X}},\boldsymbol{\Sigma}_{\mathbf{Y}}$, respectively, where (for $j,j'=1,\ldots,r$ $(r\leqslant k)$) $\boldsymbol{\Sigma}_{\mathbf{X}}=(\mathbf{R}_{jj'})$, $\boldsymbol{\Sigma}_{\mathbf{Y}}=(\mathbf{S}_{jj'})$, $\mathbf{R}_{jj'}=\mathbf{0}$, $\mathbf{S}_{jj'}=\mathbf{0}$ for $j\neq j'$, $\mathbf{R}_{jj}$ is $a_j\times a_j$, $\mathbf{S}_{jj}$ is $b_j\times b_j$ and the elements of $\mathbf{R}_{jj}$ and $\mathbf{S}_{jj}$ are one on the diagonal and ρ $(\rho\geqslant 0)$ otherwise. Show that if $(a_1,\ldots,a_r)'\succ(b_1,\ldots,b_r)'$, then for every Borel-measurable set $A\subset\mathfrak{R}^1$ we have
$$P\left[\bigcap_{i=1}^{k}\{X_i\in A\}\right]\geqslant P\left[\bigcap_{i=1}^{k}\{Y_i\in A\}\right]$$
(Tong, 1977).
13. (Continuation) In Problem 12, if $\mathbf{X}$ and $\mathbf{Y}$ are multivariate t variables as defined in (3.1.1) with correlation matrices $\boldsymbol{\Sigma}_{\mathbf{X}}$ and $\boldsymbol{\Sigma}_{\mathbf{Y}}$, respectively, and degrees of freedom n, then the same inequality holds (Tong, 1977).
14. Let $\mathbf{X}=(X_1,\ldots,X_k)'$ and $\mathbf{Y}=(Y_1,\ldots,Y_k)'$ be two random variables. Find a counterexample to the following false statement: X_i is stochastically larger than Y_i for each i implies that $\mathbf{X}$ is stochastically larger than $\mathbf{Y}$.

15. Show that $\mathbf{X}$ is stochastically larger than $\mathbf{Y}$ if and only if $P[\mathbf{X}\in A]\geqslant P[\mathbf{Y}\in A]$ holds for every increasing Borel-measurable set A (Lehmann, 1955).

16. It is immediate that if $(X_1,\ldots,X_k)'$ is stochastically larger than $(Y_1,\ldots,Y_k)'$, then $P[\cap_{i=1}^k\{X_i\leqslant a_i\}]\leqslant P[\cap_{i=1}^k\{Y_i\leqslant a_i\}]$ holds for all $(a_1,\ldots,a_k)'$. Show that the converse is false by giving a counterexample (Marshall and Olkin, 1979 p. 482).

17. Show that if $\mathbf{X}_{\boldsymbol{\xi}}\overset{\text{st}}{\succ}\mathbf{X}_{\boldsymbol{\theta}}$, then the distribution of $\Sigma_{i=1}^k X_i$ remains the same under $\boldsymbol{\xi}$ or $\boldsymbol{\theta}$ (Nevius, Proschan, and Sethuraman, 1977a). Hence the same thing can be said for $U=h(\Sigma_{i=1}^k X_i)$ where $h\colon\mathfrak{R}^1\to\mathfrak{R}^1$ is any Borel-measurable function.

18. Show that if $\mathbf{X}_{\boldsymbol{\xi}}\overset{\text{st}}{\succ\succ}\mathbf{X}_{\boldsymbol{\theta}}$, then $\mathbf{X}_{\boldsymbol{\xi}}\overset{\text{st}}{\geqslant}\mathbf{X}_{\boldsymbol{\theta}}$ (Nevius, Proschan, and Sethuraman, 1977b). Hence $\Sigma_{i=1}^k X_i$ under $\boldsymbol{\xi}$ is stochastically larger than that under $\boldsymbol{\theta}$.

19. Let $X_1,\ldots,X_k$; $Y_1,\ldots,Y_k$ be independent random variables with distributions G_x and H_y, respectively. Show that $(X_1,\ldots,X_k)'$ stochastically weakly majorizes $(Y_1,\ldots,Y_k)'$ if and only if X_1 is stochastically larger than Y_1 (i.e., $G_x(a)\leqslant H_y(a)$ holds for all a).

20. Let $X_1,\ldots,X_k$ be independent Bernoulli variables with parameters $\theta_1,\ldots,\theta_k$, respectively, and let $S_k=\Sigma_{i=1}^k X_i$. Show that the variance of S_k is Schur-concave in $\boldsymbol{\theta}$.

21. Let $X_1,\ldots,X_k$ be independent random variables with continuous distributions $F_i(x)$, $i=1,\ldots,k$. Show that for fixed i and $F_i(x)$ the probability

$$P_{(F_1,\ldots,F_k)}\Big[X_i=\max_{1\leqslant j\leqslant k}X_j\Big]$$

is bounded above when the distributions F_j, $j\neq i$, are replaced by a common distribution

$$\bar{F}(x)=\frac{1}{k-1}\sum_{j\neq i}F_j(x).$$

REFERENCES

Eaton, M. L., and Perlman, M. D. (1977). Reflection groups, generalized Schur functions and the geometry of majorization. *Ann. Probab.* **5**, 829–860.

Gleser, L. J. (1975). On the distribution of the number of successes in independent trials. *Ann. Probab.* **3**, 182–188.

Hardy, G. H., Littlewood, J. E., and Pólya, G. (1959). *Inequalities*, 2nd ed. Cambridge Univ. Press, London and New York.

Hoeffding, W. (1956). On the distribution of the number of successes in independent trials. *Ann. Math. Statist.* **27**, 713–721.

Jogdeo, K. (1977). Association and probability inequalities. *Ann. Statist.* **5**, 495–504.

Lehmann, E. L. (1955). Ordered families of distributions. *Ann. Math. Statist.* **26**, 399–419.

Marshall, A. W., and Olkin, I. (1974). Majorization in multivariate distributions. *Ann. Statist.* **2**, 1189–1200.

Marshall, A. W., and Olkin, I. (1979). *Inequalities: Theory of Majorization and Its Applications*. Academic Press, New York.

Marshall, A. W., and Proschan, F. (1965). An inequality for convex functions involving majorization. *J. Math. Anal. Appl.* **12**, 87–90.

Mudholkar, G. S. (1966). The integral of an invariant unimodal function over an invariant convex set—an inequality and applications. *Proc. Amer. Math. Soc.* **17**, 1327–1333.

Mudholkar, G. S. (1969). A generalized monotone character of d.f.'s and moments of statistics from some well-known populations. *Ann. Inst. Statist. Math.* **21**, 277–285.

Nevius, S. E., Proschan, F., and Sethuraman, J. (1977a). Schur functions in statistics. II. Stochastic majorization. *Ann. Statist.* **5**, 263–273.

Nevius, S. E., Proschan, F., and Sethuraman, J. (1977b). A stochastic version of weak majorization, with applications. In *Statistical Decision Theory and Related Topics* (S. S. Gupta and D. S. Moore, eds.), Vol. II, pp. 281–296. Academic Press, New York.

Olkin, I. (1972). Monotonicity properties of Dirichlet integrals with applications to the multinomial distribution and the analysis of variance. *Biometrika* **59**, 303–307.

Olkin, I., Sobel, M., and Tong, Y. L. (1976). Estimating the true probability of correct selection for location and scale parameter families. Tech. Rep. No. 110, Department of Statistics, Stanford Univ., Stanford, California.

Pledger, G., and Proschan, F. (1971). Comparisons of order statistics and of spacings from heterogeneous distributions. In *Optimizing Methods in Statistics* (J. S. Rustagi, ed.), pp. 89–113. Academic Press, New York.

Proschan, F., and Sethuraman, J. (1974). Schur functions in statistics. I: The preservation theorem. Tech. Rep. No. M254-RR, Department of Statistics, Florida State Univ., Tallahassee, Florida.

Proschan, F., and Sethuraman, J. (1977). Schur functions in Statistics. I. The preservation theorem. *Ann. Statist.* **5**, 256–262.

Rinott, Y. (1973). Multivariate majorization and rearrangement inequalities with some applications to probability and statistics. *Israel J. Math.* **15**, 60–67.

Samuels, S. M. (1965). On the number of successes of independent trials. *Ann. Math. Statist.* **36**, 1272–1278.

Sen, P. K. (1970). A note on order statistics for heterogeneous distributions. *Ann. Math. Statist.* **41**, 2137–2139.

Šidák, Z. (1973). A chain of inequalities for some types of multivariate distributions, with nine special cases. *Apl. Mat.* **18**, 110–118.

Tong, Y. L. (1977). An ordering theorem for conditionally independent and identically distributed random variables. *Ann. Statist.* **5**, 274–277.

Tong, Y. L. (1978). An adaptive solution to ranking and selection problems. *Ann. Statist.* **6**, 658–672.

CHAPTER

7

Distribution-Free Inequalities

In previous chapters we studied probability inequalities in multivariate distributions under various conditions. Those conditions range from very specific (e.g., the assumption of normality in Chapter 2) to very general, but basically they all depend on the properties of the density or the distribution of the random variable involved. The nature of the inequalities to be studied in this chapter is quite different. They are distribution-free in the following sense: Either they are completely independent of the distributions, or they depend only on the moments. Hence they are applicable in most cases. But on the other hand, because there is more ground to cover, in a given problem they might not be as sharp if compared with some of the other inequalities already given in previous chapters.

The first group of inequalities to be studied in this chapter is of Bonferroni type. The original Bonferroni inequality is well known, and its applications have appeared frequently in the statistical literature. We shall see several versions of its refinements in Section 7.1. Section 7.2 concerns multivariate versions of the Chebyshev inequality, and the lower bounds for the probabilities of intersection of events involve only the first two moments of the random variables. In Section 7.3 we shall study a few probability inequalities in connection with an application of the Kolmogorov inequality for independent random variables.

7.1. BONFERRONI-TYPE INEQUALITIES

Let $\mathbf{X}=(X_1,\ldots,X_k)'$ denote a k-dimensional random variable. The following theorem gives a universal lower bound for the probability of the intersection of events.

Theorem 7.1.1. Let $A_1,\ldots,A_k$ be Borel-measurable subsets of the real line, and denote

$$p_i = P[X_i \in A_i], \qquad i = 1,\ldots,k. \tag{7.1.1}$$

Then

$$P\left[\bigcap_{i=1}^{k} \{X_i \in A_i\}\right] \geqslant 1 - \sum_{i=1}^{k} (1-p_i). \tag{7.1.2}$$

The proof follows immediately from the facts that $\cap_{i=1}^{k} A_i = (\cup_{i=1}^{k} A_i^c)^c$, where A^c denotes the complement of A, and that $P[\cup_{i=1}^{k}\{X_i \in A_i^c\}] \leqslant \Sigma_{i=1}^{k} P[X_i \in A_i^c]$. This inequality is nontrivial when the lower bound is positive. Since it involves only the marginal probabilities $p_1,\ldots,p_k$, we may therefore identify it as the Bonferroni inequality of degree one.

A Bonferroni-type inequality of degree two can be derived from a result of Chung and Erdös (1952), which was obtained in connection with their problem concerning the Borel–Cantelli lemma. That inequality provides a lower bound for the probability of the union of events, and the bound depends only on the marginal probabilities and the probabilities of the pairwise intersections. For notational convenience let us define the events

$$B_i = [X_i \notin A_i], \qquad i = 1,\ldots,k, \tag{7.1.3}$$

and define

$$q_i = P(B_i) = 1 - p_i, \qquad q_{ij} = P(B_i \cap B_j), \qquad i,j = 1,\ldots,k. \tag{7.1.4}$$

Theorem 7.1.2. Let $\mathbf{X}$ be a k-dimensional random variable and $A_1,\ldots,A_k$ be Borel-measurable subsets of the real line. Then

$$1 - Q_1 \leqslant P\left[\bigcap_{i=1}^{k} \{X_i \in A_i\}\right] \leqslant 1 - \frac{Q_1^2}{Q_1 + 2Q_2}, \tag{7.1.5}$$

where

$$Q_1 = \sum_{i=1}^{k} q_i, \qquad Q_2 = \sum_{i=2}^{k} \sum_{j=1}^{i-1} q_{ij}. \tag{7.1.6}$$

Proof. The lower bound is a restatement of Theorem 7.1.1, which is given here for the purpose of comparison. The upper bound follows immediately from

$$P\left(\bigcup_{i=1}^{k} B_i\right) \geqslant \frac{Q_1^2}{Q_1 + 2Q_2}, \tag{7.1.7}$$

which was given by Chung and Erdös (1952). To prove (7.1.7) let $\chi_i(\omega)$, $\chi_{ij}(\omega)$ denote the indicator functions of B_i and $B_i \cap B_j$, respectively. Taking expectations on both sides of the identity

$$2\sum_{i=2}^{k} \sum_{j=1}^{i-1} \chi_{ij}(\omega) = \left[\sum_{i=1}^{k} \chi_i(\omega)\right]^2 - \sum_{i=1}^{k} \chi_i^2(\omega)$$

yields

$$2Q_2 = E\left[\sum_{i=1}^{k} \chi_i(\omega)\right]^2 - \sum_{i=1}^{k} E\chi_i^2(\omega).$$

Now let us define

$$U = \begin{cases} 1 & \text{if } \displaystyle\sum_{i=1}^{k} \chi_i(\omega) > 0, \\ 0 & \text{otherwise,} \end{cases}$$

and $V = \sum_{i=1}^{k} \chi_i(\omega)$. From $[E(UV)]^2 = (EV)^2$ and the Cauchy–Schwarz inequality we obtain

$$\left[E\sum_{i=1}^{k} \chi_i(\omega)\right]^2 \leqslant P\left[\sum_{i=1}^{k} \chi_i(\omega) > 0\right] E\left[\sum_{i=1}^{k} \chi_i(\omega)\right]^2. \tag{7.1.8}$$

This together with the facts that $E\chi_i(\omega) = E\chi_i^2(\omega) = P(B_i)$ and $P[\sum_{i=1}^{k} \chi_i(\omega) > 0] = P(\cup_{i=1}^{k} B_i)$ gives the inequality

$$2Q_2 \geqslant Q_1^2 / P\left(\bigcup_{i=1}^{k} B_i\right) - Q_1;$$

hence (7.1.7) follows. ■

We note that Q_2 in (7.1.6) is the sum of the probabilities of all pairwise intersections, and the denominator in the upper bound in (7.1.5) is simply $\Sigma_{i=1}^k \Sigma_{j=1}^k q_{ij}$ ($q_{ii} = q_i$ for all i). Also note that if $q_{ij} = 0$ for all $i \neq j$, then both the upper bound and the lower bound in (7.1.5) are $1 - Q_1$.

Another Bonferroni-type inequality of order two can be derived from a result of Dawson and Sankoff (1967), which says that

$$P\left(\bigcup_{i=1}^{k} B_i\right) \geqslant \frac{\lambda Q_1^2}{(2-\lambda)Q_1 + 2Q_2} + \frac{(1-\lambda)Q_1^2}{(1-\lambda)Q_1 + 2Q_2}, \tag{7.1.9}$$

where $\lambda = (2Q_2/Q_1) - [2Q_2/Q_1]$ is in $[0, 1)$ and $[x]$ indicates the largest integer less than or equal to x. Note that the result of Chung and Erdös was stated incorrectly as $P(\cup_{i=1}^k B_i) \geqslant (Q_1^2 - Q_1)/2Q_2$ in the paper of Dawson and Sankoff.

An inequality given by Gallot (1966), which represents a refinement of the inequality of Chung and Erdös (1952) and of a result of Whittle (1959), can be given in an elegant quadratic form. Let us define a $k \times 1$ vector Λ_1 and a $k \times k$ matrix Λ_2 by

$$\Lambda_1 = (q_1, \ldots, q_k)', \qquad \Lambda_2 = (q_{ij}), \tag{7.1.10}$$

where q_i and q_{ij} are as defined in (7.1.4). Gallot's result says that

Lemma 7.1.1. If Λ_2 is nonsingular, then for every real nonzero vector $\mathbf{c} = (c_1, \ldots, c_k)'$ we have

$$P\left(\bigcup_{i=1}^{k} B_i\right) \geqslant \frac{\mathbf{c}'\Lambda_1\Lambda_1'\mathbf{c}}{\mathbf{c}'\Lambda_2\mathbf{c}}. \tag{7.1.11}$$

Hence we have

$$P\left(\bigcup_{i=1}^{k} B_i\right) \geqslant \sup_{\mathbf{c}} \frac{\mathbf{c}'\Lambda_1\Lambda_1'\mathbf{c}}{\mathbf{c}'\Lambda_2\mathbf{c}} = \Lambda_1'\Lambda_2^{-1}\Lambda_1. \tag{7.1.12}$$

Proof. For fixed $\mathbf{c}$ let $V = \Sigma_{i=1}^k c_i \chi_i(\omega)$, where χ_i is the indicator function of B_i, and define

$$U = \begin{cases} 1 & \text{if} \quad V \neq 0, \\ 0 & \text{otherwise.} \end{cases}$$

Let

$$v_1 = EV = \sum_{i=1}^{k} c_i q_i = \mathbf{c}'\boldsymbol{\Lambda}_1,$$

$$v_2 = EV^2 = \sum_{i=1}^{k} \sum_{j=1}^{k} c_i c_j q_{ij} = \mathbf{c}'\boldsymbol{\Lambda}_2\mathbf{c}$$

be the first two moments of V. Then from $[E(UV)]^2 = (EV)^2$ and the Cauchy–Schwarz inequality we have

$$v_1^2 \leqslant v_2 P[V \neq 0]. \tag{7.1.13}$$

On the other hand, it is easy to see that $(\cup_{i=1}^{k} B_i)^c \subset [V=0]$. This together with (7.1.13) yields

$$P\left(\bigcup_{i=1}^{k} B_i\right) \geqslant P[V \neq 0] \geqslant \frac{v_1^2}{v_2},$$

which is the inequality in (7.1.11).

Now suppose that

$$\sup_{\mathbf{c}} \frac{\mathbf{c}'\boldsymbol{\Lambda}_1\boldsymbol{\Lambda}_1'\mathbf{c}}{\mathbf{c}'\boldsymbol{\Lambda}_2\mathbf{c}} = \lambda.$$

Then we may simultaneously diagonalize $\boldsymbol{\Lambda}_1\boldsymbol{\Lambda}_1'$ and $\boldsymbol{\Lambda}_2$, thus obtaining λ as the unique nonzero root of the equation $|\lambda\boldsymbol{\Lambda}_2 - \boldsymbol{\Lambda}_1\boldsymbol{\Lambda}_1'| = 0$; that is, $\lambda = \boldsymbol{\Lambda}_1'\boldsymbol{\Lambda}_2^{-1}\boldsymbol{\Lambda}_1$ as given in (7.1.12). ■

The proof given above is slightly different from Gallot's original proof (note that Eq. (3) in his proof contains an error). Furthermore, when we choose $\mathbf{c} = (1, \ldots, 1)'$, then the lower bound in (7.1.11) reduces to $Q_1^2/(Q_1 + 2Q_2)$, which is the inequality given by Chung and Erdös (1952). Therefore the bound given in (7.1.12) is sharper than the bound in (7.1.7).

The nonsingularity condition on $\boldsymbol{\Lambda}_2$ was later removed by Kounias (1968). Therefore we obtain

Theorem 7.1.3. Let $\mathbf{X}$ be a k-dimensional random variable and $A_1, \ldots, A_k$ be Borel-measurable subsets of the real line. Then for q_i, q_{ij}, $\boldsymbol{\Lambda}_1$, and $\boldsymbol{\Lambda}_2$ as defined in (7.1.4) and (7.1.10), respectively, $(i,j = 1, \ldots, k)$,

$$P\left[\bigcap_{i=1}^{k} \{X_i \in A_i\}\right] \leqslant 1 - \boldsymbol{\Lambda}_1'\boldsymbol{\Lambda}_2^{-}\boldsymbol{\Lambda}_1, \tag{7.1.14}$$

where $\boldsymbol{\Lambda}_2^{-}$ is a generalized inverse of $\boldsymbol{\Lambda}_2$ satisfying $\boldsymbol{\Lambda}_2\boldsymbol{\Lambda}_2^{-}\boldsymbol{\Lambda}_2 = \boldsymbol{\Lambda}_2$.

This theorem gives an upper bound for the joint probability in multivariate distributions when the marginal probabilities $P[X_i \in A_i]$ and $P[\{X_i \in A_i\} \cap \{X_j \in A_j\}]$ are known for all i,j (in this case q_i, q_{ij} are uniquely determined).

From an inequality given by Kounias (1968) we can obtain another theorem. Note that the lower bound here is an improvement of that stated in Theorem 7.1.1.

Theorem 7.1.4. Let $\mathbf{X}$, A_i, q_i, and q_{ij} $(i,j=1,\ldots,k)$ be defined as in Theorem 7.1.3. Then

$$1 - Q_1 + \max_{1 \leqslant l \leqslant k} \sum_{i \neq l} q_{il} \leqslant P\left[\bigcap_{i=1}^{k} \{X_i \in A_i\}\right] \leqslant 1 - Q_1 + Q_2. \tag{7.1.15}$$

Proof. This is equivalent to

$$Q_1 - Q_2 \leqslant P\left(\bigcup_{i=1}^{k} B_i\right) \leqslant Q_1 - \max_{1 \leqslant l \leqslant k} \sum_{i \neq l} q_{il}, \tag{7.1.16}$$

which was given by Kounias. To prove (7.1.16) we realize that the inequalities

$$\sum_{i=1}^{k} \chi_i(\omega) - \sum_{i=2}^{k} \sum_{j=1}^{i-1} \chi_i(\omega)\chi_j(\omega) \leqslant \max_{1 \leqslant i \leqslant k} \chi_i(\omega)$$

$$\leqslant (1-\chi_l(\omega)) \sum_{i=1}^{k} \chi_i(\omega) + \chi_l(\omega)$$

hold for every l and every ω in the sample space, where $\chi_i(\omega)$ is the indicator function of B_i. Hence (7.1.16) follows by taking expectations. ■

The bounds given in Theorems 7.1.2–7.1.4 are Bonferroni-type of degree two, and their applications require only knowledge of the probabilities of the pairwise intersections. Extensions to higher degrees appear possible, but are in general tedious unless the events are exchangeable. With this additional condition of exchangeability the probability becomes permutation invariant and the bounds become more workable by applying certain combinatorial arguments. This problem was studied by Sobel and Uppuluri (1972).

To see their result let us first introduce some notation. Let $\mathbf{X}$ be a k-dimensional random variable such that its distribution is permutation invariant. Let A be a Borel-measurable subset of the real line. Then the events $[X_1 \in A],\ldots,[X_k \in A]$ are exchangeable. Let us define

$$p^{(m)} = P\left[\bigcap_{i=1}^{m} \{X_i \in A\}\right], \qquad m = 1,\ldots,k, \tag{7.1.17}$$

and for $\nu \leqslant k$ define

$$M_{\nu-j}^{(\nu)} = \sum_{l=0}^{\nu-j} (-1)^l \binom{k-\nu-1+l}{l} \binom{k-j}{\nu-j-l} p^{(\nu-l)}, \qquad j = 0,\ldots,\nu. \tag{7.1.18}$$

Note that the $M_{\nu-j}^{(\nu)}$ $(j=0,1,\ldots,\nu)$ are all of degree ν (i.e., they all depend on $p^{(\nu)}$). They proved the following.

Theorem 7.1.5. Let the distribution of $\mathbf{X}=(X_1,\ldots,X_k)'$ be permutation invariant and let $A \subset \mathfrak{R}^1$ be Borel-measurable. Then for all $\nu \leqslant k$ and all odd (even) nonnegative integers $\nu_0 \leqslant \nu$ $(\nu_e \leqslant \nu)$

$$\max\{M_1^{(\nu)}, M_3^{(\nu)},\ldots,M_{\nu_0}^{(\nu)}\} \leqslant P\left[\bigcap_{i=1}^{k} \{X_i \in A\}\right] \leqslant \min\{M_0^{(\nu)}, M_2^{(\nu)},\ldots,M_{\nu_e}^{(\nu)}\}. \tag{7.1.19}$$

The proof of this theorem depends on another inequality for the union of exchangeable events, which in turn depends on an expression in terms of the indicator functions of the events, an application of the inclusion–exclusion formula, and certain combinatorial arguments. The details are not given here.

For $\nu=2$, 3, and 4 the expressions for $M_{\nu-j}^{(\nu)}$ were simplified by Sobel and Uppuluri:

(a) $\nu=2$

$$M_1^{(2)} = (k-1)p^{(2)} - (k-2)p^{(1)},$$

$$M_0^{(2)} = p^{(2)},$$

$$M_2^{(2)} = \binom{k}{2}p^{(2)} - k(k-2)p^{(1)} + \binom{k-1}{2}.$$

(b) $\nu=3$

$$M_1^{(3)} = (k-2)p^{(3)} - (k-3)p^{(2)},$$

$$M_3^{(3)} = \binom{k}{3}p^{(3)} - (k-3)\binom{k}{2}p^{(2)} + k\binom{k-2}{2}p^{(1)} - \binom{k-1}{3},$$

$$M_0^{(3)} = p^{(3)},$$

$$M_2^{(3)} = \binom{k-1}{2}p^{(3)} - (k-1)(k-3)p^{(2)} + \binom{k-2}{2}p^{(1)}.$$

(c) $\nu=4$

$$M_1^{(4)} = (k-3)p^{(4)} - (k-4)p^{(3)},$$

$$M_3^{(4)} = \binom{k-1}{3}p^{(4)} - (k-4)\binom{k-1}{2}p^{(3)} + (k-1)\binom{k-3}{2}p^{(2)} - \binom{k-2}{3}p^{(1)},$$

$$M_0^{(4)} = p^{(4)},$$

$$M_2^{(4)} = \binom{k-2}{2}p^{(4)} - (k-2)(k-4)p^{(3)} + \binom{k-3}{2}p^{(2)},$$

$$M_4^{(4)} = \binom{k}{4}p^{(4)} - (k-4)\binom{k}{3}p^{(3)} + \binom{k}{2}\binom{k-3}{2}p^{(2)} - k\binom{k-2}{3}p^{(1)} + \binom{k-1}{4}.$$

When ν_0 and ν_e are taken to be the largest integers less than or equal to ν, then the bounds given in Theorem 7.1.5 are sharper when ν increases. For $\nu=k$, the $(k+1)$ $M_{k-j}^{(k)}$ values are equal ($j=0,1,\ldots,k$), and the inequalities become equalities.

This theorem applies to all exchangeable random variables. Hence in particular it applies to random variables that are positively dependent by mixture (see Section 5.3 and the examples in Section 3.3). In this case lower and upper bounds for the joint probability may be obtained when the marginal probabilities with degree ν ($\nu<k$) are known. We note that even if the random variables under consideration are not exchangeable, it may still be possible first to obtain a bound that involves only exchangeable random variables (e.g., this is possible through the applications of certain theorems given in Chapter 6) and then to apply Theorem 7.1.5 to obtain another bound for this joint probability of events of exchangeable random variables (e.g., see Problem 3).

7.2. CHEBYSHEV-TYPE INEQUALITIES

The univariate version of the Chebyshev inequality gives a lower bound for the probability that a random variable X will take values within certain units of the standard deviation from its mean. Specifically, it says that if $EX=\mu$ and $E(X-\mu)^2=\sigma^2<\infty$, then

$$P\left[|X-\mu|\leqslant a\sigma\right] \geqslant 1-1/a^2 \tag{7.2.1}$$

holds for all $a>0$ (for the inequality to be nontrivial we require $a>1$). The inequality depends only on the first two moments of X.

In this section we study some multivariate extensions of this inequality. Let us first see two trivial extensions: If $X_1,\ldots,X_k$ are independent random variables, then clearly from (7.2.1) we have

$$P\left[\bigcap_{i=1}^{k} |X_i-\mu_i|\leqslant a_i\sigma_i\right] \geqslant \prod_{i=1}^{k}\left(1-\frac{1}{a_i^2}\right). \tag{7.2.2}$$

For the general case in which $X_1,\ldots,X_k$ are not necessarily independent, by combining (7.2.1) with Bonferroni's inequality of degree one (Theorem 7.1.1) we can obtain

$$P\left[\bigcap_{i=1}^{k} |X_i-\mu_i|\leqslant a_i\sigma_i\right] \geqslant 1-\sum_{i=1}^{k}\frac{1}{a_i^2}. \tag{7.2.3}$$

The inequality in (7.2.3) does not depend on the correlations of the X_is. Hence if the random variables are highly correlated, then it may become too conservative. This suggests that improvement can be made when the bounds are also functions of the correlation coefficients.

Berge (1937) obtained such an inequality (given below) for bivariate random variables and for the special case $a_1=a_2=a$.

Theorem 7.2.1. Let $\mathbf{X}=(X_1,X_2)'$ have a mean vector $(\mu_1,\mu_2)'$, variances σ_1^2,σ_2^2, and a correlation ρ. Then

$$P\left[\bigcap_{i=1}^{2}\{|X_i-\mu_i|\leqslant a\sigma_i\}\right] \geqslant 1-\left[1+(1-\rho^2)^{1/2}\right]/a^2 \tag{7.2.4}$$

holds for all $a>0$.

Proof. To prove this theorem, Berge adopted an argument first used by Pearson (1918). Denote $\mathbf{Y}=(Y_1, Y_2)'$, where

$$Y_i = (X_i - \mu_i)/a\sigma_i, \qquad i = 1,2,$$

then the variances of Y_1, Y_2 are $1/a^2$ and their covariance is ρ/a^2. Let A denote the subset in $\mathscr{R}^2$ given by

$$A = \{\mathbf{y}=(y_1,y_2)' \mid |y_i| \leqslant 1, i=1,2\},$$

and for arbitrary but fixed $c \in (-1,1)$ define a 2×2 positive definite matrix

$$\mathbf{C} = \begin{pmatrix} 1 & c \\ c & 1 \end{pmatrix}^{-1} = \frac{1}{(1-c^2)}\begin{pmatrix} 1 & -c \\ -c & 1 \end{pmatrix}.$$

Then we have

$$E(\mathbf{Y}'\mathbf{C}\mathbf{Y}) = \frac{1}{(1-c^2)} E\left[Y_1^2 - 2cY_1Y_2 + Y_2^2\right] = \frac{2(1-c\rho)}{a^2(1-c^2)}. \tag{7.2.5}$$

On the other hand, it is easy to check that $\mathbf{y}'\mathbf{C}\mathbf{y} \geqslant 1$ holds for all $\mathbf{y} \notin A$. This together with the fact that $\mathbf{C}$ is positive definite yields the inequality

$$E(\mathbf{Y}'\mathbf{C}\mathbf{Y}) \geqslant \int_{A^c} (\mathbf{y}'\mathbf{C}\mathbf{y})\, dG(\mathbf{y}) \geqslant 1 - \int_A dG(\mathbf{y}), \tag{7.2.6}$$

where A^c denotes the complement of A and $G(\mathbf{y})$ is the distribution function of $\mathbf{Y}$. Combining (7.2.5) and (7.2.6), we may conclude that

$$P[\mathbf{Y} \in A] \geqslant 1 - 2(1-c\rho)/\{a^2(1-c^2)\}$$

holds for all $c \in (-1,1)$. This implies

$$P[\mathbf{Y} \in A] \geqslant 1 - \inf_{c \in (-1,1)} \left[2(1-c\rho)/\{a^2(1-c^2)\}\right]. \tag{7.2.7}$$

It is then straightforward to check that the right-hand side of (7.2.7) is $[1 - \{1+(1-\rho^2)^{1/2}\}/a^2]$, and the proof is complete. ■

The lower bound given in the above theorem depends on the correlation coefficient ρ, and it ranges from $(1-2/a^2)$ (when $\rho=0$) to $(1-1/a^2)$ (when $\rho=\pm 1$). An example offered by Berge shows that the equality is attainable. Note that the upper bound is strictly decreasing in $|\rho|$, and it reduces to the bound in the univariate

Chebyshev inequality when $\rho = \pm 1$. A generalization of this inequality was later made by Lal (1955). His generalization covers the case in which the a values are not necessarily equal. It says that

$$P\left[\bigcap_{i=1}^{2}\{|X_i - \mu_i| \leqslant a_i\sigma_i\}\right]$$
$$\geqslant 1 - \left[(a_1^2 + a_2^2) + \left\{(a_1^2 + a_2^2)^2 - 4\rho^2 a_1^2 a_2^2\right\}^{1/2}\right] / 2a_1^2 a_2^2. \tag{7.2.8}$$

This inequality reduces to Berge's inequality when $a_1 = a_2 = a$. Lal also obtained a result for general k, but that inequality is sharp only for $k = 2$, in which case it reduces to (7.2.8).

A multivariate generalization of Berge's inequality was obtained simultaneously by Olkin and Pratt (1958) and Whittle (1958). Their basic approach to the problem is the following: Let us define for fixed $a_i > 0$

$$Y_i = (X_i - \mu_i)/a_i\sigma_i, \qquad i = 1,\ldots,k,$$
$$A = \{\mathbf{y} \,|\, |y_i| \leqslant 1,\ i = 1,\ldots,k\}.$$

It follows that

$$P\left[\bigcap_{i=1}^{k}\{|X_i - \mu_i| \leqslant a_i\sigma_i\}\right] = P[\mathbf{Y} \in A],$$

where $\mathbf{Y} = (Y_1,\ldots,Y_k)'$. Now let $\mathbf{C} = (c_{ij})$ be any $k \times k$ matrix satisfying the condition

$$\mathbf{C} \text{ is positive definite and } \mathbf{y} \notin A \Rightarrow \mathbf{y}'\mathbf{C}\mathbf{y} \geqslant 1. \tag{7.2.9}$$

It is easy to check that $\mathbf{T} = (\tau_{ij})$, the covariance matrix of $\mathbf{Y}$, is of the form

$$\tau_{ij} = \rho_{ij}/a_i a_j \tag{7.2.10}$$

when ρ_{ij} is the correlation between X_i and X_j ($\rho_{ii} = 1$). Hence

$$E(\mathbf{Y}'\mathbf{C}\mathbf{Y}) = \sum_{i=1}^{k}\sum_{j=1}^{k} c_{ij}\tau_{ij} = \mathrm{tr}(\mathbf{CT}). \tag{7.2.11}$$

On the other hand, from the condition stated in (7.2.9) we may conclude that

$$E(\mathbf{Y}'\mathbf{C}\mathbf{Y}) \geqslant \int_{A^c} (\mathbf{y}'\mathbf{C}\mathbf{y})\,dG(\mathbf{y}) \geqslant 1 - \int_A dG(\mathbf{y}), \tag{7.2.12}$$

where A^c denotes the complement of A and $G(\mathbf{y})$ is the distribution of $\mathbf{Y}$. From (7.2.11) and (7.2.12) it follows that

$$P[\mathbf{Y} \in A] \geqslant 1 - \operatorname{tr}(\mathbf{CT}). \tag{7.2.13}$$

Therefore we obtain

Theorem 7.2.2. Let $\mathbf{X} = (X_1, \ldots, X_k)'$ have mean vector $\boldsymbol{\mu}$ and covariance matrix $\boldsymbol{\Sigma} = (\rho_{ij}\sigma_i\sigma_j)$. For $a_i > 0$ let $\mathbf{T} = (\tau_{ij})$ denote a matrix as defined in (7.2.10), and let $\mathcal{C}$ be the class of all $k \times k$ matrices satisfying (7.2.9). Then

$$P\left[\bigcap_{i=1}^{k} \{|X_i - \mu_i| \leqslant a_i\sigma_i\}\right] \geqslant 1 - \inf_{\mathbf{C} \in \mathcal{C}} \operatorname{tr}(\mathbf{CT}). \tag{7.2.14}$$

In order to obtain the sharpest bound possible one needs to maximize the right-hand side of (7.2.14). This sharpest bound was given by both Olkin and Pratt (1958) and Whittle (1958). It can be obtained by solving a certain matrix equation, and cannot be computed easily in general. It was also shown that this sharpest bound is attainable for general k. In the special case in which one considers only a certain subclass of matrices with a special form, the quantity on the right-hand side of (7.2.14) becomes more workable. We shall now state two special results as corollaries.

Corollary 1. Under the assumptions stated in Theorem 7.2.2,

$$P\left[\bigcap_{i=1}^{k} \{|X_i - \mu_i| \leqslant a_i\sigma_i\}\right] \geqslant 1 - \sum_{i=1}^{k} \frac{1}{a_i^2}.$$

This follows by letting $\mathbf{C}$ be the identity matrix in (7.2.13). Note that this inequality was already observed in (7.2.3) as a direct consequence of the Bonferroni inequality and the univariate Chebyshev inequality.

The next corollary is obtained from Theorem 7.2.2 by considering the minimization of $\operatorname{tr}(\mathbf{CT})$ over the subclass of matrices of the following form: Suppose $\mathbf{C} = (c_{ij})$ is the inverse of $\mathbf{D} = (d_{ij})$, where

$$d_{ij} = \begin{cases} 1 & \text{for } i = j, \\ c & \text{for } i \neq j, \end{cases} \tag{7.2.15}$$

and c satisfies $c \in (-1/(k-1), 1)$. Then the elements of $\mathbf{C}$ are

$$c_{ij} = \begin{cases} [1+(k-2)c]/\Delta(c) & \text{for } i=j, \\ -c/\Delta(c) & \text{for } i \neq j, \end{cases} \tag{7.2.16}$$

where $\Delta(c) = [1+(k-2)c-(k-1)c^2]$. Defining

$$s = \sum_{i=1}^{k} \sum_{j=1}^{k} \frac{\rho_{ij}}{a_i a_j}, \qquad t = \sum_{i=1}^{k} \frac{1}{a_i^2}, \tag{7.2.17}$$

we can write

$$\operatorname{tr}(\mathbf{CT}) = \frac{1}{\Delta(c)} [\{1+(k-2)c\}t - c(s-t)], \tag{7.2.18}$$

and then consider the minimization of tr(**CT**) over c. The derivative of the right-hand side of (7.2.18) equals zero at

$$c = [t \pm \{s(kt-s)/(k-1)\}^{1/2}]/[s-(k-1)t].$$

This together with the condition $c \in (-1/(k-1), 1)$ implies that the right-hand side of (7.2.18) is minimized at

$$c = [t - \{s(kt-s)/(k-1)\}^{1/2}]/[s-(k-1)t].$$

Substituting this c value in (7.2.18), we obtain

Corollary 2. Under the conditions stated in Theorem 7.2.2

$$P\left[\bigcap_{i=1}^{k} \{|X_i - \mu_i| \leq a_i \sigma_i\right] \geq 1 - \frac{[s^{1/2} + \{(k-1)(kt-s)\}^{1/2}]^2}{k^2}. \tag{7.2.19}$$

Note that for $k=2$ the right-hand side of (7.2.19) reduces to (7.2.8). For general k and $a_1 = \cdots = a_k = a$, $\rho_{ij} = \rho$ $(i \neq j)$, it reduces to

$$P\left[\bigcap_{i=1}^{k} \{|X_i - \mu_i| \leq a\sigma_i\}\right]$$
$$\geq 1 - \frac{[(k-1)(1-\rho)^{1/2} + \{1+(k-1)\rho\}^{1/2}]^2}{ka^2}. \tag{7.2.20}$$

If in addition $k=2$, then (7.2.20) reduces to Berge's inequality.

A one-sided Chebyshev-type inequality was obtained by Marshall and Olkin (1960). Their result is a generalization of the known inequality (see, e.g., Uspensky (1937)) that if X has mean 0 and variance σ^2, then

$$P[X \geqslant 1] \leqslant \sigma^2/(1+\sigma^2), \tag{7.2.21}$$

or to put it in another form

$$P[X \leqslant a\sigma] \geqslant 1 - 1/(1+a^2), \qquad a > 0. \tag{7.2.22}$$

Their generalization was done for the case in which the random variables are equally correlated, and it is stated in the following lemma.

Lemma 7.2.1. Let $\mathbf{Y}=(Y_1,\ldots,Y_k)'$ have means zero, variances σ^2, and correlations $\rho \in (-1/(k-1), 1)$, and let u denote

$$u = (k-1)(1-\rho) - 1.$$

If

(a) $1-\sigma^2 u > 0$ and
(b) $k \geqslant \sigma^2(k-1)(1+u)$,

then

$$P\left[\bigcup_{i=1}^{k} \{Y_i \geqslant 1\}\right] \leqslant \frac{k\sigma^2\left[\{(1+(k-1)\rho)(1-\sigma^2 u)\}^{1/2} + (k-1)(1-\rho)^{1/2}\right]^2}{\left[k+\sigma^2\{1+(k-1)\rho\}\right]^2} \tag{7.2.23}$$

holds.

Outline of the Proof. Let A denote the set

$$A = \{\mathbf{y} \mid y_i \leqslant 1, i=1,\ldots,k\}. \tag{7.2.24}$$

Consider a function ϕ of the form

$$\phi(\mathbf{y}) = (\mathbf{y}-\mathbf{b})'\mathbf{C}(\mathbf{y}-\mathbf{b}), \tag{7.2.25}$$

where $\mathbf{b}$ is a real vector, $\mathbf{C}$ is a $k \times k$ matrix such that $\mathbf{b}$ and $\mathbf{C}$ satisfy $\phi(\mathbf{y}) \geqslant 0$ for all $\mathbf{y}$ and $\phi(\mathbf{y}) \geqslant 1$ for all $\mathbf{y} \notin A$. Then

$$E\phi(\mathbf{Y}) = \operatorname{tr}(\mathbf{C}(\mathbf{T}+\mathbf{bb}')),$$

where $\mathbf{T}$ is the covariance matrix of $\mathbf{Y}$. On the other hand it follows that

$$E\phi(\mathbf{Y}) \geqslant \int_{A^c} \phi(\mathbf{y})\, dG(\mathbf{y}) \geqslant P[\mathbf{Y} \notin A].$$

Hence for all such $\mathbf{b}, \mathbf{C}$ we have

$$P[\mathbf{Y} \notin A] \leqslant \operatorname{tr}(\mathbf{C}(\mathbf{T}+\mathbf{bb}')). \tag{7.2.26}$$

It can be shown that under the conditions imposed on $\mathbf{b}$ and $\mathbf{C}$ the infimum of the right-hand side in (7.2.26) is the quantity given in (7.2.23). For details the reader is referred to Marshall and Olkin (1960). ■

From this lemma we can obtain the following theorem.

Theorem 7.2.3. Let $\mathbf{X}=(X_1,\ldots,X_k)'$ have a mean vector $\boldsymbol{\mu}$, variances σ_i^2 $(i=1,\ldots,k)$, and a common correlation $\rho \in [-1/(k-1), 1]$. Then for a satisfying

$$a > (k-1)[(1-\rho)/k]^{1/2} \tag{7.2.27}$$

we have

$$P\left[\bigcap_{i=1}^{k} \{X_i \leqslant \mu_i + a\sigma_i\}\right]$$

$$\geqslant 1 - \frac{\left[\{(1+(k-1)\rho)(a^2-u)\}^{1/2} + a(k-1)(1-\rho)^{1/2}\right]^2}{k\left[a^2+\{1+(k-1)\rho\}/k\right]^2}, \tag{7.2.28}$$

where u is as defined in Lemma 7.2.1.

Proof. The proof follows from Lemma 7.2.1 by letting $Y_i=(X_i-\mu_i)/a\sigma_i$, and it is easy to check that conditions (a) and (b) in the lemma are satisfied if and only if a satisfies (7.2.27). ■

This theorem provides a bound for the cumulative probability of a k-dimensional random variable when only the first two moments are known and when the correlations are equal.

A different approach to this problem was later adopted by Mudholkar and Rao (1967), and they obtained a Chebyshev-type inequality for the probability content of a region defined by a concave function.

Theorem 7.2.4. Let $\mathbf{X}=(X_1,\ldots,X_k)'$ be a k-dimensional random variable. Let ϕ: $\mathfrak{R}^k \to [0,\infty)$ be a concave function, and for fixed $a>0$ define

$$A = \{\mathbf{x} | \phi(\mathbf{x}) \leqslant a\}. \tag{7.2.29}$$

If $E\mathbf{X}=\boldsymbol{\mu}$ and $E\phi(\mathbf{X})$ exist, then

$$P[\mathbf{X} \in A] \geqslant 1 - \phi(\boldsymbol{\mu})/a. \tag{7.2.30}$$

Moreover, the equality is attainable if in addition $\mathbf{X} \geqslant \mathbf{0}$ a.s., ϕ is a homogeneous function and if $\phi(\boldsymbol{\mu}) \leqslant a$.

Proof. Since ϕ is nonnegative and concave, by Jensen's inequality we have

$$aP[\phi(\mathbf{X}) \geqslant a] \leqslant E\phi(\mathbf{X}) \leqslant \phi(\boldsymbol{\mu}),$$

which establishes (7.2.30). To show that it is attainable under the additional conditions for given ϕ and $a>0$ if we define a new random variable

$$\mathbf{X}_0 = \begin{cases} (a/\phi(\boldsymbol{\mu}))\boldsymbol{\mu} & \text{with probability} \quad \phi(\boldsymbol{\mu})/a, \\ 0 & \text{with probability} \quad 1-\phi(\boldsymbol{\mu})/a, \end{cases}$$

then $E\mathbf{X}_0=\boldsymbol{\mu}$ and from the homogeneity of ϕ

$$[\phi(\mathbf{X}_0) \geqslant a] = [\mathbf{X}_0 = (a/\phi(\boldsymbol{\mu}))\boldsymbol{\mu}],$$

which implies

$$P[\phi(\mathbf{X}_0) \geqslant a] = \phi(\boldsymbol{\mu})/a.$$

Now suppose that the inequality in (7.2.30) is strict, i.e., suppose that there does exist an $a_0>a$ such that $P[\phi(\mathbf{X})>a] \leqslant \phi(\boldsymbol{\mu})/a_0$ holds for all $\mathbf{X}$ with $E\mathbf{X}=\boldsymbol{\mu}$. Let a_1 satisfy $a_0>a_1>a$. Then similarly we would be able to construct a random variable $\mathbf{X}_1$ with mean $\boldsymbol{\mu}$ such that

$$\phi(\boldsymbol{\mu})/a_1 = P[\phi(\mathbf{X}_1) \geqslant a_1] \leqslant P[\phi(\mathbf{X}_1) > a] \leqslant \phi(\boldsymbol{\mu})/a_0.$$

This is a contradiction since $\phi(\boldsymbol{\mu})/a_0 < \phi(\boldsymbol{\mu})/a_1$. ■

An obvious advantage of this result is that it gives inequalities for regions other than the two-sided and one-sided rectangular regions considered in Theorems 7.2.1–7.2.3. Also, a particular application of the theorem, which is given below, yields an inequality involving only the first moments. Let $X_1,\ldots,X_k$ be nonnegative random variables with means $\mu_1,\ldots,\mu_k$, respectively, and let $X_{(1)} \leqslant \cdots \leqslant X_{(k)}$ denote the order statistics. Then for $c_1 \geqslant \cdots \geqslant c_k \geqslant 0$ the function $\phi(\mathbf{x}) = \sum_{i=1}^k c_i x_{(i)}$ is concave in $\mathbf{x} = (x_1, \ldots, x_k)'$. Therefore Theorem 7.2.4 implies that if $EX_{(k)}$ exists, then for $a > 0$

$$P\left[\sum_{i=1}^{k} c_i X_{(i)} \leqslant a\right] \geqslant 1 - \sum_{i=1}^{k} c_i \mu_{(i)}/a, \tag{7.2.31}$$

where $\mu_{(1)} \leqslant \cdots \leqslant \mu_{(k)}$ are the ordered means. In the special case for $c_1 = 1$ and $c_2 = \cdots = c_k = 0$ this reduces to

$$P\left[\min_{1 \leqslant i \leqslant k} X_i \leqslant a\right] \geqslant 1 - \frac{1}{a} \min_{1 \leqslant i \leqslant k} \mu_i,$$

which is equivalent to

$$P\left[\bigcap_{i=1}^{k} \{X_i > a\}\right] \leqslant \frac{1}{a} \min_{1 \leqslant i \leqslant k} \mu_i. \tag{7.2.32}$$

Note that here the independence of the random variables is not essential.

For certain other Chebyshev-type inequalities, the reader is referred to Karlin and Studden (1966, Chap. 13).

7.3. KOLMOGOROV-TYPE INEQUALITIES

The well-known Kolmogorov inequality says

Theorem 7.3.1. Let $Y_1, \ldots, Y_k$ be independent random variables with means η_i and variances τ_i^2 $(i = 1, \ldots, k)$, and let $v_k = (\sum_{i=1}^k \tau_i^2)^{1/2}$. Then for every fixed $a > 0$

$$P\left[\bigcap_{r=1}^{k} \left\{\left|\sum_{i=1}^{r} (Y_i - \eta_i)\right| \leqslant a v_k\right\}\right] \geqslant 1 - 1/a^2. \tag{7.3.1}$$

The proof of this theorem can be found in books on probability (see, e.g., Loève (1963, p. 235)). It provides a lower bound for the probability that the deviations of the partial sums from their means are simultaneously small, and it reduces to the Chebyshev inequality when $k = 1$. Note that the inequality again involves only the first two moments.

Now let us consider a random variable $\mathbf{X} = (X_1, \ldots, X_k)'$ with mean vector $\boldsymbol{\mu}$ and a covariance matrix $\boldsymbol{\Sigma} = (\sigma_{ij})$. If there exists a random variable $\mathbf{Y}=(Y_1,\ldots,Y_k)'$ with independent components such that $(Y_1,\Sigma_{i=1}^2 Y_i,\ldots,\Sigma_{i=1}^k Y_i)'$ and $\mathbf{X}$ are identically distributed, then Theorem 7.3.1 can be applied to obtain a lower bound for the probability $P[\cap_{i=1}^k |X_i-\mu_i|\leqslant a\sigma_k]$. This is illustrated in the next example.

Example 7.3.1. Let $\mathbf{X}$ be a multivariate normal variable with mean vector $\mathbf{0}$, and assume that the elements of its covariance matrix satisfy the condition

$$\sigma_{ij} = \sigma^2\min(i,j) \qquad \text{for all} \quad i,j. \tag{7.3.2}$$

Then from Theorem 7.3.1 we may conclude that

$$P\left[\bigcap_{i=1}^{k}\{|X_i|\leqslant a\sigma\sqrt{k}\,\}\right] \geqslant 1-1/a^2. \tag{7.3.3}$$

Note that this inequality applies to a Wiener process.

A generalization of Kolmogorov inequality under certain conditions was made by Hájek and Rényi (1955). The generalization provides a bound for the probability

$$P\left[\bigcap_{r=1}^{k}\left\{\left|\sum_{i=1}^{r}(Y_i-\eta_i)\right|\leqslant a_r v_k\right\}\right]$$

when the a_rs are not necessarily equal. A multivariate version of this extension was made by Sen (1971) in connection with an application to simultaneous confidence regions.

A one-sided analogue of Kolmogorov inequality was obtained by Marshall (1960). It says

Theorem 7.3.2. Let $Y_1,\ldots,Y_k$ satisfy the conditions stated in Theorem 7.3.1. Then for every fixed $a>0$,

$$P\left[\bigcap_{r=1}^{k}\left\{\sum_{i=1}^{r}(Y_i-\eta_i)\leqslant av_k\right\}\right] \geqslant a^2/(1+a^2).$$

Proof. The proof given by Marshall is quite similar to the standard proof of Kolmogorov inequality. Without loss of generality we may assume that $\eta_i=0$ for $i=1,\ldots,k$. Let $G(\mathbf{y})$ be the distribution of $\mathbf{Y}$, and denote

$$\phi(\mathbf{y}) = \{v_k^2(a^2+1)\}^{-2}\left(av_k \sum_{i=1}^{k} y_i + v_k^2\right)^2,$$

$$D_r = \left\{\mathbf{y} \,\middle|\, \sum_{i=1}^{l} y_i \leqslant av_k, l=1,\dots,r-1;\ \sum_{i=1}^{r} y_i > av_k\right\}$$

for $r=1,\dots,k$. Then we conclude

$$\begin{aligned} P\left[\bigcup_{r=1}^{k}\left\{\sum_{i=1}^{r} Y_i > av_k\right\}\right] &= \sum_{r=1}^{k} P(D_r) \\ &\leqslant \{v_k^2(a^2+1)\}^{-2}\sum_{r=1}^{k}\int_{D_r}\left(av_k\sum_{i=1}^{r} y_i + v_k^2\right)^2 dG(\mathbf{y}) \\ &\leqslant \sum_{r=1}^{k}\int_{D_r}\phi(\mathbf{y})\,dG(\mathbf{y}) \leqslant E\phi(\mathbf{Y}) = 1/(1+a^2), \end{aligned}$$

where the second inequality follows from the condition of independence. The theorem now follows immediately from this inequality. ■

Let $\mathbf{X}=(X_1,\dots,X_k)'$ be a given random variable. If the X_is can be written as partial sums of independent random variables, then Theorem 7.3.2 can be applied to obtain a lower bound for the cumulative probability of $\mathbf{X}$. For instance if $\mathbf{X}$ is the random variable considered in Example 7.3.1, then the inequality

$$P\left[\bigcap_{i=1}^{k}\{X_i - \mu_i \leqslant a\sigma\sqrt{k}\}\right] \geqslant a^2/(1+a^2)$$

holds for all $a>0$.

PROBLEMS

1. Let $\mathbf{X}$ have a distribution function $F(x_1,\dots,x_k)$, and let $F_i(x_i)$, $i=1,\dots,k$, denote the marginal distributions. Show that we always have

$$F(x_1,\dots,x_k) \leqslant \left(\prod_{i=1}^{k} F_i(x_i)\right)^{1/k}$$

(Wilks, 1962, p. 71).

2. Find a necessary and sufficient condition under which the inequality of Chung–Erdös becomes an equality.
3. Let $\mathbf{X}=(X_1,\dots,X_k)'$ have a density function $f(\mathbf{x})$ which is Schur-concave, where $k\geqslant 2$. Show that for all real numbers $a_1,\dots,a_k$ we have

$$P\left[\bigcap_{i=1}^{k}\{X_i\leqslant a_i\}\right]$$
$$\leqslant \min\left\{p^{(2)},\binom{k}{2}p^{(2)}-k(k-2)p^{(1)}+\binom{k-1}{2}\right\},$$

where

$$p^{(1)}=P\left[X_1\leqslant \bar a\right],\qquad p^{(2)}=P\left[\bigcap_{i=1}^{2}\{X_i\leqslant \bar a\}\right],$$

and

$$\bar a=\frac{1}{k}\sum_{i=1}^{k}a_i.$$

4. Show that if $X_1,\dots,X_k$ have means zero, variances σ_i^2 ($i=1,\dots,k$) and if $|X_1|,\dots,|X_k|$ are associated (see Definition 5.2.1), then

$$P\left[\bigcap_{i=1}^{k}\{|X_i|\leqslant a_i\}\right]\geqslant \prod_{i=1}^{k}\left[1-\sigma_i^2/a_i^2\right].$$

5. Construct an example to show that the bound in the inequality of Berge is attainable.
6. Find a counterexample to illustrate the fact that if $\mathbf{C}$ satisfies the condition in (7.2.9), this does not imply that $\mathbf{C}^{-1}$ also satisfies that condition.
7. Find the value for $\rho=\rho_M$ ($\rho=\rho_m$) such that the right-hand side of (7.2.20) is maximized (minimized).
8. Let $\mathbf{X}_1,\dots,\mathbf{X}_n$ be a sequence of independent k-dimensional random variables with a common mean vector $\boldsymbol{\mu}$ and a common covariance matrix $\boldsymbol{\Sigma}$ (finite), and let $\overline{\mathbf{X}}_n=(1/n)\Sigma_{i=1}^n\mathbf{X}_i$. Show that there exists a real number c that depends only on k and the elements of $\boldsymbol{\Sigma}$ such that

$$P\left[\|\overline{\mathbf{X}}_n-\boldsymbol{\mu}\|\leqslant a\right]\geqslant 1-c/na^2,$$

where $\|\bar{\mathbf{x}}-\boldsymbol{\mu}\|$ denotes the distance $[\Sigma_{i=1}^k(\bar x_i-\mu_i)^2]^{1/2}$.

9. Let $X_1,\ldots,X_k$ be independent random variables with means zero and variances σ_i^2. Show that for $a_r>0$

$$P\left[\bigcap_{r=1}^{k}\left\{\sum_{i=1}^{r} X_i \leqslant a_r\right\}\right] \geqslant \prod_{r=1}^{k}\left[a_r^2\Big/\left\{a_r^2+\sum_{i=1}^{r}\sigma_i^2\right\}\right].$$

10. Let $\{X_t : t>0\}$ be a Wiener process with covariance kernel of the form

$$E(X_{t_1}X_{t_2}) = \sigma^2 \min(t_1,t_2)$$

for some $\sigma^2>0$. Let $0<t_1<\cdots<t_k$ be arbitrary but fixed. Apply Theorem 7.3.1 to obtain a lower bound for the probability $P[\max_{1\leqslant i\leqslant k}|X_{t_i}|\leqslant a]$, $a>0$.

11. For the process $\{X_t : t>0\}$ given in Problem 10, apply Theorem 7.3.2 to obtain a lower bound for the probability $P[\max_{1\leqslant i\leqslant k} X_{t_i} \leqslant a]$, $a>0$.

REFERENCES

Berge, P. O. (1937). A note on a form of Tchebycheff's theorem for two variables. *Biometrika* **29**, 405–406.

Chung, K. L., and Erdös, P. (1952). On the application of the Borel–Cantelli lemma. *Trans. Amer. Math. Soc.* **72**, 179–186.

Dawson, D. A., and Sankoff, D. (1967). An inequality for probabilities. *Proc. Amer. Math. Soc.* **18**, 504–507.

Gallot, S. (1966). A bound for the maximum of a number of random variables. *J. Appl. Probab.* **3**, 556–558.

Hájek, J., and Rényi, A. (1955). Generalization of an inequality of Kolmogorov. *Acta Math. Acad. Sci. Hungar.* **6**, 281–283.

Karlin, S., and Studden, W. J. (1966). *Tchebycheff Systems with Applications in Analysis and Statistics*. Wiley (Interscience), New York.

Kounias, E. G. (1968). Bounds for the probability of a union, with applications. *Ann. Math. Statist.* **39**, 2154–2158.

Lal, D. N. (1955). A note on a form of Tchebycheff's inequality for two or more variables. *Sankhyā* **15**, 317–320.

Loève, M. (1963). *Probability Theory*, 3rd ed. Van Nostrand-Reinhold, New York.

Marshall, A. W. (1960). A one-sided analog of Kolmogoroff's inequality. *Ann. Math. Statist.* **31**, 483–487.

Marshall, A. W., and Olkin, I. (1960). Multivariate Chebyshev inequalities. *Ann. Math. Statist.* **31**, 1001–1014.

Mudholkar, G. S., and Rao, P. S. R. S. (1967). Some sharp multivariate Tchebycheff inequalities. *Ann. Math. Statist.* **38**, 393–400.

Olkin, I., and Pratt, J. W. (1958). A multivariate Tchebycheff inequality. *Ann. Math. Statist.* **29**, 226–234.

Pearson, K. (1918). On generalized Tchebycheff theorems in the mathematical theory of statistics. *Biometrika* **12**, 284–296.

Sen, P. K. (1971). A Hájek–Rényi type inequality for stochastic vectors with applications to simultaneous confidence regions. *Ann. Math. Statist.* **42**, 1132–1134.

Sobel, M., and Uppuluri, V. R. R. (1972). On Bonferroni-type inequalities of the same degree for the probability of unions and intersections. *Ann. Math. Statist.* **43**, 1549–1558.

Uspensky, J. V. (1937). *Introduction to Mathematical Probability*. McGraw-Hill, New York.

Whittle, P. (1958). A multivariate generalization of Tchebychev's inequality. *Quart. J. Math. Oxford Ser.* (2) **9**, 232–240.

Whittle, P. (1959). Sur la distribution du maximum d'un polynôme trigonométrique à coefficients aléatoires. *Le Calcul des Probabilités et ses Applications*, p. 173. CNRS, Paris.

Wilks, S. S. (1962). *Mathematical Statistics*. Wiley, New York.

CHAPTER

8

Some Applications

In Chapters 2–7 we studied probability inequalities in multivariate distributions from a theoretical point of view and did not describe their applications. The purpose of this chapter is to concentrate on the applications of certain inequalities in the areas of estimation, hypothesis testing, simultaneous comparisons, ranking and selection, and reliability theory. (There are a number of other areas in statistics (e.g., Bayesian inference and nonparametric inference) in which such inequalities are also useful, but these will not be discussed here.) Most applications described in this chapter have already appeared in the literature, and it is likely that many new applications will be made available in the future through a wider use of the theory.

It should be emphasized here that we do not try to include all possible applications in this chapter. Instead we hope to demonstrate the main point that the inequalities, although developed in a theoretical way, are important and powerful tools for solving many real-life problems in statistics.

8.1. SIMULTANEOUS CONFIDENCE REGIONS

The problem of obtaining a simultaneous confidence region for a parameter vector when the random variables are dependent is perhaps a major source leading to the development of certain probability inequalities. In this section we shall see applications of the inequalities to this problem for normal means, variances, regression coefficients, and variance ratios and variance components in analysis-of-variance. Most applications that have already appeared in the literature require the assumption of normality. However, in view of the more general inequalities that have become available only recently the solutions of these problems under more general conditions are now possible.

8.1.1. Normal Means

Suppose that there are k normal variables ($k \geqslant 2$) under consideration, and that μ_i and σ_i^2 are the mean and variance of the ith variable ($i=1,\ldots,k$). Our problem is to obtain a confidence region for the mean vector $\boldsymbol{\mu}=(\mu_1,\ldots,\mu_k)'$. To formulate this problem mathematically we shall let $\mathbf{X}=(X_1,\ldots,X_k)'$ denote the normal variable with mean vector $\boldsymbol{\mu}$ and covariance matrix $\boldsymbol{\Sigma}=(\rho_{ij}\sigma_i\sigma_j)$. Let $\mathbf{X},\mathbf{X}_1,\ldots,\mathbf{X}_n$ be independent and identically distributed, where $\mathbf{X}_m=(X_{1m},\ldots,X_{km})'$. We define the sample means by

$$\overline{X}_i = \frac{1}{n}\sum_{m=1}^{n} X_{im}, \qquad i = 1,\ldots,k.$$

Then $\overline{\mathbf{X}}=(\overline{X}_1,\ldots,\overline{X}_k)'$ is a multivariate normal variable with mean $\boldsymbol{\mu}$ and covariance matrix $\mathbf{T}$ that has elements

$$\tau_{ij} = \rho_{ij}\sigma_i\sigma_j/n, \qquad 1 \leqslant i,j \leqslant k.$$

Now let $A \subset \mathfrak{R}^k$ denote a given convex set that is symmetric about the origin, and let us consider the confidence region $A+\overline{\mathbf{X}}$, which is a convex set centered at $\overline{\mathbf{X}}$. It is easy to see that the confidence probability is (as a function of the population covariance matrix $\boldsymbol{\Sigma}$)

$$\gamma_0(\boldsymbol{\Sigma}) = P_{\mathbf{T}}\left[\left(\overline{\mathbf{X}}-\boldsymbol{\mu}\right)\in A\right]. \tag{8.1.1}$$

This is the probability that a multivariate normal variable with mean $\mathbf{0}$ and covariance matrix $\mathbf{T}$ will take values in A, and it depends on $\boldsymbol{\Sigma}$ only through $\mathbf{T}$. Let $\mathbf{T}_1$, $\mathbf{T}_2$ denote two possible covariance matrices of $\overline{\mathbf{X}}$ when the true population matrix $\boldsymbol{\Sigma}$ is either $\boldsymbol{\Sigma}_1$ or $\boldsymbol{\Sigma}_2$. It follows that $(\mathbf{T}_2-\mathbf{T}_1)$ is positive semidefinite if and only if $(\boldsymbol{\Sigma}_2-\boldsymbol{\Sigma}_1)$ is (see Problem 1). Applying either the corollary of Theorem 4.1.3 or Theorem 4.3.4, we may conclude that if $(\boldsymbol{\Sigma}_2-\boldsymbol{\Sigma}_1)$ is positive semidefinite, then $\gamma_0(\boldsymbol{\Sigma}_1)\geqslant\gamma_0(\boldsymbol{\Sigma}_2)$. (This is, of course, trivial when both $\boldsymbol{\Sigma}_1$ and $\boldsymbol{\Sigma}_2$ are diagonal matrices.) This offers an inequality for the probability content of such a confidence region through a partial ordering of the covariance matrices.

In certain applications it is customary to have a rectangular confidence region that depends on the variances or the estimates of the variances. When the population variances are known, the one-sided and two-sided rectangular confidence regions for $\boldsymbol{\mu}$ are given by (see, e.g., Dunn (1958), Khatri (1967), Scott (1967), and Šidák (1967))

$$R_1 = \{\mathbf{x}|\mathbf{x}\in\mathfrak{R}^k, x_i \leqslant \bar{X}_i + d_i\sigma_i\ i=1,\ldots,k\}, \tag{8.1.2}$$

$$R_2 = \{\mathbf{x}|\mathbf{x}\in\mathfrak{R}^k, \bar{X}_i - d_i\sigma_i \leqslant x_i \leqslant \bar{X}_i + d_i\sigma_i, i=1,\ldots,k\}, \qquad d_i > 0. \tag{8.1.3}$$

It is straightforward to verify that if the populations are independent (for the normal diatribution this is equivalent to $\rho_{ij}=0$ for $i\neq j$), then the confidence probabilities γ_1 and γ_2 for the confidence regions R_1 and R_2 are, respectively,

$$\gamma_1(\boldsymbol{\Sigma}_0) = \prod_{i=1}^{k} \Phi(a_i), \tag{8.1.4}$$

$$\gamma_2(\boldsymbol{\Sigma}_0) = \prod_{i=1}^{k} [\Phi(a_i) - \Phi(-a_i)], \tag{8.1.5}$$

where $\boldsymbol{\Sigma}_0$ denotes the diagonal matrix with diagonal elements σ_i^2 $(i=1,\ldots,k)$, Φ is the standard normal distribution function, and $a_i = \sqrt{n}\ d_i$ for $i=1,\ldots,k$.

In many applications in statistics, however, the assumption of independence is unjustifiable. For example,

> In biological research growth data are often obtained with measurements taken on n individuals at k different times; the measurements would be highly correlated. The psychologist may measure n individuals' responses to k different levels of a stimulus; again, a high degree of dependence would be expected (Dunn, 1958, p. 1096).

When those situations occur, it is easy to see that

(i) $\gamma_1(\boldsymbol{\Sigma})$ (which depends on the covariance matrix $\boldsymbol{\Sigma}$) is strictly increasing in each ρ_{ij} (Theorem 2.1.1). Hence if $\rho_{ij} \geqslant 0$ for all $i\neq j$, then $\gamma_1(\boldsymbol{\Sigma})$ is bounded below by $\gamma_1(\boldsymbol{\Sigma}_0)$. In this case if one chooses $a_1 = \cdots = a_k = a$ such that $\Phi(a) = (\gamma^*)^{1/k}$, then the true confidence probability is at least γ^*. This provides a conservative solution to the problem. If it is also true that $\rho_{ij} \geqslant \rho_0 > 0$ for all $i \neq j$ for some ρ_0, then a less conservative solution can be obtained by letting a satisfy

$$P\left[\bigcap_{i=1}^{k} \{Z_i \leqslant a\}\right] = \gamma^*,$$

where $(Z_1,\ldots,Z_k)'$ has a multivariate normal distribution with means zero, variances one, and correlations ρ_0.

(ii) It follows from Theorem 2.2.2, Theorem 2.2.5, or Theorem 4.3.1 that $\gamma_2(\boldsymbol{\Sigma}) \geqslant \gamma_2(\boldsymbol{\Sigma}_0)$ holds for all $a_1,\ldots,a_k$ and all $\boldsymbol{\Sigma}$. Therefore if one chooses $a_1 = \cdots = a_k = a$ to satisfy

$$[\Phi(a) - \Phi(-a)] = (\gamma^*)^{1/k},$$

then no matter what the true correlations are one always has $\gamma_2(\boldsymbol{\Sigma}) \geqslant \gamma^*$.

When the variances are unknown and possibly unequal, a two-sided rectangular confidence region for $\boldsymbol{\mu}$ may be of the form

$$R_3 = \{\mathbf{x}|\mathbf{x} \in \mathfrak{R}^k, \bar{X}_i - d_i S_i \leqslant x_i \leqslant \bar{X}_i + d_i S_i, i = 1,\ldots,k\}, \quad (8.1.6)$$

where $d_i > 0$ and

$$S_i^2 = \left[\frac{1}{n-1}\sum_{m=1}^{n}(X_{im} - \bar{X}_i)^2\right] \quad (8.1.7)$$

for $i = 1,\ldots,k$. A confidence region of this type was considered previously by Dunn (1958) and Scott (1967) among others. If $\rho_{ij} = 0$ for all $i \neq j$, then its confidence probability is

$$\gamma_3(\boldsymbol{\Sigma}_0) = \prod_{i=1}^{k} P[-a_i \leqslant t_i \leqslant a_i],$$

where t_i is a Student t variable with $n-1$ degrees of freedom. When the variables are dependent, however, the problem becomes complicated. Of course for general $\boldsymbol{\Sigma}$ one has (by Theorem 2.2.2 or 2.2.5)

$$\gamma_3(\boldsymbol{\Sigma}) \geqslant E\prod_{i=1}^{k}[\Phi(a_i S_i/\sigma_i) - \Phi(-a_i S_i/\sigma_i)],$$

but this in general cannot be simplified further. The difficulty here is caused by the fact that $S_1,\ldots,S_k$ are no longer independent. Scott (1967) made the claim that

$$\gamma_3(\boldsymbol{\Sigma}) \geqslant \gamma_3(\boldsymbol{\Sigma}_0) \quad (8.1.8)$$

holds for all $\boldsymbol{\Sigma}$; but unfortunately his proof was later found to be in error (see, e.g., Šidák (1971, 1975)). From a result of Šidák (1971) (see Theorem 3.1.2) if the correlation matrix $\boldsymbol{\Sigma}$ has the structure l (see Definition 2.2.1), then the inequality in (8.1.8) holds. However, the general validity of (8.1.8) still remains an open problem.

Therefore to find a rectangular confidence region for $\boldsymbol{\mu}$ when the variables are dependent we do not yet know how to use the Student t table for a general solution. Instead it can be obtained from the Bonferroni inequality (Theorem 7.1.1). It is clear that if we choose $a_1,\ldots,a_k$ to satisfy

$$\sum_{i=1}^{k} 2P[t_i \leqslant -a_i] \leqslant (1-\gamma^*),$$

then $\gamma_3(\boldsymbol{\Sigma}) \geqslant \gamma^*$ holds no matter what the true correlations are. This solution was offered by Dunn (1958).

When the variables are unknown but equal, a rectangular confidence region of the form

$$R_4 = \{\mathbf{x} | \mathbf{x} \in \Re^k, \bar{X}_i - d_i S_1 \leqslant x_i \leqslant \bar{X}_i + d_i S_1, i=1,\ldots,k\} \quad (8.1.9)$$

was proposed by Dunn (1958). It follows from Theorem 2.2.2 (or 2.2.5) and Lemma 2.2.1 that the confidence probability satisfies

$$\gamma_4(\boldsymbol{\Sigma}) \geqslant \{P[-a_i \leqslant t_1 \leqslant a_i]\}^k$$

for all $\boldsymbol{\Sigma}$, where t_1 is a Student t variable with $n-1$ degrees of freedom. Therefore a conservative confidence region for $\boldsymbol{\mu}$ may again be obtained from the Student t table.

It might be suggested that in order to fully utilize the information the polled sample standard deviation

$$S_{\mathrm{p}} = \left[\sum_{i=1}^{k} \sum_{m=1}^{n} (X_{im} - \bar{X}_i)^2 / (k(n-1)) \right]^{1/2}, \quad (8.1.10)$$

instead of S_1, should be used in (8.1.9). Although this is a very desirable thing to do, it involves a difficulty that has not yet been resolved. The difficulty is caused by the fact that when some of the correlations are nonzero, the distribution of $k(n-1)S_{\mathrm{p}}^2/\sigma^2$ depends on the correlations; hence it is no longer a chi-square variable. In this case it is not yet known how to obtain a confidence region with confidence coefficient at least γ^* (preassigned) when S_{p} is used. Of course if it is known that all the ρ_{ij}s are zero for $i \neq j$, then the difficulty does not exist. Hence under that circumstance one should use S_{p} instead of S_1 in (8.1.9), and a conservative solution can be obtained accordingly from the Student t table.

The confidence regions R_1–R_4 considered above are of rectangular type. It is well known that elliptical confidence regions for $\boldsymbol{\mu}$ may also be obtained by using the chi-square or F distribution, depending on whether the covariance matrix $\boldsymbol{\Sigma}$ is known or unknown (see, e.g., Sections 3.3 and 5.3 of Anderson (1958)). If a reduction in dimensionality for the confidence region is needed, then again the applications of several theorems in Chapters 2–4 yield a conservative solution. This is illustrated in the following: Let us consider the partition of $\overline{\mathbf{X}}$, $\boldsymbol{\mu}$, and $\boldsymbol{\Sigma}$ into the form

$$\overline{\mathbf{X}} = \begin{pmatrix} \overline{\mathbf{X}}^{(1)} \\ \overline{\mathbf{X}}^{(2)} \end{pmatrix}, \qquad \boldsymbol{\mu} = \begin{pmatrix} \boldsymbol{\mu}^{(1)} \\ \boldsymbol{\mu}^{(2)} \end{pmatrix}, \qquad \boldsymbol{\Sigma} = \begin{pmatrix} \boldsymbol{\Sigma}_{11} & \boldsymbol{\Sigma}_{12} \\ \boldsymbol{\Sigma}_{21} & \boldsymbol{\Sigma}_{22} \end{pmatrix}, \tag{8.1.11}$$

where $\overline{\mathbf{X}}^{(1)}$, $\boldsymbol{\mu}^{(1)}$ are $q \times 1$ and $\boldsymbol{\Sigma}_{11}$ is $q \times q$, $q<k$. When $\boldsymbol{\Sigma}_{11}$ and $\boldsymbol{\Sigma}_{22}$ are known,

$$A_1 = \left\{ \mathbf{x} | \mathbf{x} \in \mathcal{R}^q, n(\overline{\mathbf{X}}^{(1)} - \mathbf{x})'\boldsymbol{\Sigma}_{11}^{-1}(\overline{\mathbf{X}}^{(1)} - \mathbf{x}) \leqslant \chi^2_{\gamma_1, q} \right\}, \tag{8.1.12}$$

$$A_2 = \left\{ \mathbf{x} | \mathbf{x} \in \mathcal{R}^{k-q}, n(\overline{\mathbf{X}}^{(2)} - \mathbf{x})'\boldsymbol{\Sigma}_{22}^{-1}(\overline{\mathbf{X}}^{(2)} - \mathbf{x}) \leqslant \chi^2_{\gamma_2, k-q} \right\} \tag{8.1.13}$$

are confidence regions for $\boldsymbol{\mu}^{(1)}$ and $\boldsymbol{\mu}^{(2)}$, respectively, with confidence coefficient γ_i ($\chi^2_{\gamma,p}$ is the γth percentile of the chi-square distribution with p degrees of freedom). If the rank of $\boldsymbol{\Sigma}_{12}$ is either zero or one, then (by Theorem 2.2.2 or 2.2.3)

$$P\left[\bigcap_{i=1}^{2} \{ \boldsymbol{\mu}^{(i)} \in A_i \} \right] \geqslant \gamma_1 \gamma_2. \tag{8.1.14}$$

If $\boldsymbol{\Sigma}$ has the structure l (see Definition 2.2.1), which implies that $\boldsymbol{\Sigma}_{12}$ has rank either zero or one, then one can partition $\overline{\mathbf{X}}$ and $\boldsymbol{\mu}$ into r vectors and consider separate confidence regions $A_1, \ldots, A_r$ for $\boldsymbol{\mu}^{(1)}, \ldots, \boldsymbol{\mu}^{(r)}$ with confidence probabilities $\gamma_1, \ldots, \gamma_r$. Theorem 2.2.4 implies that

$$P\left[\bigcap_{i=1}^{r} \{ \boldsymbol{\mu}^{(i)} \in A_i \} \right] \geqslant \prod_{i=1}^{r} \gamma_i \tag{8.1.15}$$

for any $r \geqslant 2$. When $r=k$, this reduces to the inequality for the rectangular confidence region R_2 defined in (8.1.3).

8.1.2. Normal Variances

We shall continue to adopt the same notation, and let $\mathbf{X}_1, \ldots, \mathbf{X}_n$ denote independent normal variables with mean vector $\boldsymbol{\mu}$ and covariance matrix $\Sigma=(\rho_{ij}\sigma_i\sigma_j)$. Let S_i^2 be the sample variances ($i=1,\ldots,k$). To find conservative confidence regions for the population variances σ_i^2 we need to apply certain inequalities for the multivariate chi-square distribution given in Chapters 3 and 4.

For $k=2$ a two-sided confidence region for $(\sigma_1^2,\sigma_2^2)'$, as considered by Jensen and Jones (1969), can be given via an inequality obtained by Jensen (1969). For any $a>b>0$ let

$$A = \{\mathbf{x}|\mathbf{x}\in\mathfrak{R}^2,(n-1)S_i^2/a\leqslant x_i\leqslant(n-1)S_i^2/b,i=1,2\} \tag{8.1.16}$$

be a confidence region for $(\sigma_1^2,\sigma_2^2)'$. Then it follows that (see (3.2.7))

$$\begin{aligned} P\left[(\sigma_1^2,\sigma_2^2)'\in A\right] &= P\left[\bigcap_{i=1}^{2}\{b\leqslant\chi_i^2\leqslant a\}\right] \\ &\geqslant \prod_{i=1}^{2} P\left[\{b\leqslant\chi_i^2\leqslant a\}\right], \end{aligned} \tag{8.1.17}$$

where χ_i^2 is a chi-square variable with $n-1$ degrees of freedom. If b and a are chosen from the univariate chi-square table so that the right-hand side of (8.1.17) is γ^*, then the true probability of coverage is at least γ^*.

For the general case inequalities other than Bonferroni inequality for such a two-sided confidence region are not yet available. Hence only one-sided confidence regions for $\boldsymbol{\sigma}^2=(\sigma_1^2,\ldots,\sigma_k^2)'$ will be considered here. Let us define

$$A^* = \{\mathbf{x}|\mathbf{x}\in\mathfrak{R}^k,(n-1)S_i^2/a_i\leqslant x_i,i=1,\ldots,k\}, \tag{8.1.18}$$

$$A^{**} = \{\mathbf{x}|\mathbf{x}\in\mathfrak{R}^k,x_i\leqslant(n-1)S_i^2/a_i,i=1,\ldots,k\}. \tag{8.1.19}$$

Then by the first corollary of Theorem 4.3.5

$$P[\boldsymbol{\sigma}^2\in A^*] = P\left[\bigcap_{i=1}^{k}\{\chi_i^2\leqslant a_i\}\right] \geqslant \prod_{i=1}^{k} P[\chi_i^2\leqslant a_i] \tag{8.1.20}$$

holds for all Σ. If Σ has the structure l, then either from the second corollary of Theorem 4.3.5 or from Theorem 3.2.1 we have

$$P[\boldsymbol{\sigma}^2\in A^{**}] = P\left[\bigcap_{i=1}^{k}\{\chi_i^2\geqslant a_i\}\right] \geqslant \prod_{i=1}^{k} P[\chi_i^2\geqslant a_i]. \tag{8.1.21}$$

From the inequalities in (8.1.20) and (8.1.21) it is possible to obtain conservative one-sided confidence regions for the variances from the univariate chi-square table only.

8.1.3. Regression Coefficients

Let us consider a general linear model given by

$$\mathbf{Y} = \mathbf{H}\boldsymbol{\beta} + \boldsymbol{\varepsilon}, \tag{8.1.22}$$

where $\mathbf{Y}$ and $\boldsymbol{\varepsilon}$ are $n\times 1$, $\boldsymbol{\beta}$ is $k\times 1$, and $\mathbf{H}$ is $n\times k$ $(n>k)$. We make the usual assumption that $\boldsymbol{\varepsilon}$ has a multivariate normal distribution with mean vector $\mathbf{0}$ and covariance matrix $\sigma^2\mathbf{I}$ (σ^2 is unknown and $\mathbf{I}$ is the identity matrix). Hence $\mathbf{Y}$ is a normal variable with mean vector $\mathbf{H}\boldsymbol{\beta}$ and covariance matrix $\sigma^2\mathbf{I}$. If $\mathbf{H}$ has full rank, then $\mathbf{H}'\mathbf{H}$ is nonsingular and the best linear unbiased estimator of $\boldsymbol{\beta}$ is $\hat{\boldsymbol{\beta}}=(\mathbf{H}'\mathbf{H})^{-1}\mathbf{H}'\mathbf{Y}$. This estimator $\hat{\boldsymbol{\beta}}$ has a normal distribution with mean vector $\boldsymbol{\beta}$ and covariance matrix $\sigma^2(\mathbf{H}'\mathbf{H})^{-1}\equiv\sigma^2(h_{ij})$. We shall also use $\hat{\sigma}^2$ to denote the usual estimator of σ^2.

In the literature the problem of obtaining linear segment confidence bands for $\boldsymbol{\beta}$, whose elements are the regression coefficients in regression analysis, has been studied frequently (see, e.g., Hoel (1951), Roy and Bose (1953), Roy (1954), Gafarian (1964), Graybill and Bowden (1967), Folks and Antle (1967)). Since the random variable

$$(t_1,\ldots,t_k)' = \hat{\sigma}^{-1}\left((\hat{\beta}_1-\beta_1)/\sqrt{h_{11}}\,,\ldots,(\hat{\beta}_k-\beta_k)/\sqrt{h_{kk}}\,\right)' \tag{8.1.23}$$

has a multivariate t distribution with correlation matrix $(h_{ij}/(h_{ii}h_{jj})^{1/2})$ and degrees of freedom $n-k$, certain probability inequalities in Chapters 2–5 can be applied to obtain a conservative solution. We shall summarize the results in the following ($\mathbf{H}$ is assumed to have full rank):

(i) Let A denote the rectangular confidence region

$$A = \left\{\mathbf{x}\,|\,\mathbf{x}\in\Re^k, \hat{\beta}_i - d_i\hat{\sigma} \leqslant x_i \leqslant \hat{\beta}_i + d_i\hat{\sigma}, i=1,\ldots,k\right\}. \tag{8.1.24}$$

Then for $a_i=d_i/\sqrt{h_{ii}}$ $(i=1,\ldots,k)$

$$P[\boldsymbol{\beta}\in A] = P\left[\bigcap_{i=1}^{k}\{|t_i|\leqslant a_i\}\right] \geqslant \prod_{i=1}^{k} P[|t_i|\leqslant a_i] \tag{8.1.25}$$

holds for all design matrices $\mathbf{H}$. This follows from Theorems 2.2.2 (or 2.2.5) and 3.1.1 and the inequality in (3.1.5).

(ii) If $\mathbf{H}$ is chosen such that the correlation between t_i and t_j is $\lambda_i\lambda_j\rho_{ij}$ for all i, j for some correlation matrix (ρ_{ij}) and a real vector $\boldsymbol{\lambda}=(\lambda_1,\ldots,\lambda_k)'$, then for each i the confidence probability is nondecreasing (nonincreasing) in λ_i for $\lambda_i>0$ $(\lambda_i<0)$. This follows from Theorems 2.2.5 and 3.1.1 or Theorem 4.3.1.

(iii) If $\mathbf{H}$ is chosen such that $h_{ii}=h_{jj}$ for all i, j and $h_{ij}=\rho h_{11}$ for all $i\neq j$ for some $\rho\geqslant 0$, then the joint distribution of the random variables $t_1,\ldots,t_k$ defined in (8.1.23) are conditionally independent and identically distributed, or they are positively dependent by mixture (see Section 5.3 and Problem 12, Chapter 5). Hence for $a_1=\cdots=a_k=a$ the confidence probability γ is bounded below by

$$\gamma \geqslant \left(P\left[\bigcap_{i=1}^{r} \{|t_i|\leqslant a\}\right]\right)^{k/r} \tag{8.1.26}$$

for all $k\geqslant r\geqslant 1$. This follows from Theorems 2.3.4 and 3.1.1 or Theorem 3.3.1. Moreover, the lower bound is monotonically nondecreasing in r (see Problem 5). Therefore if a is chosen to satisfy

$$P\left[\bigcap_{i=1}^{r} \{|t_i|\leqslant a\}\right] = (\gamma^*)^{r/k},$$

then the true confidence probability is at least γ^*. For $r>1$ this solution is less conservative than that obtained from the inequality in (8.1.25).

8.1.4. Variance Ratios and Variance Components

The problem of obtaining conservative simultaneous confidence regions for variance ratios and variance components in analysis-of-variance problems was considered by several authors (see, e.g., Broemeling (1969, 1978), Sahai and Anderson (1973), Sahai (1974), Broemeling and Bee (1976)). This concerns basically the applications of inequalities for the multivariate F distribution. Let us consider a balanced random-effects model

$$\mathbf{Y} = \mu\mathbf{1} + \sum_{i=1}^{k} \mathbf{H}_i\mathbf{b}_i + \boldsymbol{\varepsilon}, \tag{8.1.27}$$

where $\mathbf{Y}$ is a $n\times 1$ vector of observations, μ a scalar real parameter, $\mathbf{1}$ a $n\times 1$ vector of ones, and $\mathbf{H}_i$ an $n\times r_i$ full-rank known matrix. The $r_i\times 1$ and $n\times 1$ random variables $\mathbf{b}_i$ and $\boldsymbol{\varepsilon}$ are independent normal variables which have independent components with means zero and variances σ_i^2 ($i=1,\ldots,k$) and $\sigma_0^2>0$, respectively. The parameters of interest under this random-effects model are usually the variance ratios

$$\theta_i = \sigma_i^2/\sigma_0^2, \qquad i = 1,\ldots,k. \tag{8.1.28}$$

Let $S_0^2,\ldots,S_k^2$ and $\nu_0,\ldots,\nu_k$ denote the mean squares and the degrees of freedom, which are easily obtainable from existing results in linear model theory. It is known that for this balanced random-effects model

$$\chi_0^2 = \nu_0 S_0^2/\sigma_0^2 \tag{8.1.29}$$

and

$$\chi_i^2 = \nu_i S_i^2/\tau_i^2, \qquad i = 1,\ldots,k \tag{8.1.30}$$

and independent chi-square variables with degrees of freedom $\nu_0,\ldots,\nu_k$, respectively, where

$$\tau_i^2 = \sigma_0^2\left(1+\sum_{j=1}^{k} c_{ij}\theta_j\right). \tag{8.1.31}$$

(The c_{ij}s can be determined from the design matrices $\mathbf{H}_1,\ldots,\mathbf{H}_k$; see, e.g., Hartley (1967).)

The model in (8.1.27) is quite general in application. It includes the following important special cases:

(i) The one-way random-effects model

$$Y_{ij} = \mu + u_i + \varepsilon_{ij}, \qquad i = 1,\ldots,I, \quad j = 1,\ldots,J, \tag{8.1.32}$$

where $u_1,\ldots,u_I$ and $\varepsilon_{11},\ldots,\varepsilon_{IJ}$ are independent normal variables with means zero and variances σ_1^2 and σ_0^2, respectively. The standard method in analysis-of-variance gives

$$S_0^2 = \sum_{i=1}^{I}\sum_{j=1}^{J}\left(Y_{ij}-\bar{Y}_{i\cdot}\right)^2/I(J-1),$$

$$S_1^2 = J\sum_{i=1}^{I}\left(\bar{Y}_{i\cdot}-\bar{Y}_{\cdot\cdot}\right)^2/(I-1),$$

and $\nu_0=I(J-1)$, $\nu_1=I-1$.

(ii) The two-way classification random model without interaction

$$Y_{ij} = \mu + u_i + v_j + \varepsilon_{ij}, \qquad i = 1,\ldots,I, \quad j = 1,\ldots,J. \tag{8.1.33}$$

The S_i^2 and ν_i are given by

$$S_0^2 = \sum_{i=1}^{I} \sum_{j=1}^{J} \left(Y_{ij} - \bar{Y}_{i\cdot} - \bar{Y}_{\cdot j} + \bar{Y}_{\cdot\cdot}\right)^2 / (I-1)(J-1),$$

$$S_1^2 = J \sum_{i=1}^{I} \left(Y_{i\cdot} - \bar{Y}_{\cdot\cdot}\right)^2 / (I-1),$$

$$S_2^2 = I \sum_{j=1}^{J} \left(Y_{\cdot j} - \bar{Y}_{\cdot\cdot}\right)^2 / (J-1);$$

and

$$\nu_0 = (I-1)(J-1), \qquad \nu_1 = (I-1), \qquad \nu_2 = (J-1).$$

(iii) The two-fold nested random model

$$Y_{ijm} = \mu + u_i + v_{ij} + \varepsilon_{ijm} \tag{8.1.34}$$

for $i=1,\ldots,I, j=1,\ldots,J$, and $m=1,\ldots,M$. Here σ_1^2, σ_2^2, and σ_0^2 are the variances of u_i, v_{ij}, and ε_{ijm}, respectively. Again the standard method gives

$$S_0^2 = \sum_{i=1}^{I} \sum_{j=1}^{J} \sum_{m=1}^{M} \left(Y_{ijm} - \bar{Y}_{ij\cdot}\right)^2 / IJ(M-1),$$

$$S_1^2 = JM \sum_{i=1}^{I} \left(\bar{Y}_{i\cdot\cdot} - \bar{Y}_{\cdot\cdot\cdot}\right)^2 / (I-1),$$

$$S_2^2 = M \sum_{i=1}^{I} \sum_{j=1}^{J} \left(\bar{Y}_{ij\cdot} - \bar{Y}_{i\cdot\cdot}\right)^2 / I(J-1);$$

and

$$\nu_0 = IJ(M-1), \qquad \nu_1 = (I-1), \qquad \nu_2 = I(J-1).$$

An exact solution for obtaining a simultaneous confidence region for the variance ratios $\boldsymbol{\theta}=(\theta_1,\ldots,\theta_k)'$ was given by Hartley and Rao (1967). But that solution is difficult to simplify. In the following we show how the inequalities in Chapter 3 may be applied to find a conservative solution for this problem. Clearly for

$$F_i = \left(S_i^2/\tau_i^2\right)/\left(S_0^2/\sigma_0^2\right), \qquad i = 1,\ldots,k, \tag{8.1.35}$$

the random variable $\mathbf{F}=(F_1,\ldots,F_k)'$ has a multivariate F distribution as defined in Theorem 3.2.2. Let us then consider the one-sided confidence regions

$$A^* = \{\mathbf{x}|\mathbf{x}\in\mathcal{R}^k, S_i^2/(a_iS_0^2) \leqslant x_i, i=1,\ldots,k\}, \qquad (8.1.36)$$

$$A^{**} = \{\mathbf{x}|\mathbf{x}\in\mathcal{R}^k, x_i \leqslant S_i^2/(a_iS_0^2), i=1,\ldots,k\} \qquad (8.1.37)$$

for the parameter vector

$$\boldsymbol{\tau} = (\tau_1^2/\sigma_0^2,\ldots,\tau_k^2/\sigma_0^2)' = \left(1+\sum_{j=1}^{k} c_{1j}\theta_j,\ldots,1+\sum_{j=1}^{k} c_{kj}\theta_j\right)'. \qquad (8.1.38)$$

It follows from Theorem 3.2.2 that

$$P[\boldsymbol{\tau}\in A^*] = P\left[\bigcap_{i=1}^{k}\{F_i \leqslant a_i\}\right] \geqslant \prod_{i=1}^{k} P[F_i \leqslant a_i], \qquad (8.1.39)$$

$$P[\boldsymbol{\tau}\in A^{**}] = P\left[\bigcap_{i=1}^{k}\{F_i \geqslant a_i\}\right] \geqslant \prod_{i=1}^{k} P[F_i \geqslant a_i]. \qquad (8.1.40)$$

Hence if $a_1,\ldots,a_k$ are chosen to satisfy $\Pi_{i=1}^k P[F_i \leqslant a_i]=\gamma^*$ or $\Pi_{i=1}^k P[F_i \geqslant a_i]=\gamma^*$, then the region defined by A^* or A^{**} is a conservative confidence region for the variance ratios $\boldsymbol{\theta}=(\theta_1,\ldots,\theta_k)'$, and the true confidence probability is at least γ^*. This region is the intersection of k sets, each of which is a region bounded by a hyperplane in $\mathcal{R}^k$. If the design matrices $\mathbf{H}_1,\ldots,\mathbf{H}_k$ satisfy $c_{ij}=0$ for $i\neq j$ and $c_{ii}>0$ for all i, then this region reduces to a one-sided confidence region for $\boldsymbol{\theta}$.

This approach was adopted by Broemeling (1969) and Broemeling and Bee (1976). The sharpness of the bounds in (8.1.39) and (8.1.40) was assessed by Sahai and Anderson (1973).

Broemeling (1978) made an attempt to extend his work to the two-sided case. It was claimed that for the same random variables $F_1,\ldots,F_k$ one has

$$P\left[\bigcap_{i=1}^{k}\{b_i \leqslant F_i \leqslant a_i\}\right] \geqslant \prod_{i=1}^{k} P[b_i \leqslant F_i \leqslant a_i]. \qquad (8.1.41)$$

He then proceeded to derive the main result without justifying this statement. Unfortunately this statement is false, and the following counterexample was offered by Tong (1979). It should be pointed out that the proof given below applies to a large class of mixtures of distributions.

Example 8.1.1. Consider $k=2$. For arbitrary but fixed positive real numbers a_1 and b_2, there exist a small $b_1>0$ and a large a_2 such that

$$P\left[\bigcap_{i=1}^{2}\{b_i \leqslant F_i \leqslant a_i\}\right] < \prod_{i=1}^{2} P[b_i \leqslant F_i \leqslant a_i]. \tag{8.1.42}$$

Hence in general the assertion in (8.1.41) is false.

Proof. Let us define

$$\delta(b_1,a_2) = P\left[\bigcap_{i=1}^{2}\{b_i \leqslant F_i \leqslant a_i\}\right] - \prod_{i=1}^{2} P[b_i \leqslant F_i \leqslant a_i] \tag{8.1.43}$$

as a function of b_1 and a_2 over the region $b_1 \in [0,a_1]$ and $a_2 \geqslant b_2$. Since the random variables are continuous, $\delta(b_1,a_2)$ is a continuous function of (b_1,a_2). Therefore an application of Lemma 2.2.1 yields

$$\begin{aligned}
\delta_0 &= \lim_{b_1\to 0}\lim_{a_2\to\infty} \delta(b_1,a_2)\\
&= P[F_1 \leqslant a_1, b_2 \leqslant F_2] - P[F_1 \leqslant a_1]P[b_2 \leqslant F_2]\\
&= E\{P[(S_1^2/\tau_1^2) \leqslant (a_1 S_0^2/\sigma_0^2) | S_0^2 = s_0^2]P[(b_2 S_0^2/\sigma_0^2) \leqslant (S_2^2/\tau_2^2) | S_0^2 = s_0^2]\}\\
&\qquad - P[F_1 \leqslant a_1]P[b_2 \leqslant F_2]\\
&< 0.
\end{aligned}$$

This implies that there exist a small $b_1^* > 0$ and a large a_2^* that depend on a_1, b_2, and δ_0 such that

$$|\delta(b_1^*,a_2^*) - \delta_0| \leqslant -\delta_0/2.$$

Therefore we have

$$\delta(b_1^*,a_2^*) \leqslant \delta_0/2 < 0,$$

and the inequality in (8.1.42) holds at $(b_1,a_2)=(b_1^*,a_2^*)$. ■

Because Broemeling's argument depends on the assertion in (8.1.41), which is false, the proof of his main result in his 1978 paper is in error. However, if the design matrices $\mathbf{H}_1,\ldots,\mathbf{H}_k$ in (8.1.27) satisfy the condition that the degrees of freedom for the mean squares are equal (i.e., $\nu_1 = \cdots = \nu_k$) and if a_i, b_i satisfy

$$b_1 = \cdots = b_k < a_1 = \cdots = a_k,$$

then $F_1,\ldots,F_k$ are conditionally independent and identically distributed. In this special case we know that the inequality in (8.1.41) is true (by Theorem 6.2.11). Hence his result can be justified under such additional conditions.

The problem of obtaining a conservative confidence region for the variance components σ_0^2, σ_1^2, σ_2^2 in the two-way classification and two-fold nested random models was studied by Sahai (1974). His solution also depends on an application of the inequalities for multivariate F distribution.

8.2. HYPOTHESIS-TESTING AND SIMULTANEOUS COMPARISONS

For $k>1$ let $\mathbf{X}=(X_1,\ldots,X_k)'$ denote a k-dimensional random variable which has a density function $f_{\boldsymbol{\theta}}(\mathbf{x})$, where $\boldsymbol{\theta}$ is the unknown parameter vector. Let Ω denote the parameter space. The functional form of f is assumed to be known.

8.2.1. Monotonicity of Power Functions and Conservative Critical Regions

Consider the problem of testing the hypotheses

$$H_0:\quad \boldsymbol{\theta}\in\Omega_0 \qquad \text{versus} \qquad H_1:\quad \boldsymbol{\theta}\in\Omega_1, \tag{8.2.1}$$

where Ω_0 and Ω_1 are disjoint subsets of Ω. Suppose that ϕ is a given test† defined by a certain critical region. We now illustrate how existing probability inequalities in multivariate distributions may be applied to establish the monotonicity of the power function of a test under suitable conditions.

(a) When testing the hypotheses

$$H_0:\quad \boldsymbol{\theta}=\boldsymbol{\theta}_0 \qquad \text{versus} \qquad H_1:\quad \boldsymbol{\theta}\neq\boldsymbol{\theta}_0 \tag{8.2.2}$$

for the mean vector $\boldsymbol{\theta}$ of a normal population with a known covariance matrix $\boldsymbol{\Sigma}$, $\mathbf{X}=\overline{\mathbf{X}}$ becomes the sample mean vector with sample size n. Hence $f_{\boldsymbol{\theta}}$ is a multivariate normal density function with mean $\boldsymbol{\theta}$ and covariance matrix $(1/n)\boldsymbol{\Sigma}$. The likelihood ratio test ϕ calls for the rejection of H_0 when $(\overline{\mathbf{X}}-\boldsymbol{\theta}_0)\notin A$, where

$$A=\{\mathbf{x}\mid\mathbf{x}\in\mathfrak{R}^k, n\mathbf{x}'\boldsymbol{\Sigma}^{-1}\mathbf{x}\leqslant a\} \tag{8.2.3}$$

†For notational convenience ϕ is assumed to be a nonrandomized test.

and a is the $(1-\alpha)$th percentile of a chi-square distribution with k degrees of freedom. The power function of this test is

$$\pi(\boldsymbol{\theta}) = P_{\boldsymbol{\theta}}\left[(\overline{\mathbf{X}}-\boldsymbol{\theta}_0)\notin A\right] = 1 - P_{\boldsymbol{\theta}}\left[(\overline{\mathbf{X}}-\boldsymbol{\theta})\in(A+(\boldsymbol{\theta}_0-\boldsymbol{\theta}))\right].$$

Denoting $\mathbf{y}=(\boldsymbol{\theta}-\boldsymbol{\theta}_0)$ and realizing that

$$P_{\boldsymbol{\theta}}\left[(\overline{\mathbf{X}}-\boldsymbol{\theta})\in(A+(\boldsymbol{\theta}_0-\boldsymbol{\theta}))\right] = P_{\boldsymbol{\theta}}\left[(\overline{\mathbf{X}}-\boldsymbol{\theta})\in(A-(\boldsymbol{\theta}_0-\boldsymbol{\theta}))\right].$$

we write

$$\pi(\boldsymbol{\theta}) = 1 - P_{\boldsymbol{\theta}}\left[(\overline{\mathbf{X}}-\boldsymbol{\theta})\in(A+\mathbf{y})\right] \equiv \pi^*(\mathbf{y}), \tag{8.2.4}$$

which depends on $\boldsymbol{\theta}$ through $\mathbf{y}$. From this we may conclude

(i) $\pi^*(\lambda\mathbf{y})$ is monotonically increasing in $|\lambda|$. Hence if the true mean moves away from $\boldsymbol{\theta}_0$ along a straight line with this given direction, the power of the test becomes larger. This is a consequence of Theorem 4.1.1. It is also consistent with the classical known result that the power involves the probability of a noncentral chi-square distribution with a noncentrality parameter proportional to $(\boldsymbol{\theta}-\boldsymbol{\theta}_0)'\boldsymbol{\Sigma}^{-1}(\boldsymbol{\theta}-\boldsymbol{\theta}_0)$.

(ii) When the components of $\overline{\mathbf{X}}$ have equal variances and equal covariances, then the density function $f_{\boldsymbol{\theta}=\mathbf{0}}(\mathbf{x})$ and the indicator function of A given in (8.2.3) are both Schur-concave. Hence Theorem 6.2.2 implies that $\pi^*(\mathbf{y})$ is a Schur-convex function of $\mathbf{y}$. Specifically, let $\boldsymbol{\theta}$ and $\boldsymbol{\xi}$ be two parameter vectors. If $(\boldsymbol{\xi}-\boldsymbol{\theta}_0)$ majorizes $(\boldsymbol{\theta}-\boldsymbol{\theta}_0)$, then $\pi(\boldsymbol{\xi})\geqslant\pi(\boldsymbol{\theta})$.

(b) When testing the hypothesis in (8.2.2) for the mean vector of a normal distribution with unknown covariance matrix $\boldsymbol{\Sigma}$, Hotelling's T^2 test calls for the rejection of H_0 when $(\overline{\mathbf{X}},\mathbf{S})\notin A$, where A is the acceptance region

$$A = \left\{(\mathbf{x},\mathbf{S})|\mathbf{x}\in\mathcal{R}^k, (\mathbf{x}-\boldsymbol{\theta}_0)'\mathbf{S}^{-1}(\mathbf{x}-\boldsymbol{\theta}_0)\leqslant a\right\}, \tag{8.2.5}$$

$\mathbf{S}$ is the sample covariance matrix, and a is to be chosen from the F distribution. Since $\mathbf{S}$ is positive definite with probability one, we can write, for $\mathbf{y}=\boldsymbol{\theta}-\boldsymbol{\theta}_0$

$$\begin{aligned}\pi(\boldsymbol{\theta}) &= 1 - E\left\{P_{\boldsymbol{\theta}}\left[(\overline{\mathbf{X}}-\boldsymbol{\theta}_0)'\mathbf{S}^{-1}(\overline{\mathbf{X}}-\boldsymbol{\theta}_0)\leqslant a|\mathbf{S}=\mathbf{S}_0\right]\right\}\\ &\equiv \pi^{**}(\mathbf{y}).\end{aligned} \tag{8.2.6}$$

Now from the discussion in (a) for every fixed $\mathbf{S}_0$ an inequality can be obtained through a partial ordering of $\boldsymbol{\theta}$, and this inequality is preserved after taking expectations. Hence we may conclude that in

Hotelling's T^2 test $\pi^{**}(\lambda\mathbf{y})$ is also monotonically increasing in $|\lambda|$, this is related to a known result in noncentral F distribution.

(c) A multivariate version of this problem was considered by Das Gupta, Anderson, and Mudholkar (1964). In their paper $\mathbf{X}=(\mathbf{X}^{(1)},\ldots,\mathbf{X}^{(k)})'$ is a $k\times p$ random matrix; and it was assumed that $\mathbf{X}^{(1)},\ldots,\mathbf{X}^{(k)}$ are independent $1\times p$ normal variables with means $\boldsymbol{\theta}^{(1)},\ldots,\boldsymbol{\theta}^{(k)}$ and a common covariance matrix. They showed that in testing a multivariate general linear hypothesis the power functions of a certain class of invariant tests are monotonically increasing in each of the nonzero characteristic roots of the population parameter matrix. That class of tests has the property that the acceptance region is convex and symmetric in $\mathbf{x}^{(i)}$ when the other variables are kept fixed. The proof of their result follows from a repeated application of Theorem 4.1.1, considering one variable at a time.

(d) Another monotonicity property of the power functions of a certain class of permutation invariant tests for testing a general linear hypothesis was proved by Eaton and Perlman (1974). They considered the problem of multivariate analysis-of-variance, and showed that inequalities for the power of certain tests may be obtained through the partial ordering of two vectors of the characteristic roots via majorization. The proof depends on an application of either Theorem 6.2.1 or Theorem 6.2.2. Since two vectors $\mathbf{y}_1$ and $\mathbf{y}_2$ might be comparable via majorization even if there does not exist an λ such that $\mathbf{y}_2=\lambda\mathbf{y}_1$, there are situations under which the inequality of Eaton and Perlman applies and the inequality of Das Gupta–Anderson–Mudholkar does not. In their paper Eaton and Perlman provided a comparison for the regions in which the inequalities may be applied.

Previously, certain monotonicity properties of the power functions of simultaneous analysis of variance and multivariate analysis of variance tests were also studied by Ghosh (1963) and Krishnaiah (1965, 1969). Their results depend on the monotonicity property of the noncentral F distribution.

(e) Let $(Y_1,\ldots,Y_k)'$ be a normal variable with mean vector $\boldsymbol{\theta}$, variances one, and correlations $\rho\in(-1/(k-1),1)$. We consider the noncentral chi-square and F variables defined by

$$\chi^{*2}=\sum_{i=1}^{k}Y_i^2,\qquad F^*=\sum_{i=1}^{k}Y_i^2/kS^2,$$

where S is independent of $(Y_1,\ldots,Y_k)'$ and νS^2 has a chi-square distribution with ν degrees of freedom. Marshall and Olkin (1974) showed that the distribution functions of these noncentral variables are Schur-concave functions of $\boldsymbol{\theta}$ (see Example 6.3.1 and the remark following it). Hence the comparison of the power of some standard tests, whose acceptance regions are of the forms

$$\left\{\mathbf{y}|\mathbf{y}\in\mathfrak{R}^k,\ \sum_{i=1}^{k} y_i^2 \leqslant a\right\},\qquad \left\{(\mathbf{y},s^2)|\mathbf{y}\in\mathfrak{R}^k, s^2>0,\ \sum_{i=1}^{k}(y_i^2/s^2)\leqslant a\right\}$$

can be obtained by the partial ordering of two configurations via majorization. This statement also follows from the more general result of Eaton and Perlman (1974).

(f) Let us now see some applications of inequalities via weak majorization. Suppose that the random variable $\mathbf{X}$ is either continuous or discrete, the components $X_1,\ldots,X_k$ of $\mathbf{X}$ are nonnegative, and the density of $\mathbf{X}$ is $f_{\boldsymbol{\theta}}(\mathbf{x})=\prod_{i=1}^{k} g_{\theta_i}(x_i)$, where $g_{\theta_i}(x_i)$ is the density function of X_i and $\theta_i>0$. Let g satisfy the TP_2 property and the semigroup property stated in Theorem 6.2.10. In testing the hypothesis in (8.2.1), if the critical region of a test is convex, permutation invariant, and decreasing, then its power function $\pi(\boldsymbol{\theta})$ is monotone in the following sense: $\boldsymbol{\xi}\gg\boldsymbol{\theta}$ implies $\pi(\boldsymbol{\theta})\geqslant\pi(\boldsymbol{\xi})$. This follows from Theorem 6.3.6 and the fact that the critical region is a decreasing Schur-concave set (see Problem 8, Chapter 6).

The above result applies to multiparameter hypothesis testing for binomial, Poisson, hypergeometric, and gamma families among others. In general, when testing the hypotheses

$$H_0:\quad \boldsymbol{\theta}=(a,\ldots,a)' \qquad \text{versus}\quad H_1:\quad \boldsymbol{\theta}=(b,\ldots,b)' \tag{8.2.7}$$

with $b<a$ for the exponential family, the critical region of the Neyman–Pearson test is decreasing and Schur-concave (see Problems 2 and 3).

(g) The application in (f) is valid only for nonnegative random variables. Now suppose that the density function belongs to a location parameter family, i.e., $f_{\boldsymbol{\theta}}(\mathbf{x})=f(\mathbf{x}-\boldsymbol{\theta})$, and that $f(\mathbf{x})$ is both Schur-concave and decreasing in absolute value. In testing the hypotheses

$$H_0:\quad \boldsymbol{\theta}=\mathbf{0} \qquad \text{versus} \qquad H_1:\quad \boldsymbol{\theta}\neq\mathbf{0}, \tag{8.2.8}$$

if H_0 is rejected when $\mathbf{x}\notin A$ for some acceptance region A that is Schur-concave and decreasing in absolute value, then the critical region is Schur-convex and increasing in absolute value. Theorem

6.3.9 says that $|\boldsymbol{\xi}| \succ\!\succ_w |\boldsymbol{\theta}|$ implies

$$\pi(\boldsymbol{\xi}) = 1 - P_{\boldsymbol{\xi}}[\mathbf{X} \in A] \geqslant \pi(\boldsymbol{\theta}). \tag{8.2.9}$$

Therefore a monotonicity property of the power function can be obtained via weak majorization even when the random variables may take negative values.

In the special case in which $f_{\boldsymbol{\theta}}(\mathbf{x})$ is the multivariate normal density with mean vector $\boldsymbol{\theta}$, known variances σ^2, and correlations zero (i.e., $\mathbf{X}$ has independent components) the likelihood ratio test for testing the hypotheses in (8.2.8) yields an acceptance region of the form

$$A = \left\{ \mathbf{x} | \mathbf{x} \in \mathcal{R}^k, \sum_{i=1}^{k} x_i^2 \leqslant a \right\}.$$

It is known that the power function $\pi(\boldsymbol{\theta})$ of this test involves a noncentral chi-square probability with degrees of freedom k and a noncentrality parameter proportional to $\sum_{i=1}^{k} \theta_i^2$. Hence in this special case the classical known result is stronger than the inequality that $|\boldsymbol{\xi}| \succ\!\succ_w |\boldsymbol{\theta}|$ implies $\pi(\boldsymbol{\xi}) \geqslant \pi(\boldsymbol{\theta})$ (see Problem 4).

In addition to establishing the monotonicity properties of the power functions, probability inequalities can also be applied to derive conservative critical regions. Consider a test for testing

$$H_0: \quad \boldsymbol{\theta} = (\theta_1, \ldots, \theta_k)' = (\theta_1^0, \ldots, \theta_k^0) = \boldsymbol{\theta}_0 \tag{8.2.10}$$

versus the alternative hypothesis

$$H_1: \quad \theta_i > \theta_i^0 \qquad \text{for some} \quad i. \tag{8.2.11}$$

Suppose that H_0 is rejected if and only if $\mathbf{X} \notin A$, where

$$A = \{ \mathbf{x} | \mathbf{x} \in \mathcal{R}^k, x_i \leqslant a_i, i = 1, \ldots, k \}. \tag{8.2.12}$$

This type of test may be regarded as compound decision rules in a multiple decision problem. Specifically, suppose that we have k component decision problems for testing k sets of hypotheses

$$H_0^{(i)}: \quad \theta_i = \theta_i^{(0)} \qquad \text{versus} \qquad H_1^{(i)}: \quad \theta_i > \theta_i^{(0)} \tag{8.2.13}$$

for $i = 1, \ldots, k$. Let $\phi^{(i)}(x_i)$ be the test for $H_0^{(i)}$ versus $H_1^{(i)}$ such that $\phi^{(i)}$ depends on $\mathbf{x}$ only through x_i and $H_0^{(i)}$ is rejected if and only if $x_i > a_i$. Then the test ϕ given by

$$\phi(\mathbf{x}) = \begin{cases} 0 & \text{for} \quad \mathbf{x} \in A, \\ 1 & \text{otherwise} \end{cases} \tag{8.2.14}$$

is a test for the hypotheses

$$H_0\colon \bigcap_{i=1}^{k} H_0^{(i)} \qquad \text{versus} \qquad H_1\colon \bigcup_{i=1}^{k} H_1^{(i)}. \tag{8.2.15}$$

A typical application of such a test is to the slippage problem (see, e.g., Lehmann (1966) for some references) in which one wishes to conclude whether all the parameters of k populations are equal to a specific value or at least one of the populations has a larger parameter. Lehmann (1957) studied the properties of this type of compound decision rules. He showed that under reasonable conditions on the loss functions they are optimal among a certain class of decision rules if each component decision rule is optimal.

In many applications the random variables $X_1,\ldots,X_k$ (the components of $\mathbf{X}$) are associated. (This includes the special case in which they are independent.) In this case for the test ϕ defined in (8.2.14) we have immediately

$$P[\phi(\mathbf{X})=0] = P[\mathbf{X}\in A] \geqslant \prod_{i=1}^{k} P[X_i \leqslant a_i].$$

Hence if we choose $a_1,\ldots,a_k$ to satisfy

$$P[X_i > a_i] \leqslant \alpha_i, \qquad i = 1,\ldots,k,$$

then the type I error α of the test ϕ is bounded above by

$$\alpha \leqslant 1 - \prod_{i=1}^{k} (1-\alpha_i). \tag{8.2.16}$$

Therefore a conservative critical region for the test ϕ can be obtained based on the marginal distributions only.

In an analysis-of-variance problem Kimball (1951) considered a test of this type. Suppose that in a complete two-way layout S_1^2, S_2^2, and S_0^2 are the mean squares of the treatment, the block, and the error, respectively. Then the two F ratios

$$F_1 = S_1^2/S_0^2, \qquad F_2 = S_2^2/S_0^2$$

are associated (see Problem 8, Chapter 5). To test the null hypothesis that the treatment effect and the block effect are both zero, one may choose a_1, a_2 from the F table and reject the null hypothesis if $F_1 > a_1$ or $F_2 > a_2$. For this test the type I error is bounded above by $1-\prod_{i=1}^{2} P[F_i \leqslant a_i]$.

Olkin (1972) applied another inequality to solve the above analysis-of-variance problem. Under the additional condition that the degrees of freedom for S_1^2, S_2^2 be equal, the density function of $(F_1,F_2)'$ is Schur-concave. Hence for all $a>0$ we have (see (6.2.21))

$$P\left[\bigcap_{i=1}^{2}\{F_i>a\}\right] \geqslant P[F_1>2a].$$

This together with the identity

$$P\left[\bigcap_{i=1}^{2}\{F_i\leqslant a_i\}\right] = P[F_1\leqslant a_1]+P[F_2\leqslant a_2] + P\left[\bigcap_{i=1}^{2}\{F_i>a_i\}\right]-1$$

implies that

$$P\left[\bigcap_{i=1}^{2}\{F_i\leqslant a\}\right] \geqslant 2P[F_1\leqslant a]-P[F_1\leqslant 2a].$$

The type I error of the test is then bounded above by

$$\alpha \leqslant 1-2P[F_1\leqslant a]+P[F_1\leqslant 2a], \tag{8.2.17}$$

which again depends only on the univariate F distribution. A comparison of the two solutions given by Olkin shows that for moderate a values the inequality in (8.2.17) is sharper.

Probability inequalities have also been found useful in various problems concerning simultaneous test procedures. These procedures were developed in the spirit of Roy's (1953) union–intersection principle, and a survey of them can be found in Krishnaiah (1979). Under such test procedures, the test usually is of the form given in (8.2.14). Hence to obtain a bound for the cutoff point of such a test one needs to apply inequalities. When testing the means, a numerical comparison for the sharpness of different bounds was given recently by Cox, Krishnaiah, Lee, Reising, and Schuurmann (1979).

8.2.2. Simultaneous Comparisons

Let us now see the applications of certain probability inequalities in simultaneous comparisons. For the purpose of simplicity let us consider only normal variables and assume that $X_1,\ldots,X_k$ have means θ_i and variances σ_i^2 $(i=1,\ldots,k)$. In hypothesis-testing problems the concern is either to accept or to reject the null hypothesis that $\boldsymbol{\theta}=\mathbf{0}$, where $\boldsymbol{\theta}$ is the mean of the normal variable

$\mathbf{X}=(X_1,\ldots,X_k)'$. In simultaneous comparisons, however, one is also interested in knowing which one of the components of $\mathbf{X}$ is responsible for the rejection of the null hypothesis if it is rejected. For example, in the bivariate case one "needs a decision not merely between H_0 and *not* H_0," but among

$$H_0:\quad \theta_1 = 0, \qquad \theta_2 = 0, \tag{8.2.18}$$

$$H_1:\quad \theta_1 \neq 0, \qquad \theta_2 = 0, \tag{8.2.19}$$

$$H_2:\quad \theta_1 = 0, \qquad \theta_2 \neq 0, \tag{8.2.20}$$

$$H_3:\quad \theta_1 \neq 0, \qquad \theta_2 \neq 0. \tag{8.2.21}$$

(See Miller (1966, p. 3).) A complete discussion of this approach and the references in this area can be found in Miller's book.

For many well-known procedures in simultaneous comparisons the acceptance region for the hypothesis H_0 is of the form

$$A_0 = \left\{\mathbf{x}|\mathbf{x}\in\mathfrak{R}^k, |x_i| \leqslant a_i, i=1,\ldots,k\right\}. \tag{8.2.22}$$

(The values of a_i are usually equal when the variances $\sigma_1^2,\ldots,\sigma_k^2$ are equal.) A problem of concern is to choose $a_1,\ldots,a_k$ such that the probability of rejecting H_0 when H_0 is true is at most α (preassigned). When the variances are known and when $X_1,\ldots,X_k$ are independent, this can be done easily by choosing $a_1,\ldots,a_k$ to satisfy

$$\prod_{i=1}^{k}\left[\Phi(a_i/\sigma_i)-\Phi(-a_i/\sigma_i)\right] = 1-\alpha. \tag{8.2.23}$$

But in reality the X_is are not always independent. Suppose that the variances of the X_is are known but their correlation matrix is either unknown or very complicated. Then from Theorem 2.2.2 or 2.2.5 or 4.3.1 we know that the solution given in (8.2.23) (which is easy to obtain) is a conservative one, *no matter what the true correlations are*. If the Bonferroni inequality is used instead, then the solution is overprotective.

When the variances are unknown but equal to σ^2 and the X_is are independent, then the a_is in A_0 may depend on S, an estimate of the common variance σ^2 such that S and $(X_1,\ldots,X_k)'$ are independent and $\nu S^2/\sigma^2$ has a chi-square distribution with ν degrees of freedom. In this case a solution similar to (8.2.23) can be obtained from the Student t table for a given α. This again is a conservative solution because no matter what the true correlations are the true probability of rejecting H_0 when it is true is at most α (by Theorem 3.1.1 and the inequality in (3.1.5)).

In some special cases inequalities for the probabilities of accepting the other hypotheses may also be of interest. Let us consider the case of $k=2$. For testing hypotheses H_0–H_3 stated in (8.2.18)–(8.2.21) let us assume that the following decision rule is adopted:

$$\begin{aligned}&\text{accept } H_0 \text{ if } |X_1| \leqslant a_1, |X_2| \leqslant a_2,\\ &\text{accept } H_1 \text{ if } |X_1| > a_1, |X_2| \leqslant a_2,\\ &\text{accept } H_2 \text{ if } |X_1| \leqslant a_1, |X_2| > a_2,\\ &\text{accept } H_3 \text{ if } |X_1| > a_1, |X_2| > a_2.\end{aligned}$$

Then under the assumption that the means of X_1 and X_2 are zero and their correlation is ρ, we may conclude that

(i) The probability of accepting H_0 or H_3 is decreasing in ρ for $\rho<0$ and increasing in ρ for $\rho>0$. Hence it is minimized when X_1, X_2 are independent (Theorems 2.2.5 and 2.3.3).

(ii) The probability of accepting H_1 or H_2 is increasing in ρ for $\rho<0$ and decreasing in ρ for $\rho>0$. Hence it is maximized when X_1, X_2 are independent. This follows from Theorem 2.3.3 and the identity

$$P_\rho\left[|X_i|>a_i, |X_j|\leqslant a_j\right] = P\left[|X_i|>a_i\right] - P_\rho\left[|X_i|>a_i, |X_j|>a_j\right]$$

for $i \neq j$.

The above discussions show how the probability of accepting H_0 or the other hypotheses depends on the correlations when the means are zero. Let us also illustrate how this probability depends on the means when the X_is are independent and have equal variances. For this application we consider a region A_0 in $\mathfrak{R}^k$ for accepting H_0 that is Schur-concave and decreasing in absolute value. It can be seen in Miller's book that this condition is satisfied under many existing procedures in simultaneous comparisons. Since the distribution of $(X_1,\ldots,X_k)'$ is Schur-concave and decreasing in absolute value, Theorem 6.3.9 says that if $|\boldsymbol{\xi}| \succ\succ |\boldsymbol{\theta}|$, then

$$P_{\boldsymbol{\theta}}\left[\text{accepting } H_0\right] \geqslant P_{\boldsymbol{\xi}}\left[\text{accepting } H_0\right].$$

Therefore inequalities for this probability can be given through a partial ordering of the mean vector via weak majorization.

8.2.3. Simultaneous Comparisons with a Control

A particular real-life application of simultaneous inference is the problem of comparing k treatments simultaneously with a control. Let $\Pi_0, \Pi_1, \ldots, \Pi_k$ denote $k+1$ independent normal populations with means μ_i and variances σ_i^2 $(i=0,1,\ldots,k)$. Here Π_0 denotes the control population, and it is to be compared simultaneously with the experimental populations $\Pi_1,\ldots,\Pi_k$. This approach was due to Dunnett (1955, 1964), and it was called a "many–one" comparison in Miller (1966, Sections 2.5, 4.1, 4.3).

For $i=1,\ldots,k$ let

$$\theta_1 = \mu_i - \mu_0 \tag{8.2.24}$$

denote the difference of the means between Π_i and Π_0. Suppose that $\bar{X}_i$ is the sample mean with sample size n_i from the ith population Π_i and that the inference on $\boldsymbol{\theta}=(\theta_1,\ldots,\theta_k)'$ is to be based on $\bar{X}_i-\bar{X}_0$, $i=1,\ldots,k$. When we standardize the differences of the sample means by defining

$$X_i = (\bar{X}_i - \bar{X}_0)\Big/\left(\frac{\sigma_i^2}{n_i}+\frac{\sigma_0^2}{n_0}\right)^{1/2}, \tag{8.2.25}$$

the new random variables X_i have means θ_i and variances one $(i=1,\ldots,k)$. For notational convenience we write

$$\tau_i^2 = \frac{\sigma_i^2}{n_i}, \qquad i = 0,1,\ldots,k, \tag{8.2.26}$$

which is the variance of $\bar{X}_i$. It is easy to check that $\mathbf{X}=(X_1,\ldots,X_k)'$ is a multivariate normal variable with mean vector $\boldsymbol{\theta}$, variances one, and correlations

$$\operatorname{cor}(X_i,X_j) = \frac{\tau_0^2}{\left[(\tau_i^2+\tau_0^2)(\tau_j^2+\tau_0^2)\right]^{1/2}} \equiv \lambda_i\lambda_j. \tag{8.2.27}$$

Thus the covariance matrix of $\mathbf{X}$ has the structure l (see Definition 2.2.1) for all $\sigma_0^2,\ldots,\sigma_k^2$ and $n_0,\ldots,n_k$. Moreover, from (8.2.27) we note that

$$\lambda_i = \frac{\tau_0}{(\tau_i^2+\tau_0^2)^{1/2}} \tag{8.2.28}$$

is positive, decreasing in τ_i^2 for fixed τ_0^2, and increasing in τ_0^2 for fixed τ_i^2.

(a) When the variances are known, then under the simultaneous comparison procedure Dunnett (1955, 1964) proposed, one accepts the null hypothesis H_0 that none of the experimental populations is different from the control if and only if

$$\mathbf{X}\in A \equiv \left\{\mathbf{x}\,|\,\mathbf{x}\in\mathcal{R}^k, |x_i|\leqslant a_i, i=1,\ldots,k\right\}.$$

We now see how the probability of accepting H_0 depends on the sample sizes for fixed variances $\sigma_0^2,\ldots,\sigma_k^2$ under the assumption that $\boldsymbol{\theta}=\mathbf{0}$. Since A is convex and symmetric about the origin, combining (8.2.26) and (8.2.27) with Corollary 1 of Theorem 2.2.5, we know that

(i) For fixed n_j, $j\neq i$, the probability is increasing in n_i for each $i=1,\ldots,k$.

(ii) For fixed $n_1,\ldots,n_k$ the probability is decreasing in n_0.

(b) In some other applications a one-sided comparison may be of interest. Such a comparison serves the purpose of testing whether or not any of the experimental populations is better than the control. According to Dunnet's (1955, 1964) procedure, one concludes that none is better than the control if

$$\mathbf{X}\in A\equiv\{\mathbf{x}|\mathbf{x}\in\mathcal{R}^k, x_i\leqslant a_i, i=1,\ldots,k\}.$$

Under this procedure the probability of accepting this null hypothesis is a cumulative multivariate normal probability. Therefore from Slepian's inequality (Theorem 2.1.1) this probability is again increasing in n_i for fixed n_j ($j\neq i$, $i=1,\ldots,k$) and decreasing in n_0 for fixed $n_1,\ldots,n_k$.

It seems natural to apply these inequalities to solve the problem of optimal allocation of observations when comparing several treatments simultaneously with a control. But it appears that this tool has not yet been fully employed. Dunnett (1955) originally considered the problem of optimal allocation of observations for the one-sided case in his paper. The method he used is basically that of numerical calculation and differentiation.

(c) When the variances are unknown but equal, the common variance σ^2 may be estimated by a polled sample variance S^2, where $\nu S^2/\sigma^2$ has a chi-square distribution with ν degrees of freedom and $\mathbf{X}$ and S^2 are independent. If one takes an equal number of observations from the experimental populations with $n_1=\cdots=n_k$, then the joint distribution of the random variables

$$t_i=\{(\bar{X}_i-\bar{X}_0)-\theta_i\}/S\left(\frac{1}{n_1}+\frac{1}{n_0}\right)^{1/2},\quad i=1,\ldots,k,\tag{8.2.29}$$

is multivariate t with ν degrees of freedom, and the elements of the correlation matrix are

$$\rho_{ij}=\begin{cases}1 & \text{for}\quad i=j,\\ n_1/(n_0+n_1) & \text{for}\quad i\neq j.\end{cases}\tag{8.2.30}$$

Therefore $t_1,\ldots,t_k$ are positively dependent by mixture; in other words, they are conditionally independent and identically distributed random variables (see Example 3.3.2). From Theorem 6.2.11 we know that for $r \leqslant k$

$$P\left[\bigcap_{i=1}^{k}\{|t_i|\leqslant a\}\right] \geqslant \left(P\left[\bigcap_{i=1}^{r}\{|t_i|\leqslant a\}\right]\right)^{k/r}, \qquad (8.2.31)$$

$$P\left[\bigcap_{i=1}^{k}\{t_i\leqslant a\}\right] \geqslant \left(P\left[\bigcap_{i=1}^{r}\{t_i\leqslant a\}\right]\right)^{k/r}, \qquad (8.2.32)$$

and the bounds in (8.2.31) and (8.2.32) are monotonically nondecreasing in r (see Problem 5). From these inequalities one can obtain from existing univariate or multivariate t tables the sample sizes n_0, n_1 that are required to guarantee that the probability in question is at least γ^* (preassigned).

Nonparametric procedures for comparing several treatments with a control were considered by Steel (1959a, b). Suppose that $F_0, F_1, \ldots, F_k$ are the continuous distribution functions of $Y_0, Y_1, \ldots, Y_k$. For $i=1,\ldots,k$ let $X_i = Y_i - Y_0$, and denote by θ_i the median of X_i. In testing the null hypothesis

$$H_0:\quad \theta_i = 0 \qquad \text{for} \quad i = 1,\ldots,k$$

versus the alternative hypothesis

$$H_1:\quad \theta_i \neq 0 \qquad \text{for some} \quad i,$$

the sign test Steel proposed may be described as the following: For $i=0,1,\ldots,k$ take observations Y_{im}, $m=1,\ldots,n$. Let S_i^+ (S_i^-) be the number of positive (negative) observations in the n observations $(Y_{im}-Y_{0m})$, $m=1,\ldots,n$ (note that $S_i^- = n - S_i^+$), and let $S_i = \max(S_i^+, S_i^-)$. The procedure calls for the rejection of H_0 if $\max_{1\leqslant i\leqslant k} S_i > a$ for some a. The rank sum test Steel proposed is quite similar. In that test each experimental population is to be compared with the control, and the decision rule depends on the rank sums. (For details, see Problem 8, Chapter 3.)

The exact distributions of the testing statistics are complicated. However it is clear that under both the sign test and the rank sum test the statistic S_i is of the form

$$S_i = \psi(X_{01},\ldots,X_{0n}; X_{i1},\ldots,X_{in}), \qquad i = 1,\ldots,k$$

for a Borel-measurable function ψ. Therefore if $F_1 = \cdots = F_k$ holds (which is implied by H_0), then $S_1,\ldots,S_k$ are conditionally independent and identically distributed random variables, and their joint distribution is a mixture (see Section 5.3 and Theorem 6.2.11). Hence for the sign test (under H_0)

$$P[H_0 \text{ is accepted}] = P\left[\bigcap_{i=1}^{k} \{S_i \leqslant a\}\right]$$

$$\geqslant \left(P\left[\bigcap_{i=1}^{r} \{S_i \leqslant a\}\right]\right)^{k/r} \tag{8.2.33}$$

holds for all $r \leqslant k$, and the bound is sharper when r is closer to k (see Problem 5). A conservative critical region can thus be constructed for any large k because tables for lower dimensions are already available (e.g., for $r=2,\dots,9$, see Miller (1966, Table VI, p. 248)). For Steel's rank sum test exactly the same conclusion can be drawn, and tables for $r=2,\dots,9$ can also be found in Miller (1966, Table VIII, p. 251). In particular, a solution can be obtained from the table for the Wilcoxon–Mann–Whitney test by taking $r=1$. The same thing can be said for the one-sided test.

As a final note we remark that Theorem 6.2.11 applies to many other existing procedures in simultaneous comparisons. Hence the theorem provides a conservative solution to those problems for which the value of k is beyond the capacity of existing statistical tables made for such purpose.

8.3. RANKING AND SELECTION PROBLEMS

Suppose that $\mathcal{F} = \{F_\theta(y) : \theta \in \Lambda\}$ is a stochastically increasing family of distribution functions (for the definition see Section 5.1). For $i=1,\dots,k$ let $Y_{i1},\dots,Y_{in}$ be an independent random sample of size n from the ith population with distribution $F_{\theta_i}(y) \in \mathcal{F}$, and let

$$\theta_{[1]} \leqslant \cdots \leqslant \theta_{[k]} \tag{8.3.1}$$

denote the ordered parameters. Then the population associated with the largest parameter $\theta_{[k]}$ is called the best population, and the problem under consideration is to select that population.

For $i=1,\dots,k$ let

$$X_i = T(Y_{i1},\dots,Y_{in}), \tag{8.3.2}$$

where T is an appropriate statistic. Let $G_{n,\theta_i}(x)$ denote the distribution function of X_i. We shall assume that the family of distributions $\mathcal{G}=\{G_{n,\theta}(x):\theta\in\Lambda\}$ is also stochastically increasing.† Thus the order is preserved by T, and we may consider the problem of selecting the population with distribution $G_{n,\theta_{[k]}}(x)$ based on $X_1,\ldots,X_k$ only.

The area of ranking and selection problems originated from a paper on selecting the largest normal mean considered by Bechhofer (1954). During the past two decades this area has grown considerably. For a general discussion of the existing ranking procedures and a bibliography the reader is referred to Gibbons, Olkin, and Sobel (1977).

8.3.1. Conservative Solutions under Indifference-Zone and Subset Formulations

Under the indifference-zone formulation a distance function δ: $\Lambda\times\Lambda\to[0,\infty)$ is adopted such that the following conditions are satisfied: (i) $\delta(\theta,\theta^*)\geqslant 0$ for all $\theta,\theta^*\in\Lambda$, $\delta(\theta,\theta^*)=\delta(\theta^*,\theta)$, and $\delta(\theta,\theta^*)=0$ if and only if $\theta=\theta^*$; and (ii) $\delta(\theta,\theta^*)$ is strictly increasing (decreasing) in θ for fixed θ^* when $\theta>\theta^*$ $(\theta<\theta^*)$ (see Bechhofer, Kiefer, and Sobel (1968, p. 37)). For a preassigned $\delta^*>0$ let Ω^* be the subset of the product parameter space such that

$$\Omega^* = \{\boldsymbol{\theta}=(\theta_1,\ldots,\theta_k)'|\delta(\theta_{[k]},\theta_{[k-1]})\geqslant\delta^*\}. \tag{8.3.3}$$

Then for every $\boldsymbol{\theta}\in\Omega^*$ the parameters $\theta_{[1]},\ldots,\theta_{[k-1]}$ are considered to be significantly distinct from $\theta_{[k]}$. If the natural decision rule "always choose the population with the largest X value" is used, the probability of a correct selection (CS) for every $\boldsymbol{\theta}$ and every n is‡

$$P_{\boldsymbol{\theta}}(\mathrm{CS}) = \int\prod_{i=1}^{q} G_{n,\theta_{[i]}}(x)\,dG_{n,\theta^*}(x), \tag{8.3.4}$$

where $q=k-1$ and θ^* is used to denote $\theta_{[k]}$ for notational convenience. Our statistical problem now is to find the smallest n satisfying $\inf_{\boldsymbol{\theta}\in\Omega^*}P_{\boldsymbol{\theta}}(\mathrm{CS})\geqslant P^*$ for a preassigned $P^*>1/k$. Since $\mathcal{G}$ is

†Tong (1972) showed that if $\mathcal{F}$ is a stochastically increasing family and if T is nondecreasing in each argument, then $\mathcal{G}$ is a stochastically increasing family.

‡For notational convenience we consider here only nonrandomized selection procedures. For procedures with randomization similar results follow with the obvious changes.

a stochastically increasing family, it then follows that

$$\inf_{\boldsymbol{\theta} \in \Omega^*} P_{\boldsymbol{\theta}}(\mathrm{CS}) = \inf_{\theta^* \in \Lambda} \int G^q_{n,\theta_*}(x)\, dG_{n,\theta^*}(x), \tag{8.3.5}$$

where for every fixed $\theta^* \in \Lambda$, θ_* satisfies $\theta_* < \theta^*$ and $\delta(\theta^*, \theta_*) = \delta^*$.

Under the subset formulation of the problem a nonempty subset of the k populations is to be selected, and it is required that the probability of a correct selection (i.e., that the best population with parameter $\theta_{[k]}$ is contained in the subset selected) is at least P^*, which is preassigned. Let us consider the decision rule R_h: Include the ith population in the selected subset if and only if $h(X_i) \geqslant \max_{1 \leqslant j \leqslant k} X_j$. Let the function $h(x) = h_{c,d}(x)$ possess the following properties: (i) $h_{c,d}(x) \geqslant x$, $h_{1,0}(x) = x$, and $h_{c,d}(x)$ is continuous in c,d for all x; (ii) $\lim_{d \to \infty} h_{c,d}(x) = \infty$, c is fixed and/or $\lim_{c \to \infty} h_{c,d}(x) = \infty$, d is fixed, $x \neq 0$; and (iii) for fixed c,d, $h_{c,d}(x)$ is monotonically increasing in x (Gupta and Panchapakesan, 1972). It can be shown that for every fixed $\theta^* \in \Lambda$ and every n the probability of a correct selection satisfies

$$P_{\boldsymbol{\theta}}(\mathrm{CS}) \geqslant \int G^q_{n,\theta^*}(h(x))\, dG_{n,\theta^*}(x). \tag{8.3.6}$$

Hence it is uniformly bounded below by

$$P_{\boldsymbol{\theta}}(\mathrm{CS}) \geqslant \inf_{\theta^* \in \Lambda} \int G^q_{n,\theta^*}(h(x))\, dG_{n,\theta^*}(x). \tag{8.3.7}$$

The preceding discussion illustrates the fact that under both approaches to the problem the probability of a correct selection is a cumulative probability of conditionally independent and identically distributed random variables. If there exists a $\theta'' \in \Lambda$ such that the values on the right-hand sides in (8.3.5) and (8.3.7) are attainable at $\theta^* = \theta''$ (this is the case when Λ is a compact set and the probability function is continuous in $\boldsymbol{\theta}$ or when the probability function is invariant under location or scale translation), then the right-hand sides in (8.3.5) and (8.3.7) may be written, respectively, as

$$\beta_q = \int G^q_{n,\theta_*}(x)\, dG_{n,\theta''}(x) = E_{\theta''} G^q_{n,\theta_*}(X), \tag{8.3.8}$$

$$\gamma_q = \int G^q_{n,\theta''}(h(x))\, dG_{n,\theta''}(x) = E_{\theta''} G^q_{n,\theta''}(h(X)), \tag{8.3.9}$$

where θ_* satisfies $\theta_* < \theta''$ and $\delta(\theta'', \theta_*) = \delta^*$. The problem now is to determine the value of n for given δ^* (under the indifference-zone formulation) and the values of c,d for fixed n (under the subset formulation) so that β_q and γ_q are both at least P^*.

To find an exact solution for this problem one needs to calculate the values of the above integrals, which may be either inconvenient or difficult to achieve to a certain accuracy (because of significant round-off errors). Therefore probability inequalities that yield lower bounds for these integrals are useful in obtaining conservative solutions. Applying Theorem 6.2.11, if a_j and b_j are nonnegative integers satisfying $\mathbf{a} \succ \mathbf{b}$, then (see Tong (1977))

$$\beta_q \geqslant \prod_{j=1}^{r} \int G_{n,\theta^*}^{a_j}(x)\,dG_{n,\theta''}(x)$$

$$\geqslant \prod_{j=1}^{r} \int G_{n,\theta^*}^{b_j}(x)\,dG_{n,\theta''}(x)$$

$$\geqslant \left\{ \int G_{n,\theta^*}(x)\,dG_{n,\theta''}(x) \right\}^q, \tag{8.3.10}$$

$$\gamma_q \geqslant \prod_{j=1}^{r} \int G_{n,\theta''}^{a_j}(h(x))\,dG_{n,\theta''}(x)$$

$$\geqslant \prod_{j=1}^{r} \int G_{n,\theta''}^{b_j}(h(x))\,dG_{n,\theta''}(x)$$

$$\geqslant \left\{ \int G_{n,\theta''}(h(x))\,dG_{n,\theta''}(x) \right\}^q \tag{8.3.11}$$

hold. For fixed k $(k=q+1)$ and P^* let $n^*(k,P^*)$, $(c^*(k,P^*), d^*(k,P^*))$ be the n (c,d) values required to guarantee the probability requirements $\beta_q \geqslant P^*$, $\gamma_q \geqslant P^*$, respectively. We then conclude that

(i) For every $q'' < q$ let n'' be the sample size required to satisfy $\beta_{q''} \geqslant (P^*)^{q''/q}$; then we must have $n^* \leqslant n''$. Hence n'' is an upper bound on n^*. This follows from the fact that

$$\beta_q \geqslant (\beta_{q''})^{q/q''}, \tag{8.3.12}$$

which in turn follows from (8.3.10).

(ii) For some fixed $q'' < q$ let us consider only $q''+1$ populations. Let (c'',d'') be the parameter values needed in the decision rule R_h under the subset formulation such that $\gamma_{q''} \geqslant (P^*)^{q''/q}$. Then (c'',d'') is a conservative solution for (c^*,d^*) in the following sense: When letting $(c^*,d^*)=(c'',d'')$ in the decision problem for $k=q+1$ populations, the probability of correct selection satisfies $\gamma_q \geqslant P^*$.

The above result says that, when the value of k is beyond the capacity of existing statistical tables, we can always use the existing tables to find a conservative solution for the sample size required under the indifference-zone formulation or for the parameter values required in the decision rule under the subset formulation. From Theorem 6.2.11 we also know that the bounds are sharper when the value of q'' is closer to q (see Problem 5). This implies that when using existing tables in ranking and selection problems to find such a conservative solution, one should always go to the table with the highest dimension available.

For selecting the largest mean of k normal populations with a common known variance σ^2 the probability of correct selection is a multivariate normal probability with correlation coefficients $\frac{1}{2}$. The following example illustrates how the bounds for this probability behave as a function of the ratio q/q''.

Example 8.3.1 (Tong, 1970). For $q=8$ let $(X_1,\ldots,X_8)'$ be a multivariate normal variable with means zero, variances one, and correlations $\frac{1}{2}$. Let us denote

$$\beta_q = P\left[\bigcap_{i=1}^{q} \{X_i \leqslant 1.5\}\right], \qquad q = 1,\ldots,8.$$

From existing statistical tables (e.g., Gupta (1963)) and elementary calculation we have

$$\beta_8 = 0.7265, \qquad (\beta_7)^{8/7} = 0.7141, \qquad (\beta_5)^{8/5} = 0.6836,$$
$$(\beta_3)^{8/3} = 0.2971, \qquad (\beta_1)^8 = 0.2511.$$

In the normal case the problem of finding an approximation to, or a bound for, the sample size required has been considered by several authors (e.g., Dudewicz (1969), Dudewicz and Zaino (1971), McDonald (1971), Ramberg (1972), Tong (1970, 1977)). The bound given by Ramberg was obtained via Slepian's inequality, which is simply $\beta_q \geqslant (\beta_1)^q$. The above example illustrates that it is too conservative. Since the multivariate normal probabilities have already been tabulated by Gupta (1963) for $q'' \leqslant 12$, it is clear that, based on that table, the best bound available numerically can be obtained from $(\beta_{12})^{q/12}$ for $q > 12$.

8.3.2. Bounds for the True Probability of Correct Selection

The classical indifference-zone approach certainly has its mathematical elegance, and it has played a predominant role in the area of ranking and selection problems. However it appears that the solution of the problem under this approach is rather "artificial" because of the lack of connection between the value of δ^* and the true parameter $\boldsymbol{\theta}$. Since δ^* is preassigned and the probability is guaranteed to be at least P^* only when $\delta(\theta_{[k]},\theta_{[k-1]})\geqslant\delta^*$, in reality this true distance is unknown. Hence under that approach there is no knowledge regarding the *true* probability of correct selection in a given real-life situation. In view of this difficulty it is natural to consider more realistic approaches as alternatives.

In a technical report Olkin, Sobel, and Tong (1976) considered the problem of estimating the true probability of correct selection by estimating the true configuration involved when the sample size n is fixed. The problem was solved for location and scale parameter families. For location parameter families the distribution of X_i defined in (8.3.2) is of the form $G_{n,\theta_i}(x)=G_n(x-\theta_i)$. Let $X_{(i)}$ be the X variable with distribution function $G_n(x-\theta_{[i]})$. Then the probability of a correct selection is

$$\beta(\boldsymbol{\theta}) = P\left[\bigcap_{i=1}^{q}\{X_{(i)}\leqslant X_{(k)}\}\right] = \int\prod_{i=1}^{q}G_n(x+\delta_i)\,dG_n(x)$$
$$\equiv\beta^*(\boldsymbol{\delta}), \tag{8.3.13}$$

where

$$\delta_i = \theta_{[k]}-\theta_{[i]}, \qquad i=1,\ldots,q, \tag{8.3.14}$$

are the spacings between the best population and the ith population. Note that β depends on $\boldsymbol{\theta}$ only through the spacing vector $\boldsymbol{\delta}=(\delta_1,\ldots,\delta_q)'$.

The behavior of the function $\beta^*(\boldsymbol{\delta})$ was fully studied by Olkin, Sobel, and Tong via majorization. Let $\boldsymbol{\delta}$ and $\boldsymbol{\xi}$ be two spacing vectors such that $\Sigma_{i=1}^q\delta_i=\Sigma_{i=1}^q\xi_i$. They showed that if $G_n(x)$ is log-concave, then

$$\boldsymbol{\xi}\succ\boldsymbol{\delta} \qquad \text{implies} \qquad \beta^*(\boldsymbol{\delta})\geqslant\beta^*(\boldsymbol{\xi}). \tag{8.3.15}$$

Hence bounds for the probability function of a correct selection can be obtained through a partial ordering of the spacing vector. In particular, for fixed $\Sigma_{i=1}^q\delta_i$ the probability is maximized when $\delta_1=\cdots=\delta_q=\bar{\delta}$ (that is, $\theta_{[1]}=\cdots=\theta_{[k-1]}=\theta_{[k]}-\bar{\delta}$).

Now let us consider the estimation aspect of the problem. Let $X_{[1]}\leqslant\cdots\leqslant X_{[k]}$ denote the ordered X values, and suppose that an appropriate estimator of δ_i is

$$\hat{\delta}_i = X_{[k]} - X_{[i]}, \qquad i = 1,\ldots,k. \tag{8.3.16}$$

Then we can estimate the true probability of correct selection $\beta^*(\boldsymbol{\delta})$ by

$$\hat{\beta}^*(\boldsymbol{\delta}) = \beta^*(\hat{\boldsymbol{\delta}}) = \int \prod_{i=1}^{q} G_n(x+\hat{\delta}_i)\, dG_n(x), \tag{8.3.17}$$

where $\hat{\boldsymbol{\delta}}=(\hat{\delta}_1,\ldots,\hat{\delta}_q)'$. This estimator is not always easy to evaluate numerically. Hence bounds for the value of this estimator become useful. It then follows that $\hat{\boldsymbol{\xi}} \succ \hat{\boldsymbol{\delta}}$ a.s. implies $\beta^*(\hat{\boldsymbol{\delta}}) \geqslant \beta^*(\hat{\boldsymbol{\xi}})$ a.s. if $G_n(x)$ is log-concave. Now suppose that for r, $r^* \geqslant 1$

$$B = \{b_1,\ldots,b_r\}, \qquad C = \{c_1,\ldots,c_{r*}\} \tag{8.3.18}$$

are two sets of nonnegative integers such that

$$0 = b_1 < \cdots < b_r = q, \qquad 0 = c_1 < \cdots < c_{r*} = q.$$

Let us define

$$\hat{\boldsymbol{\delta}}(B) = \left(\hat{\delta}_1(B),\ldots,\hat{\delta}_q(B)\right)', \tag{8.3.19}$$

where for all i satisfying $b_j < i \leqslant b_{j+1}$, $\hat{\delta}_i(B)$ is given by

$$\hat{\delta}_i(B) = \left[\sum_{l=b_j+1}^{b_{j+1}} \hat{\delta}_l\right] \Big/ (b_{j+1} - b_j), \tag{8.3.20}$$

which is simply the "cluster" average of the elements $\hat{\delta}_{b_j+1},\ldots,\hat{\delta}_{b_{j+1}}$. Olkin, Sobel, and Tong showed that if $G_n(x)$ is log-concave and if B is a subset of C, then the inequality

$$\beta^*(\hat{\boldsymbol{\delta}}(B)) \geqslant \beta^*(\hat{\boldsymbol{\delta}}(C)) \tag{8.3.21}$$

holds a.s. In particular, by choosing $B=\{0,q\}$ we have

$$\beta^*(\hat{\bar{\boldsymbol{\delta}}}) > \beta^*(\hat{\boldsymbol{\delta}}) \quad \text{a.s.} \tag{8.3.22}$$

for all $\hat{\boldsymbol{\delta}}$, where the vector $\hat{\bar{\boldsymbol{\delta}}}=(\hat{\bar{\delta}},\ldots,\hat{\bar{\delta}})'$ is such that

$$\hat{\bar{\delta}} = \frac{1}{q}\sum_{i=1}^{q} \hat{\delta}_i.$$

The statement in (8.3.21) provides a chain of inequalities, and it can be used to obtain approximations based on existing statistical tables for exchangeable events. For details see Olkin, Sobel, and Tong (1976, Section 4).

For scale parameter families with a distribution $G_{n,\theta}(x)=G_n(x/\theta)$ such that $G_n(0)=0$ we define and use geometric means instead of arithmetic means. Then a similar result follows with the obvious changes.

8.3.3. Other Applications

A few other applications of probability inequalities to ranking and selection problems have appeared in the literature. Those include the following:

(a) An inequality for the probability of correct ordering. Let $X_1,\ldots,X_k$ denote independent normal variables with means θ_i and a common variance σ^2/n. Let $X_{(i)}$ have the mean $\theta_{[i]}$ $(i=1,\ldots,k)$. Then a complete ordering is said to be correct if $X_{(1)} \leqslant X_{(2)} \leqslant \cdots \leqslant X_{(k)}$. This problem was first considered by Bechhofer (1954). In a recent technical report Olkin, Sobel, and Tong (1979) obtained an inequality for this probability as a function of the configuration. After expressing this probability in a multivariate normal probability with means zero, variances one, and correlations

$$\rho_{ij} = \begin{cases} -\frac{1}{2} & \text{for} \quad |i-j|=1, \\ 0 & \text{otherwise,} \end{cases}$$

they showed that for $k=3$ the probability function is Schur-concave in $(\theta_{[3]}-\theta_{[2]},\theta_{[2]}-\theta_{[1]})'$. Hence for fixed $(\theta_{[3]}-\theta_{[1]})$ it is maximized when $\theta_{[2]}$ is the midpoint. For general k certain partial results were also given; but a complete generalization seems impossible because the probability function is not permutation invariant.

Under the configuration $\theta_{[i+1]}-\theta_{[i]}=\delta$ $(i=1,\ldots,k-1)$, the probability of correct ordering was tabulated for selected values of $\sqrt{n}\,\delta/\sigma$ by Bechhofer (1954, Table II) and approximated by Carroll and Gupta (1977, Table Ia). Therefore these tables may be used to obtain an upper bound for the probability of correct ordering over all configurations $\boldsymbol{\theta}$ satisfying $\theta_{[3]}-\theta_{[1]}=2\delta$ for fixed $\sigma/\sqrt{n}$ and $\delta>0$ when $k=3$.

(b) Least favorable configuration for selecting the largest interaction. Bechhofer, Santner, and Turnbull (1977) considered the problem of selecting the combination of treatment and block that yields the largest interaction in a two-way analysis-of-variance problem. The problem was solved under the indifference-zone formulation. In order to find the sample size required to guarantee a given probability of correct selection one needs to obtain the least favorable configuration. Under the assumption of normality this least favorable configuration was found through a probability inequality via majorization when the number of blocks and the number of treatments take certain specific values. The solution for the general case remains an open problem. Their proof depends on an interesting application of an inequality of Prékopa (1971), which says that the Gaussian measure is log-concave (see the second footnote on p. 52).

(c) Slepian's inequality and certain other inequalities for densities with the TP_2 property have frequently been applied for obtaining bounds for the probability of correct selection and for the expected subset size (under the subset formulation). (See, e.g., the works of Gupta (1966), Gupta and Panchapakesan (1969), Gupta and Studden (1970).)

(d) Bonferroni's inequality and Slepian's inequality have been applied to obtain bounds for the probability of correct selection under certain multistage selection procedures when the observations come from normal populations. This includes the works of Dudewicz and Dalal (1975), Tamhane and Bechhofer (1977, 1979), Rinott (1978), and others.

(e) Tong and Wetzell (1979) applied Slepian's inequality to prove a positive result and a negative result for the probability function when selecting the largest normal mean under single-stage selection procedures.

Suppose that there are k normal populations with means θ_i and variances σ_i^2. n_i observations are to be taken from the ith population $(i=1,\ldots,k)$. The question is the following: If one increases n_i and keeps n_j fixed $(j \neq i)$, would this increase the probability of a correct selection? This was actually assumed to be true and used in a paper by Ofosu (1973), and later it was pointed out by Bechhofer (see the correction note of Ofosu (1975)) that it "is not easily proved and indeed is quite possibly false." Bechhofer also indicated that earlier attempts to investigate this problem foundered on this point.

To solve this problem, Tong and Wetzell (1979) showed that

(i) Increasing the sample size from any population other than the best one always increases the probability of a correct selection.

(ii) When $k > 2$ and all the δ_i values are small enough, increasing the sample size from the best population results in a decrease in the probability of a correct selection.

Tong and Wetzell's proofs depend on an application of Slepian's inequality and the fact that the multivariate normal distribution function is continuous. The same negative result in (ii) was also obtained independently by Rinott (1978); his proof is similar and it was given under the least favorable configuration.

8.4. RELIABILITY AND LIFE TESTING

In addition to simultaneous inference and multiple decision problems, probability inequalities have been found useful in the area of reliability theory and life testing. Workers in this area have been discovering new inequalities—most of them are of a general nature—and have been applying them to solve many practical problems. For details the reader is referred to the book by Barlow and Proschan (1975). A particularly useful application is for the following situation: Under certain circumstances it is desirable or necessary to check whether or not the reliability of a system meets a given specification when the reliability of each component is known. The evaluation of the reliability of a complex system is not always feasible, especially when the components are heterogeneous and/or dependent. But in many cases a bound for this system reliability can be obtained, and the bound is usually easier to evaluate. If a lower bound already meets or exceeds the specification, then one knows for sure that the system meets the specification. This is a reason why probability inequalities are useful in reliability theory.

In this section we shall consider only one particular application for the purpose of illustration. A system or structure consisting of k components is called an r-out-of-k system if it functions when and only when at least r of the k components function ($k \geqslant r \geqslant 1$). Such systems are frequently encountered in practice. Let $X_1, \ldots, X_k$ denote the life lengths of the components (which are nonnegative) and $F_1(x_1), \ldots, F_k(x_k)$ their marginal distribution functions. If the components are connected in parallel (in series), then the system is one-out-of-k (k-out-of-k), and the life length of the system equals the maximum (the minimum) of $X_1, \ldots, X_k$.

For $\mathbf{X}=(X_1,\ldots,X_k)'$ and for every fixed $t>0$ if we consider the following random variables:

$$Z_i = Z_i(X_i) = \begin{cases} 1 & \text{for } X_i \geqslant t, \\ 0 & \text{otherwise,} \end{cases} \tag{8.4.1}$$

$$S_k(\mathbf{X}) = \sum_{i=1}^{k} Z_i, \tag{8.4.2}$$

and the structure function

$$\phi(\mathbf{X}) = \begin{cases} 1 & \text{for } S_k(\mathbf{X}) \geqslant r, \\ 0 & \text{otherwise,} \end{cases} \tag{8.4.3}$$

then for any such r-out-of-k system the reliability of the system is simply the expected value of $\phi(\mathbf{X})$, which is nondecreasing in each argument.

8.4.1. Systems of Heterogeneous Components

Let us consider the situation in which the components in an r-out-of-k system are independent but heterogeneous, i.e., their component reliability functions $\bar{F}_i(x_i)=1-F_i(x_i)$ are not necessarily the same. Clearly the reliability of such a system is,

$$E\phi(\mathbf{X}) = 1 - P_{\boldsymbol{\theta}}[S_k \leqslant r-1] \equiv \beta_r(\boldsymbol{\theta}), \tag{8.4.4}$$

where S_k is the sum of k independent Bernoulli variables with parameters

$$\theta_i = \bar{F}_i(t), \qquad i = 1,\ldots,k. \tag{8.4.5}$$

The quantity on the right-hand side of (8.4.4) is tedious to evaluate, especially for large k.

(a) Suppose that we replace the θ_is by their average $\bar{\theta}=\sum_{i=1}^{k}\theta_i/k$; then by Theorem 6.4.1

$$\beta_r(\bar{\boldsymbol{\theta}}) = \sum_{j=r}^{k} \binom{k}{j} \bar{\theta}^j (1-\bar{\theta})^{k-j} \tag{8.4.6}$$

is a lower bound for the system reliability $\beta(\boldsymbol{\theta})$ if $r \leqslant k\bar{\theta}$, and is an upper bound on $\beta_r(\boldsymbol{\theta})$ if $r \geqslant k\bar{\theta}+1$. This bound is a binomial probability; hence it is easy to evaluate.

(b) Suppose that there are two possible sets of component reliabilities $(\theta_1,\ldots,\theta_k)'=\boldsymbol{\theta}$ and $(\xi_1,\ldots,\xi_k)'=\boldsymbol{\xi}$, respectively, such that

$\Sigma_{i=1}^{k}\theta_i=\Sigma_{i=1}^{k}\xi_i$. Then from the concept of majorization a system with component reliabilities $\theta_1,\ldots,\theta_k$ is less diverse if $\boldsymbol{\xi}\succ\boldsymbol{\theta}$. In this case the true system reliability under $\boldsymbol{\theta}$ is closer to the bound stated in (a) for $r\leqslant[k\bar{\theta}-1]$ or $r\geqslant[k\bar{\theta}+3]$, where $[x]$ indicates the largest integer less than or equal to x. This is a consequence of Theorem 6.4.2. In the special case in which the components are homogeneous with a common $\bar{\theta}$ value, of course the bound becomes the true reliability of the system.

(c) A result for comparing system reliabilities of an r-out-of-k system was given by Pledger and Proschan (1971). Let $\theta_1,\ldots,\theta_k$ denote the component reliabilities, and

$$R_i = -\ln\theta_i, \qquad i = 1,\ldots,k, \tag{8.4.7}$$

the corresponding component hazards. They showed that if

$$(-\ln\xi_1,\ldots,-\ln\xi_k)' \succ (-\ln\theta_1,\ldots,-\ln\theta_k)', \tag{8.4.8}$$

then the reliability function satisfies

$$\beta_r(\boldsymbol{\xi}) \geqslant \beta_r(\boldsymbol{\theta}) \qquad \text{for} \quad r \leqslant k-1, \tag{8.4.9}$$

$$\beta_k(\boldsymbol{\xi}) = \beta_k(\boldsymbol{\theta}). \tag{8.4.10}$$

Therefore a comparison of system reliabilities can be made by the partial ordering of the component hazard vector via majorization. This result follows from Theorem 6.4.5.

8.4.2. Systems of Dependent Components

In many real-life situations the life-length variables $X_1,\ldots,X_k$ of the components are associated random variables (for the definition, see Section 5.2). This may happen, for example, when the system involves one or several common units or components, or when the components are subject to the same set of stresses (see Barlow and Proschan (1975, p. 29)). Let us use $h_1(t)$ and $h_k(t)$, respectively, to denote the reliability functions of a one-out-of-k system and a k-out-of-k system. When the components are associated (i.e., their life-length variables are associated), then an application of Theorems

5.2.2 and 5.2.4 yields for all $t>0$

$$h_1(t) = P\left[\bigcup_{i=1}^{k} \{X_i \geqslant t\}\right] = 1 - P\left[\bigcap_{i=1}^{k} \{X_i < t\}\right]$$

$$\leqslant 1 - \prod_{i=1}^{k} P[X_i < t], \tag{8.4.11}$$

$$h_k(t) = P\left[\bigcap_{i=1}^{k} \{X_i \geqslant t\}\right] \geqslant \prod_{i=1}^{k} P[X_i \geqslant t]. \tag{8.4.12}$$

Here the bounds on the right-hand sides of (8.4.11) and (8.4.12) are attainable when the components of the system are actually independent.

A real-life application of such a system was considered by Marshall and Olkin (1967). Suppose that a system involves three units functioning independently. Those three units u_0, u_1, and u_2 have life lengths U_0, U_1, and U_2, respectively. They form two components in the following way: Component one consists of u_0 and u_1 connected in series, and component two consists of u_0 and u_2 connected in series. The system consists of those two components connected in parallel. Such systems are encountered in shock-model problems. Let us define

$$X_i = \min(U_i, U_0), \qquad i = 1,2. \tag{8.4.13}$$

If U_0, U_1, U_2 are independent exponential variables with parameters λ_0, λ_1, and λ_2, respectively, then the marginal distribution of X_i is exponential with parameter $\theta_i = \lambda_0 + \lambda_i$ $(i=1,2)$, and $\mathbf{X} = (X_1, X_2)'$ has a bivariate exponential distribution. By Theorem 5.2.2 the random variables X_1, X_2 are associated (this is intuitively clear because both components involve the same unit u_0). Since the system is one-out-of-two, by (8.4.11) we immediately have

$$h_1(t) \leqslant 1 - \prod_{i=1}^{2} \left[1 - \exp(-(\lambda_0 + \lambda_i)t)\right]$$

$$= \exp(-(\lambda_0+\lambda_1)t) + \exp(-(\lambda_0+\lambda_2)t) - \exp(-(2\lambda_0+\lambda_1+\lambda_2)t) \tag{8.4.14}$$

for all $t>0$.

Probability inequalities through other types of positive dependence of the random variables have also been found useful in reliability theory. In certain cases the components are subject to a common handling, treatment, or environment. If in a given environment the components function independently and if the environmental condition is random, then the life lengths of the components are positively dependent by mixture. Applications of probability inequalities to this type of problem have also appeared in the literature.

PROBLEMS

1. Show that if a $k \times k$ matrix $\boldsymbol{\Sigma} = (\sigma_{ij})$ is positive semidefinite, then for an arbitrary real vector $\boldsymbol{\lambda} = (\lambda_1, \ldots, \lambda_k)'$ the matrix $\mathbf{T} = (\lambda_i \lambda_j \sigma_{ij})$ is also positive semidefinite.
2. Let $X_1, \ldots, X_k$ be independent and identically distributed random variables with density $g_\theta(x)$ that has the TP_2 property. Show that if for $\theta_1 < \theta_0$ the function $h(x) = \ln(g_{\theta_1}(x)/g_{\theta_0}(x))$ is concave in x, then the critical region of the Neyman–Pearson test for testing
$$H_0\colon\ \theta = \theta_0 \qquad \text{versus} \qquad H_1\colon\ \theta = \theta_1$$
is decreasing and Schur-concave.
3. Show that in Problem 2 if $g_\theta(x)$ belongs to the exponential family with the form
$$g_\theta(x) = P(x) Q(\theta) \exp(xR(\theta)),$$
where R is strictly increasing, then the critical region of the Neyman–Pearson test is decreasing and Schur-concave.
4. Let $X_1, \ldots, X_k$ be independent normal variables with means θ_i and variances one. It is known that
   (a) The probability $P_{\boldsymbol{\theta}}[\Sigma_{i=1}^k X_i^2 > a]$ is monotonically increasing in $\Sigma_{i=1}^k \theta_i^2$.
   (b) $|\boldsymbol{\xi}| \succ\succ |\boldsymbol{\theta}|$ implies
$$P_{\boldsymbol{\xi}}\left[\sum_{i=1}^{k} X_i^2 > a\right] \geqslant P_{\boldsymbol{\theta}}\left[\sum_{i=1}^{k} X_i^2 > a\right].$$

   Show that the statement given in (a) is a stronger result by showing that if $|\boldsymbol{\xi}| \succ\succ |\boldsymbol{\theta}|$, then $\Sigma_{i=1}^k \xi_i^2 \geqslant \Sigma_{i=1}^k \theta_i^2$.

5. For fixed k let $X_1,\ldots,X_k$ be conditionally independent and identically distributed random variables, i.e., there exist distribution functions G and H such that the distribution of $\mathbf{X}=(X_1,\ldots,X_k)'$ is of the form

$$F(x_1,\ldots,x_k) = \int \prod_{i=1}^{k} G_u(x_i)\,dH(u).$$

Let A be any Borel-measurable set and denote

$$\beta(r) = \left(P\left[\bigcap_{i=1}^{r}\{X_i \in A\}\right]\right)^{k/r}, \qquad r = 1,\ldots,k.$$

Show that $\beta(r)$ is nondecreasing in r; hence $\beta(r)$ is a lower bound on $\beta(k)$, and the bound becomes sharper when r is closer to k.

6. Let $Y_0, Y_1,\ldots, Y_k$ be independent random variables with distributions $F_{\theta_i}(y)=F(y-\theta_i)$ for $i=0,1,\ldots,k$ (a location parameter family). For comparing the k experimental populations with the control let us define

$$X_i = Y_i - Y_0, \qquad i = 1,\ldots,k.$$

Then $X_1,\ldots,X_k$ are associated random variables; hence

$$P\left[\bigcap_{i=1}^{k}\{X_i \leqslant a_i\}\right] \geqslant \prod_{i=1}^{k} P[X_i \leqslant a_i],$$

$$P\left[\bigcap_{i=1}^{k}\{X_i > a_i\}\right] \geqslant \prod_{i=1}^{k} P[X_i > a_i].$$

7. Show that in Problem 6 if $F_{\theta_i}(y)=F(y/\theta_i)$, $\theta_i>0$, $F(0)=0$, and $X_i=Y_i/Y_0$ for all i, then $X_1,\ldots,X_k$ are again associated, and the same inequalities hold.

8. Show that in (8.3.15) if we replace "$\boldsymbol{\xi} \succ \boldsymbol{\delta}$" by the new condition "$-\boldsymbol{\xi} \succ\succ -\boldsymbol{\delta}$," then the same implication follows.

9. Consider a k-out-of-k system with independent components. For arbitrary but fixed $t>0$, let $\theta_i=\bar{F}_i(t)$ $(i=1,\ldots,k)$ denote the component reliabilities. Show that for all $\theta_1,\ldots,\theta_k$ such that $\Sigma_{i=1}^{k}\theta_i$ is a fixed constant the system reliability is maximized when the θ_is are equal.

10. For $k \geqslant r \geqslant 1$ consider an r-out-of-k system with component reliabilities θ_i $(i=1,\ldots,k)$. For $\beta_r(\boldsymbol{\theta})$ given in (8.4.4) let

$\Delta_1(\boldsymbol{\theta}) = |\beta_r(\boldsymbol{\theta}) - \beta_r(\bar{\boldsymbol{\theta}})|$, where

$$\bar{\theta} = \frac{1}{k}\sum_{i=1}^{k}\theta_i \qquad \text{and} \qquad \bar{\boldsymbol{\theta}} = (\bar{\theta},\ldots,\bar{\theta})'.$$

Show that if $\boldsymbol{\xi} \succ \boldsymbol{\theta}$, then $\Delta_1(\boldsymbol{\xi}) \geqslant \Delta_1(\boldsymbol{\theta})$ holds for $r \leqslant [k\bar{\theta} - 1]$ and $r \geqslant [k\bar{\theta} + 3]$.

11. In Problem 10 define $\bar{\theta}^* = (\prod_{i=1}^{k}\theta_i)^{1/k}$, $\bar{\boldsymbol{\theta}}^* = (\bar{\theta}^*,\ldots,\bar{\theta}^*)'$, and $\Delta_2(\boldsymbol{\theta}) = \beta_r(\boldsymbol{\theta}) - \beta_r(\bar{\boldsymbol{\theta}}^*)$. Show that for every r, $\Delta_2(\boldsymbol{\theta}) \geqslant 0$ always holds and that $\Delta_2(\boldsymbol{\xi}) \geqslant \Delta_2(\boldsymbol{\theta})$ holds when (8.4.8) is satisfied.

REFERENCES

Anderson, T. W. (1958). *An Introduction to Multivariate Statistical Analysis*. Wiley, New York.

Barlow, R. E., and Proschan, F. (1975). *Statistical Theory of Reliability and Life Testing*. Holt, New York.

Bechhofer, R. E. (1954). A single-stage multiple decision procedure for ranking means of normal populations with known variances. *Ann. Math. Statist.* **25**, 16–39.

Bechhofer, R. E., Kiefer, J., and Sobel, M. (1968). *Sequential Identification and Ranking Procedures*. Chicago Univ. Press, Chicago, Illinois.

Bechhofer, R. E., Santner, T. J. and Turnbull, B. W. (1977). Selecting the largest intersection in a two-factor experiment. In *Statistical Decision Theory and Related Topics* (S. S. Gupta and D. S. Moore, eds.), Vol. II, pp. 1–18. Academic Press, New York.

Broemeling, L. D. (1969). Confidence regions for variance ratios of random models. *J. Amer. Statist. Assoc.* **64**, 660–664.

Broemeling, L. D. (1978). Simultaneous inferences for variance ratios of some mixed linear models. *Comm. Statist. A—Theory Methods* **7**, 297–305.

Broemeling, L. D., and Bee, D. E. (1976). Simultaneous confidence intervals for parameters of a balanced incomplete block. *J. Amer. Statist. Assoc.* **71**, 425–428.

Carroll, R. J., and Gupta, S. S. (1977). On the probabilities of rankings of k populations with applications. *J. Statist. Comput. Simulation* **5**, 145–157.

Cox, C. M., Krishnaiah, P. R., Lee, J. C., Reising, J., and Schuurmann, F. J. (1979). A Study on finite intersection tests for multiple comparisons of means. In *Multivariate Analysis* (P. R. Krishnaiah, ed.), Vol. V. North-Holland Publ., Amsterdam (to appear).

Das Gupta, S., Anderson, T. W., and Mudholkar, G. S. (1964). Monotonicity of the power functions of some tests of the multivariate linear hypothesis. *Ann. Math. Statist.* **35**, 200–205.

Dudewicz, E. J. (1969). An approximation to the sample size in selection problems. *Ann. Math. Statist.* **40**, 492–497.

Dudewicz, E. J., and Dalal, S. R. (1975). Allocation of observations in ranking and selection with unequal variances. *Sankhyā Ser. B* **37**, 28–78.

Dudewicz, E. J., and Zaino, N. A., Jr. (1971). Sample size for selection. In *Statistical Decision Theory and Related Topics* (S. S. Gupta and J. Yackel, eds.), pp. 347–362. Academic Press, New York.

Dunn, O. J. (1958). Estimation of the means of dependent variables. *Ann. Math. Statist.* **29**, 1095–1111.

Dunnett, C. W. (1955). A multiple comparison procedure for comparing several treatments with a control. *J. Amer. Statist. Assoc.* **50**, 1096–1121.

Dunnett, C. W. (1964). New tables for multiple comparisons with a control. *Biometrics* **20**, 482–491.

Eaton, M. L., and Perlman, M. D. (1974). A monotonicity property of the power functions of some invariant tests for MANOVA. *Ann. Statist.* **2**, 1022–1028.

Folks, J. L., and Antle, C. E. (1967). Straight line confidence regions for linear models. *J. Amer. Statist. Assoc.* **62**, 1365–1374.

Gafarian, A. V. (1964). Confidence bands in straight line regression. *J. Amer. Statist. Assoc.* **59**, 182–213.

Ghosh, M. N. (1963). Hotelling's generalized T^2 in the multivariate analysis of variance. *J. Roy. Statist. Soc. Ser. B* **25**, 358–367.

Gibbons, J. D., Olkin, I., and Sobel, M. (1977). *Selecting and Ordering Populations: A New Statistical Methodology*. Wiley, New York.

Graybill, F. A., and Bowden, D. C. (1967). Linear segment confidence bands for simple linear models. *J. Amer. Statist. Assoc.* **62**, 403–408.

Gupta, S. S. (1963). Probability integrals of multivariate normal and multivariate t. *Ann. Math. Statist.* **34**, 792–828.

Gupta, S. S. (1966). On some selection and ranking procedures for multivariate normal populations using distance functions. In *Multivariate Analysis* (P. R. Krishnaiah, ed.), pp. 457–475. Academic Press, New York.

Gupta, S. S., and Panchapakesan, S. (1969). Some selection and ranking procedures for multivariate normal populations. In *Multivariate Analysis* (P. R. Krishnaiah, ed.), Vol. II, pp. 475–505. Academic Press, New York.

Gupta, S. S., and Panchapakesan, S. (1972). On a class of subset selection procedures. *Ann. Math. Statist.* **43**, 814–822.

Gupta, S. S., and Panchapakesan, S. (1979). *Multiple Decision Procedures: Theory and Methodology of Selecting and Ranking Populations*. Wiley, New York.

Gupta, S. S., and Studden, W. J. (1970). On some selection and ranking procedures with applications to multivariate normal populations. In *Essays in Probability and Statistics* (R. C. Bose, I. M. Chakravarti, P. C. Mahalanobis, C. R. Rao, and K. J. C. Smith, eds.), pp. 327–338. Univ. of North Carolina Press, Chapel Hill, North Carolina.

Hartley, H. O. (1967). Expectations, variances, and covariances of ANOVA mean squares by "synthesis." *Biometrics* **23**, 105–114.

Hartley, H. O., and Rao, J. N. K. (1967). Maximum likelihood estimation for the mixed analysis of variance model. *Biometrika* **54**, 93–107.

Hoel, P. G. (1951). Confidence regions for linear regression. *Proc. Second Berkeley Symp. Math. Statist. Probab.* (J. Neyman, ed.), pp. 75–81. Univ. of California Press, Berkeley, California.

Jensen, D. R. (1969). An inequality for a class of bivariate chi-square distributions. *J. Amer. Statist. Assoc.* **64**, 333–336.

Jensen, D. R., and Jones, M. Q. (1969). Simultaneous confidence intervals for variances. *J. Amer. Statist. Assoc.* **64**, 324–332.

Khatri, C. G. (1967). On certain inequalities for normal distributions and their applications to simultaneous confidence bounds. *Ann Math. Statist.* **38**, 1853–1867.

Kimball, A. W. (1951). On dependent tests of significance in the analysis of variance. *Ann. Math. Statist.* **22**, 600–602.

Krishnaiah, P. R. (1965). On the simultaneous ANOVA and MANOVA tests. *Ann. Inst. Statist. Math.* **17**, 35–53.

Krishnaiah, P. R. (1969). Simultaneous test procedures under general MANOVA models. In *Multivariate Analysis* (P. R. Krishnaiah, ed.), Vol II, pp. 121–143. Academic Press, New York.

Krishnaiah, P. R. (1979). Some developments on simultaneous test procedures. In *Developments in Statistics* (P. R. Krishnaiah, ed.), Vol. 2, pp. 157–201. Academic Press, New York.

Lehmann, E. L. (1957). A theory of some multiple decision problems, I. *Ann. Math. Statist.* **28**, 1–25; II. *Ann. Math. Statist.* **28**, 547–572.

Lehmann, E. L. (1966). Some concepts of dependence. *Ann. Math. Statist.* **37**, 1137–1153.

Marshall, A. W., and Olkin, I. (1967). A multivariate exponential distribution. *J. Amer. Statist. Assoc.* **62**, 30–44.

Marshall, A. W., and Olkin, I. (1974). Majorization in multivariate distributions. *Ann. Statist.* **2**, 1189–1200.

McDonald, G. C. (1971). On approximating constants required to implement a selection procedure based on ranks. In *Statistical Decision Theory and Related Topics* (S. S. Gupta and J. Yackel, eds.), pp. 299–312. Academic Press, New York.

Miller, R. G., Jr. (1966). *Simultaneous Statistical Inference*. McGraw-Hill, New York.

Ofosu, J. B. (1973). A two-sample procedure for selecting the population with the largest mean from several normal populations with unknown variances. *Biometrika* **60**, 117–124 [Corrigenda (1975). *Biometrika* **62**, 221.]

Olkin, I. (1972). Monotonicity properties of Dirichlet integrals with applications to the multinomial distribution and the analysis of variance. *Biometrika* **59**, 303–307.

Olkin, I., Sobel, M., and Tong, Y. L. (1976). Estimating the true probability of correct selection for location and scale parameter families. Tech. Rep. No. 110, Department of Statistics, Stanford Univ., Stanford, California.

Olkin, I., Sobel, M., and Tong, Y. L. (1979). Bounds for a k-fold integral for location and scale parameter models with applications to statistical ranking and selection problems. Tech. Rep. No. 141, Department of Statistics, Stanford Univ., Stanford, California.

Pledger, G., and Proschan, F. (1971). Comparisons of order statistics and of spacings from heterogeneous distributions. In *Optimizing Methods in Statistics* (J. S. Rustagi, ed.), pp. 89–113. Academic Press, New York.

Prékopa, A. (1971). Logarithmic concave measures with applications. *Acta Sci. Math.* **32**, 301–316.

Ramberg, J. S. (1972). Selection sample size approximations. *Ann. Math. Statist.* **43**, 1977–1980.

Rinott, Y. (1978). On two-stage selection procedures and related probability inequalities. *Comm. Statist. A—Theory Methods* **7**, 799–811.

Roy, S. N. (1953). On a heuristic method of test and its use in multivariate analysis. *Ann. Math. Statist.* **24**, 220–238.

Roy, S. N. (1954). Some further results in simultaneous confidence interval estimation. *Ann. Math. Statist.* **25**, 752–761.

Roy, S. N., and Bose, R. C. (1953). Simultaneous confidence interval estimation. *Ann. Math. Statist.* **24**, 513–536.

Sahai, H. (1974). Simultaneous confidence intervals for variance components in some balanced random effects models. *Sankhyā Ser. B* **36**, 278–287.

Sahai, H., and Anderson, R. L. (1973). Confidence regions for variance ratios of random models for balanced data. *J. Amer. Statist. Assoc.* **68**, 951–952.

Scott, A. (1967). A note on conservative confidence regions for the mean of a multivariate normal. *Ann. Math. Statist.* **38**, 278–280. [Corrigenda (1968). *Ann. Math. Statist.* **39**, 2161.]

Šidák, Z. (1967). Rectangular confidence regions for the means of multivariate normal distributions. *J. Amer. Statist. Assoc.* **62**, 626–633.

Šidák, Z. (1971). On probabilities of rectangles in multivariate Student distributions: Their dependence on correlations. *Ann. Math. Statist.* **42**, 169–175.

Šidák, Z. (1975). A note on C. G. Khatri and A. Scott's papers on multivariate normal distributions. *Ann. Inst. Statist. Math.* **27**, 181–184.

Steel, R. G. D. (1959a). A multiple comparison rank sum test: Treatments versus control. *Biometrics* **15**, 560–572.

Steel, R. G. D. (1959b). A multiple comparison sign test: Treatments versus control. *J. Amer. Statist. Assoc.* **54**, 767–775.

Tamhane, A. C., and Bechhofer, R. E. (1977). A two-stage minimax procedure with screening for selecting the largest normal mean. *Comm. Statist. A—Theory Methods* **6**, 1003–1033.

Tamhane, A. C., and Bechhofer, R. E. (1979). A two-stage minimax procedure with screening for selecting the largest normal mean (II): An improved PCS lower bound and associated tables. *Comm. Statist. A—Theory Methods* **8**, 337–358.

Tong, Y. L. (1970). Some probability inequalities of multivariate normal and multivariate *t*. *J. Amer. Statist. Assoc.* **65**, 1243–1247.

Tong, Y. L. (1972). On the consistency of single-stage ranking procedures. *Ann. Inst. Statist. Math.* **24**, 271–284.

Tong, Y. L. (1977). Applications of a probability inequality to ranking and selection and other related problems. *Comm. Statist. A—Theory Methods* **6**, 1105–1120.

Tong, Y. L. (1979). Counterexamples to a result of Broemeling on simultaneous inferences for variance ratios of some mixed linear models. *Comm. Statist. A—Theory Methods* **8**, 1197–1204.

Tong, Y. L., and Wetzell, D. E. (1979). On the behavior of the probability function for selecting the best normal population. *Biometrika* **66**, 174–176.

Bibliography

This bibliography contains mostly references published during the past two decades, before June 1979. Some of them are relevant to more than one section and have been cross-listed accordingly. The journal abbreviations used here are from *Mathematical Reviews*.

A. BOOKS

Anderson, T. W. (1958). *An Introduction to Multivariate Statistical Analysis*. Wiley, New York.

Barlow, R. E., and Proschan, F. (1975). *Statistical Theory of Reliability and Life Testing*. Holt, New York.

Beckenbach, E. F., and Bellman, R. (1965). *Inequalities*. Springer-Verlag, Berlin and New York.

Hardy, G. H., Littlewood, J. E., and Pólya, G. (1959). *Inequalities*, 2nd ed. Cambridge Univ. Press, London and New York.

Johnson, N. L., and Kotz, S. (1972). *Distributions in Statistics: Continuous Multivariate Distributions*. Wiley, New York.

Karlin, S. (1968). *Total Positivity*, Vol. I. Stanford Univ. Press, Stanford, California.

Karlin, S., and Studden, W. J. (1966). *Tchebycheff Systems with Applications in Analysis and Statistics*. Wiley (Interscience), New York.

Marshall, A. W., and Olkin, I. (1979). *Inequalities: Theory of Majorization and Its Applications*. Academic Press, New York.

Miller, R. G., Jr. (1966). *Simultaneous Statistical Inference*. McGraw-Hill, New York.

Mitrinović, D. S. (1970). *Analytic Inequalities*. Springer-Verlag, Berlin and New York.

Stoyan, D. (1977). *Qualitative Eigenschaften und Abschätzungen Stochastischer Modelle*. Akademie-Verlag, Berlin.

B. INEQUALITIES FOR MULTIVARIATE NORMAL DISTRIBUTION†

Abdel-Hameed, M., and Sampson, A. R. (1978). Positive dependence of the bivariate and trivariate absolute normal, t, χ^2 and F distributions. *Ann. Statist.* **6**, 1360–1368.

Brascamp, H. J., and Lieb, E. H. (1975). Some inequalities for Gaussian measures and the long-range order of the one-dimensional plasma. In *Functional Integration and Its Applications* (A. M. Arthurs, ed.), pp. 1–14. Oxford Univ. Press (Clarendon), London and New York.

Chartres, B. (1963). A geometrical proof of a theorem due to D. Slepian. *SIAM Rev.* **5**, 335–341.

Chover, J. (1962). Certain convexity conditions on matrices with applications to Gaussian processes. *Duke Math. J.* **29**, 141–150.

Das Gupta, S. (1968). Some inequalities for multivariate normal distribution. *Calcutta Statist. Assoc. Bull.* **18**, 179–180.

Das Gupta, S. (1974). Probability inequalities and errors in classification. *Ann. Statist.* **2**, 751–762.

Das Gupta, S. (1976). On a probability inequality for multivariate normal distribution. *Apl. Mat.* **21**, 1–4.

Das Gupta, S., Eaton, M. L., Olkin, I., Perlman, M. D., Savage, L. J., and Sobel, M. (1972). Inequalities on the probability content of convex regions for elliptically contoured distributions. *Proc. Sixth Berkeley Symp. Math. Statist. Probab.*, **2** (L. M. LeCam, J. Neyman, and E. L. Scott, eds.), 241–265. Univ. of California Press, Berkeley, California.

David, F. N. (1953). A note on the evaluation of the multivariate normal integral. *Biometrika* **40**, 458–459.

David, H. T. (1963) The sample mean among the extreme normal order statistics. *Ann. Math. Statist.* **34**, 33–55.

Dunn, O. J. (1958). Estimation of the means of dependent variables. *Ann. Math. Statist.* **29**, 1095–1111.

Dunnett, C. W., and Sobel, M. (1955). Approximations to the probability integral and certain percentage points to a multivariate analogue of Student's t-distribution. *Biometrika* **42**, 258–260.

Dykstra, R. L. (1979). Product inequalities involving the multivariate normal distribution. Tech. Rep. No. 85, Department of Statistics, Univ. of Missouri, Columbia, Missouri.

Eaton, M. L., and Perlman, M. D. (1975). Monotonicity of some multivariate normal probabilities. *Canad. J. Statist.* **3**, 165–173.

Gupta, S. S. (1963). Bibliography on the multivariate normal integrals and related topics. *Ann. Math. Statist.* **34**, 829–838.

Gupta, S. S., Pillai, K. C. S., and Steck, G. P. (1964). On the distribution of linear functions and ratios of linear functions of ordered correlated normal random variables with emphasis on range. *Biometrika* **51**, 143–151.

Hall, R., Kanter, M., and Perlman, M. D. (1979). Inequalities for the probability content of a rotated square and related convolutions. *Ann. Probab.* (to appear).

Jogdeo, K. (1970). A simple proof of an inequality for multivariate normal probabilities of rectangles. *Ann. Math. Statist.* **41**, 1357–1359.

†See also the references listed in Gupta (1963).

Jogdeo, K. (1977). Association and Probability inequalities. *Ann. Statist.* **5**, 495–504.

Khatri, C. G. (1967). On certain inequalities for normal distributions and their applications to simultaneous confidence bounds. *Ann. Math. Statist.* **38**, 1853–1867.

Khatri, C. G. (1970). Further contributions to some inequalities for normal distributions and their applications to simultaneous confidence bounds. *Ann. Inst. Statist. Math.* **22**, 451–458.

Khatri, C. G. (1976). A note on an inequality for a multivariate normal distribution. *Gujarat Statist. Rev.* **3**, 1–12.

Pitt, L. D. (1977). A Gaussian correlation inequality for symmetric convex sets. *Ann. Probab.* **5**, 470–474.

Plackett, R. L. (1954). A reduction formula for normal multivariate integrals. *Biometrika* **41**, 351–360.

Pólya, G. (1949). Remarks on computing the probability integral in one and two dimensions. *Proc. First Berkeley Symp. Math. Statist. Probab.* (J. Neyman, ed.), pp. 63–78. Univ. of California Press, Berkeley, California.

Rinott, Y., and Santner, T. J. (1977). An inequality for multivariate normal probabilities with application to a design problem. *Ann. Statist.* **5**, 1228–1234.

Ruben, H. (1960). Probability content of regions under spherical normal distributions, I. *Ann. Math. Statist.* **31**, 598–618.

Ruben, H. (1960). Probability content of regions under spherical normal distributions, II. The distribution of the range in normal samples. *Ann. Math. Statist.* **31**, 1113–1121.

Ruben, H. (1961). Probability content of regions under spherical normal distributions, III: The bivariate normal integral. *Ann. Math. Statist.* **32**, 171–186.

Savage, I. R. (1962). Mills' ratio for multivariate normal distributions. *J. Res. Nat. Bur. Standards Sect. B* **66B**, 93–96.

Scott, A. (1967). A note on conservative confidence regions for the mean of a multivariate normal. *Ann. Math. Statist.* **38**, 278–280. [Corrigenda (1968). *Ann. Math. Statist.* **39**, 2161.]

Šidák, Z. (1967). Rectangular confidence regions for the means of multivariate normal distributions. *J. Amer. Statist. Assoc.* **62**, 626–633.

Šidák, Z. (1968). On multivariate normal probabilities of rectangles: Their dependence on correlations. *Ann. Math. Statist.* **39**, 1425–1434.

Šidák, Z. (1971). On probabilities of rectangles in multivariate Student distributions: Their dependence on correlations. *Ann. Math. Statist.* **42**, 169–175.

Šidák, Z. (1973). A chain of inequalities for some types of multivariate distributions, with nine special cases. *Apl. Mat.* **18**, 110–118.

Šidák, Z. (1973). On probabilities in certain multivariate distributions: Their dependence on correlations. *Apl. Mat.* **18**, 128–135.

Šidák, Z. (1975). A note on C. G. Khatri and A. Scott's papers on multivariate normal distributions. *Ann. Inst. Statist. Math.* **27**, 181–184.

Slepian, D. (1962). The one-sided barrier problem for Gaussian noise. *Bell System Tech. J.* **41**, 463–501.

Steck, G. P. (1962). Orthant probabilities for the equicorrelated multivariate normal distribution. *Biometrika* **49**, 433–445.

Steck, G. P. (1979). Lower bounds for the multivariate normal Mills' ratio. *Ann. Probab.* **7**, 547–551.

Tong, Y. L. (1970). Some probability inequalities of multivariate normal and multivariate *t*. *J. Amer. Statist. Assoc.* **65**, 1243–1247.

Tong, Y. L. (1977). An ordering theorem for conditionally independent and identically distributed random variables. *Ann. Statist.* **5**, 274–277.

Wynn, H. P. (1975). Integrals for one-sided confidence bounds: A general result. *Biometrika* **62**, 393–396.

C. INEQUALITIES FOR MULTIVARIATE *t*, CHI-SQUARE, *F*, AND OTHER WELL-KNOWN DISTRIBUTIONS

Abdel-Hameed, M., and Sampson, A. R. (1978). Positive dependence of the bivariate and trivariate absolute normal, t, χ^2 and F distributions. *Ann. Statist.* **6**, 1360–1368.

Alam, K. (1970). Monotonicity properties of the multinomial distribution. *Ann. Math. Statist.* **41**, 315–317.

Das Gupta, S., Eaton, M. L., Olkin, I., Perlman, M. D., Savage, L. J., and Sobel, M. (1972). Inequalities on the probability content of convex regions for elliptically contoured distributions. *Proc. Sixth Berkeley Symp. Math. Statist. Probab.* **2** (L. M. LeCam, J. Neyman and E. L. Scott, eds.), 241–265. Univ. of California Press, Berkeley, California.

Dunn, O. J. (1958). Estimation of the means of dependent variables. *Ann. Math. Statist.* **29**, 1095–1111.

Dunn, O. J. (1965). A property of the multivariate *t*-distribution. *Ann. Math. Statist.* **36**, 712–714.

Dunnett, C. W. (1955). A multiple comparison procedure for comparing several treatments with a control. *J. Amer. Statist. Assoc.* **50**, 1096–1121.

Dunnett, C. W., and Sobel, M. (1954). A bivariate generalization of Student's *t*-distribution with tables for special cases. *Biometrika* **41**, 153–169.

Dunnett, C. W., and Sobel, M. (1955). Approximations to the probability integral and certain percentage points to a multivariate analogue of Student's *t*-distribution. *Biometrika* **42**, 258–260.

Gupta, S. S. (1963). Bibliography on the multivariate normal integrals and related topics. *Ann. Math. Statist.* **34**, 829–838.

Gupta, S. S., and Sobel, M. (1962). On the smallest of several correlated *F* statistics. *Biometrika* **49**, 509–523.

Halperin, M. (1967). An inequality on a bivariate Student's "*t*" distribution. *J. Amer. Statist. Assoc.* **62**, 603–606.

Hewett, J. E., and Bulgren, W. G. (1971). Inequalities for some multivariate *F*-distributions with applications. *Technometrics* **13**, 397–402.

Jensen, D. R. (1969). An inequality for a class of bivariate chi-square distributions. *J. Amer. Statist. Assoc.* **64**, 333–336.

Jensen, D. R. (1970). The joint distribution of traces of Wishart matrices and some applications. *Ann. Math. Statist.* **41**, 133–145.

Jogdeo, K. (1977). Association and Probability inequalities. *Ann. Statist.* **5**, 495–504.

Kadiyala, K. R. (1968). An inequality for the ratio of two quadratic forms in normal variate. *Ann. Math. Statist.* **39**, 1762–1763.

Khatri, C. G. (1967). On certain inequalities for normal distributions and their applications to simultaneous confidence bounds. *Ann. Math. Statist.* **38**, 1853–1867.

Khatri, C. G. (1970). Further contributions to some inequalities for normal distributions and their applications to simultaneous confidence bounds. *Ann. Inst. Statist. Math.* **22**, 451–458.

Khatri, C. G. (1974). Review of a paper by Z. Šidák. *Math. Rev.* **47**, 479.

Kimball, A. W. (1951). On dependent tests of significance in the analysis of variance. *Ann. Math. Statist.* **22**, 600–602.

Mallows, C. L. (1968). An inequality involving multinomial probabilities. *Biometrika* **55**, 422–424.

Olkin, I. (1972). Monotonicity properties of Dirichlet integrals with applications to the multinomial distribution and the analysis of variance. *Biometrika* **59**, 303–307.

Scott, A. (1967). A note on conservative confidence regions for the mean of a multivariate normal. *Ann. Math. Statist.* **38**, 278–280. [Corrigenda (1968). *Ann. Math. Statist.* **39**, 2161.]

Šidák, Z. (1971). On probabilities of rectangles in multivariate Student distributions: Their dependence on correlations. *Ann. Math. Statist.* **42**, 169–175.

Šidák, Z. (1973). A chain of inequalities for some types of multivariate distributions, with nine special cases. *Apl. Mat.* **18**, 110–118.

Šidák, Z. (1973). On probabilities in certain multivariate distributions: Their dependence on correlations. *Apl. Mat.* **18**, 128–135.

Šidák, Z. (1975). A note on C. G. Khatri and A. Scott's papers on multivariate normal distributions. *Ann. Inst. Statist. Math.* **27**, 181–184.

Tong, Y. L. (1970). Some probability inequalities of multivariate normal and multivariate *t*. *J. Amer. Statist. Assoc.* **65**, 1243–1247.

Tong, Y. L. (1977). An ordering theorem for conditionally independent and identically distributed random variables. *Ann. Statist.* **5**, 274–277.

D. INTEGRAL INEQUALITIES OVER A SYMMETRIC CONVEX SET

Anderson, T. W. (1955). The integral of a symmetric unimodal function over a symmetric convex set and some probability inequalities. *Proc. Amer. Math. Soc.* **6**, 170–176.

Birnbaum, Z. W. (1948). On random variables with comparable peakedness. *Ann. Math. Statist.* **19**, 76–81.

Borell, C. (1974). Convex measures on locally convex spaces. *Ark. Mat.* **12**, 239–252.

Borell, C. (1975). Convex set functions in *d*-space. *Period. Math. Hungar.* **6**, 111–136.

Brascamp, H. J., and Lieb, E. H. (1974). On extensions of the Brunn–Minkowski and Prékopa–Leindler theorems, including inequalities for log concave functions, and with an application to the diffusion equation. *J. Funct. Anal.* **22**, 366–389.

Brascamp, H. J., and Lieb, E. H. (1975). Some inequalities for Gaussian measures and the long-range order of the one-dimensional plasma. In *Functional Integration and its Applications* (A. M. Arthurs, ed.), pp. 1–14. Oxford Univ. Press (Clarendon), London and New York.

Chung, K. L. (1953). Sur les lois de probabilité unimodales. *C. R. Séances Acad. Sci.* **236**, 583–584.

Das Gupta, S. (1976). A generalization of Anderson's theorem on unimodal distributions. *Proc. Amer. Math. Soc.* **60**, 85–91.

Das Gupta, S. (1976). *S*-Unimodal function: Related inequalities and statistical applications. *Sankhyā Ser. B* **38**, 301–314.

Das Gupta, S. (1978). Brunn–Minkowski inequality and its aftermath. Tech. Rep. No. 310, School of Statistics, Univ. of Minnesota, Minneapolis, Minnesota.

Das Gupta, S., Eaton, M. L., Olkin, I., Perlman, M. D., Savage, L. J., and Sobel, M. (1972). Inequalities on the probability content of convex regions for elliptically contoured distributions. *Proc. Sixth Berkeley Symp. Math. Statist. Probab.* **2** (L. M. LeCam, J. Neyman, and E. L. Scott, eds.), 241–265. Univ. of California Press, Berkeley, California.

Davidovič, Ju. S., Korenbljum, B. I., and Hacet, B. I. (1969). A property of logarithmically concave functions. *Soviet Math. Dokl.* **10**, 477–480.

Dharmadhikari, S. W., and Jogdeo, K. (1976). Multivariate unimodality. *Ann. Statist.* **4**, 607–613.

Eaton, M. L., and Perlman, M. D. (1974). A Monotonicity property of the power functions of some invariant tests for MANOVA. *Ann. Statist.* **5**, 1022–1028.

Fefferman, C., Jodeit, M., Jr., and Perlman, M. D. (1972). A spherical surface measure inequality for convex sets. *Proc. Amer. Math. Soc.* **33**, 114–119.

Gnedenko, B. V., and Kolmogorov, A. N. (1968). *Limit Distributions for Sums of Independent Random Variables* (K. L. Chung, translator), rev. ed. Addison-Wesley, Reading, Massachusetts.

Hall, R., Kanter, M., and Perlman, M. D. (1979). Inequalities for the probability content of a rotated square and related convolutions. *Ann. Probab.* (to appear).

Khintchine, A. Y. (1938). On unimodal distributions. *Izv. Nauchno. Issled. Inst. Mat. Mech. Tomsk. Gos. Univ.* **2**, 1–7.

Mudholkar, G. S. (1966). The integral of an invariant unimodal function over an invariant convex set—an inequality and applications. *Proc. Amer. Math. Soc.* **17**, 1327–1333.

Mudholkar, G. S. (1972). *G*-peakedness comparisons for random vectors. *Ann. Inst. Statist. Math.* **24**, 127–135.

Mudholkar, G. S., and Dalal, S. R. (1977). Some bounds on the distribution functions of linear combinations and applications. *Ann. Inst. Statist. Math.* **29**, *A*, 89–100.

Olkin, I., and Sobel, M. (1970). An inequality on the probability content of parallelepipeds based on circular symmetry. Tech. Rep. No. 43, Department of Statistics, Stanford Univ., Stanford, California.

Prékopa, A. (1971). Logarithmic concave measures with applications. *Acta Sci. Math.* **32**, 301–316.

Prékopa, A. (1973). On logarithmic concave measures and functions. *Acta Sci. Math.* **34**, 335–343.

Rinott, Y. (1976). On convexity of measures. *Ann. Probab.* **4**, 1020–1026.

Sherman, S. (1955). A theorem on convex sets with applications. *Ann. Math. Statist.* **26**, 763–766.

Wells, D. R. (1978). A result in multivariate unimodality. *Ann. Statist.* **6**, 926–931.

Wintner, A. (1938). *Asymptotic Distributions and Infinite Convolutions.* Edward Brothers, Ann Arbor, Michigan.

E. INEQUALITIES VIA DEPENDENCE, ASSOCIATION, AND MIXTURE

Abdel-Hameed, M., and Sampson, A. R. (1978). Positive dependence of the bivariate and trivariate absolute normal, t, χ^2 and F distributions. *Ann. Statist.* **6**, 1360–1368.

Ahmed, A. H. N., León, R., and Proschan, F. (1978). Generalization of associated random variables, with applications. Tech. Rep. No. M468. Department of Statistics, Florida State Univ., Tallahassee, Florida.

Ahmed, A. H. N., Langberg, N. A., León, R. V., and Proschan, F. (1978). Two concepts of positive dependence, with applications in multivariate analysis. Tech. Rep. No. M486, Department of Statistics, Florida State Univ., Tallahassee, Florida.

Ahmed, A. H. N., Langberg, N. A., León, R. V., and Proschan, F. (1979). Partial ordering of positive quadrant dependence, with applications. Tech. Rep. No. M482, Department of Statistics, Florida State Univ., Tallahassee, Florida.

Alam, K., and Wallenius, K. T. (1976). Positive dependence and monotonicity in conditional distributions. *Comm. Statist. A—Theory Methods* **5**, 525–534.

Barlow, R. E., and Proschan, F. (1975). *Statistical Theory of Reliability and Life Testing*. Holt, New York.

Barlow, R. E., Marshall, A. W., and Proschan, F. (1963). Properties of probability distributions with monotone hazard rate. *Ann. Math. Statist.* **34**, 375–389.

Bergmann, R. (1978). Some classes of semi-ordering relations for random vectors and their use for comparing covariances. *Math. Nachr.* **82**, 103–114.

Bergmann, R., and Stoyan, D. (1978). Monotonicity properties of second order characteristics of stochastically monotone Markov chains. *Math. Nachr.* **82**, 99–102.

Bickel, P. J. (1967). Some contributions to the theory of order statistics. *Proc. Fifth Berkeley Symp. Math. Statist. Probab.* **1** (L. M. LeCam, and J. Neyman, eds.), 575–591. Univ. of California Press, Berkeley, California.

Brindley, E. C., Jr., and Thompson, W. A., Jr. (1972). Dependence and aging aspects of multivariate survival. *J. Amer. Statist. Assoc.* **67**, 822–830.

Cambanis, S., Simons, G., and Stout, W. (1976). Inequalities for $Ek(X, Y)$ when the marginals are fixed. *Z. Wahrsch. Verw. Gebiete* **36**, 285–294.

Dykstra, R. L., and Hewett, J. E. (1978). Positive dependence of the roots of a Wishart matrix. *Ann. Statist.* **6**, 235–238.

Dykstra, R. L., Hewett, J. E., and Thompson, W. A., Jr. (1973). Events which are almost independent. *Ann. Statist.* **1**, 674–681.

Eaton, M. L. (1967). Some optimum properties of ranking procedures. *Ann. Math. Statist.* **38**, 124–137.

Edwards, D. A. (1978). On the Holley–Preston inequalities. *Proc. Roy. Soc. Edinburgh Sect. A* **78**, 265–272.

Esary, J. D., and Marshall, A. W. (1974). Multivariate distributions with exponential minimums. *Ann. Statist.* **1**, 84–98.

Esary, J. D., and Proschan, F. (1972). Relationships among some concepts of bivariate dependence. *Ann. Math. Statist.* **43**, 651–655.

Esary, J. D., Proschan, F., and Walkup, D. W. (1967). Association of random variables, with applications. *Ann. Math. Statist.* **38**, 1466–1474.

Farlie, D. J. C. (1960). The performance of some correlation coefficients for a general bivariate distribution. *Biometrika* **47**, 307–323.

Fortuin, C. M., Kastelyn, P. W., and Ginibre, J. (1971). Correlation inequalities on some partially ordered sets. *Comm. Math. Phys.* **22**, 89–103.

Gaver, D. P. (1970). Multivariate gamma distributions generated by mixture. *Sankhyā Ser. A* **32**, 123–126.

Hall, W. J. (1970). On characterizing dependence in joint distributions. In *Essays in Probability and Statistics* (R. C. Bose, I. M. Chakravarti, P. C. Mahalanobis, C. R. Rao, and K. J. C. Smith, eds.), pp. 339–376. Univ. of North Carolina Press, Chapel Hill, North Carolina.

Harris, R. (1970). A multivariate definition for increasing hazard rate distribution functions. *Ann. Math. Statist.* **41**, 713–717.

Hollander, M., Proschan, F., and Sethuraman, J. (1979). Functions decreasing in transposition with applications to shock, damage and down time. Tech. Rep. No. M494, Department of Statistics, Florida State Univ., Tallahassee, Florida.

Holley, R. (1974). Remarks on the FKG inequalities. *Comm. Math. Phys.* **36**, 227–231.

Jensen, D. R. (1971). A note of positive dependence and the structure of bivariate distributions. *SIAM J. Appl. Math.* **20**, 749–753.

Jogdeo, K. (1968). Characterizations of independence in certain families of bivariate and multivariate distributions. *Ann. Math. Statist.* **39**, 433–441.

Jogdeo, K. (1977). Association and probability inequalities. *Ann. Statist.* **5**, 495–504.

Johnson, N. L., and Kotz, S. (1972). *Distributions in Statistics: Continuous Multivariate Distributions*. Wiley, New York.

Karlin, S. (1968). *Total Positivity*, Vol. 1. Stanford Univ. Press, Stanford, California.

Kemperman, J. H. B. (1977). On the FKG-inequality for measures on a partially ordered space. *Indag. Math.* **39**, 313–331.

Kimball, A. W. (1951). On dependent tests of significance in the analysis of variance. *Ann. Math. Statist.* **22**, 600–602.

Kingman, J. F. C. (1978). Uses of exchangeability. *Ann. Probab.* **6**, 183–197.

Konijn, H. S. (1959). Positive and negative dependence of two random variables. *Sankhyā* **21**, 269–280.

Kowalczyk, T., and Pleszczyńska, E. (1977). Monotonic dependence functions of bivariate distributions. *Ann. Statist.* **6**, 1221–1227.

Kruskal, W. H. (1958). Ordinal measures of association. *J. Amer. Statist. Assoc.* **53**, 814–864.

Lancaster, H. O. (1958). The structure of bivariate distributions. *Ann. Math. Statist.* **29**, 719–736.

Lehmann, E. L. (1955). Ordered families of distributions. *Ann. Math. Statist.* **26**, 399–419.

Lehmann, E. L. (1959). *Testing Statistical Hypotheses*. Wiley, New York.

Lehmann, E. L. (1966). Some concepts of dependence. *Ann. Math. Statist.* **37**, 1137–1153.

Mallows, C. L. (1968). An inequality in constrained random variables. *Ann. Math. Statist.* **39**, 1080–1082.

Mardia, K. V. (1970). *Families of Bivariate Distributions*. Charles Griffin, London.

Marshall, A. W. (1975). Multivariate distributions with monotone hazard rate. In *Reliability and Fault Tree Analysis* (R. E. Barlow, J. B. Fussell, and N. D. Singpurwalla, eds.), pp. 259–284. SIAM, Philadelphia, Pennsylvania.

Olshen, R. (1974). A note on exchangeable sequences. *Z. Wahrsch. Verw. Gebiete* **28**, 317–321.

Perlman, M. D. (1979). Unbiasedness of multivariate tests: Recent results. In *Multivariate Analysis* (P. R. Krishnaiah, ed.), Vol. V. North-Holland Publ., Amsterdam (to appear).

Perlman, M. D., and Olkin, I. (1980). Unbiasedness of invariant tests for MANOVA and other multivariate problems. *Ann. Statist.* (to appear).

Plackett, R. L. (1965). A class of bivariate distributions. *J. Amer. Statist. Assoc.* **60**, 516–522.

Preston, C. J. (1974). A generalization of the FKG inequalities. *Comm. Math. Phys.* **36**, 233–241.

Saw, J. G. (1977). On inequalities in constrained random variables. *Comm. Statist. A—Theory Methods* **6**, 1301–1304.

Shaked, M. (1977). A family of concepts of dependence for bivariate distributions. *J. Amer. Statist. Assoc.* **72**, 642–654.

Shaked, M. (1977). A concept of positive dependence for exchangeable random variables. *Ann. Statist.* **5**, 505–515.

Shaked, M. (1979). Some concepts of positive dependence for bivariate exchangeable distributions. *Ann. Inst. Statist. Math. A* **31**, 67–84.

Šidák, Z. (1973). A chain of inequalities for some types of multivariate distributions, with nine special cases. *Apl. Mat.* **18**, 110–118.

Stoyan, D. (1977). Bounds and approximations in queuing through monotonicity and continuity. *Oper. Res.* **25**, 851–863.

Strassen, V. (1965). The existence of probability measures with given marginals. *Ann. Math. Statist.* **36**, 423–439.

Tong, Y. L. (1977). An ordering theorem for conditionally independent and identically distributed random variables. *Ann. Statist.* **5**, 274–277.

Whitt, W. (1976). Bivariate distributions with given marginals. *Ann. Statist.* **4**, 1280–1289.

Wynn, H. P. (1977). An inequality for certain bivariate probability integrals. *Biometrika* **64**, 411–414.

Yanagimoto, T. (1972). Families of positively dependent random variables. *Ann. Inst. Statist. Math.* **24**, 559–573.

F. INEQUALITIES VIA MAJORIZATION AND WEAK MAJORIZATION†

Anderson, T. W., and Samuels, S. M. (1967). Some inequalities among binomial and Poisson probabilities. *Proc. Fifth Berkeley Symp. Math. Statist. Probab.* **1** (L. M. LeCam, and J. Neyman, eds.), 1–12. Univ. of California Press, Berkeley, California.

Barlow, R. E., and Proschan, F. (1966). Inequalities for linear combinations of order statistics from restricted families. *Ann. Math. Statist.* **37**, 1574–1592.

Beckenbach, E. F., and Bellman, R. (1965). *Inequalities*. Springer-Verlag, Berlin and New York.

†See also the references listed in Marshall and Olkin (1979).

Conlon, J. C., León, R., Proschan, F., and Sethuraman, J. (1977). *G*-ordered functions, with applications in statistics, I. Theory. Tech. Rep. No. M432, Department of Statistics, Florida State Univ., Tallahassee, Florida.

Conlon, J. C., León, R., Proschan, F., and Sethuraman, J. (1977). *G*-ordered functions, with applications in statistics, II. Applications. Tech. Rep. No. M433, Department of Statistics, Florida State Univ., Tallahassee, Florida.

Dalal, S. R., and Fortini, P. (1979). An inequality comparing sums and maxima with application to Behrens–Fisher type problem. Tech. Rep., Department of Statistics, Rutgers Univ., New Brunswick, New Jersey.

Eaton, M. L. (1970). A note on symmetric Bernoulli random variables. *Ann. Math. Statist.* **41**, 1223–1226.

Eaton, M. L., and Olshen, R. A. (1972). Random quotients and the Behrens–Fisher problem. *Ann. Math. Statist.* **43**, 1852–1860.

Eaton, M. L., and Perlman, M. D. (1974). A monotonicity property of the power functions of some invariant tests for MANOVA. *Ann. Statist.* **2**, 1022–1028.

Eaton, M. L., and Perlman, M. D. (1977). Reflection groups, generalized Schur functions and the geometry of majorization. *Ann. Probab.* **5**, 829–860.

Gleser, L. J. (1975). On the distribution of the number of successes in independent trials. *Ann. Probab.* **3**, 182–188.

Hardy, G. H., Littlewood, J. E., and Pólya, G. (1959). *Inequalities*, 2nd ed. Cambridge Univ. Press, London and New York.

Hoeffding, W. (1956). On the distribution of the number of successes in independent trials. *Ann. Math. Statist.* **27**, 713–721.

Hollander, M., Proschan, F., and Sethuraman, J. (1977). Functions decreasing in transposition and their applications in ranking problems. *Ann. Statist.* **5**, 722–733.

Jogdeo, K. (1978). On a probability bound of Marshall and Olkin. *Ann. Statist.* **6**, 232–234.

Kanter, M. (1976). Probability inequalities for convex sets and multidimensional concentration functions. *J. Multivariate Anal.* **6**, 222–236.

Karlin, S., and Novikoff, A. (1963). Generalized convex inequalities. *Pacific J. Math.* **13**, 1251–1279.

Khatri, C. G., and Srivastava, M. S. (1975). Some probability inequalities connected with Schur functions. *J. Multivariate Anal.* **5**, 480–486.

Marshall, A. W., and Olkin, I. (1974). Majorization in multivariate distributions. *Ann. Statist.* **2**, 1189–1200.

Marshall, A. W., and Olkin, I. (1979). *Inequalities: Theory of Majorization and Its Applications*. Academic Press, New York.

Marshall, A. W., and Proschan, F. (1965). An inequality for convex functions involving majorization. *J. Math. Anal. Appl.* **12**, 87–90.

Marshall, A. W., Olkin, I., and Proschan, F. (1967). Monotonicity of ratios of means and other applications of majorization. In *Inequalities* (O. Shisha, ed.), Vol. I, pp. 177–203. Academic Press, New York.

Marshall, A. W., Walkup, D. W., and Wets, R. J.-B. (1967). Order-preserving functions; applications to majorization and order statistics. *Pacific J. Math.* **23**, 569–584.

Mudholkar, G. S. (1966). The integral of an invariant unimodal function over an invariant convex set—an inequality and applications. *Proc. Amer. Math. Soc.* **17**, 1327–1333.

Mudholkar, G. S. (1969). A generalized monotone character of d.f.'s and moments of statistics from some well-known populations. *Ann. Inst. Statist. Math.* **21**, 277–285.

Nevius, S. E., Proschan, F., and Sethuraman, J. (1977). Schur functions in statistics. II. Stochastic majorization. *Ann. Statist.* **5**, 263–273.

Nevius, S. E., Proschan, F., and Sethuraman, J. (1977). A stochastic version of weak majorization, with applications. In *Statistical Decision Theory and Related Topics* (S. S. Gupta and D. S. Moore, eds.), Vol. II, pp. 281–296. Academic Press, New York.

Olkin, I. (1972). Monotonicity properties of Dirichlet integrals with applications to the multinomial distribution and the analysis of variance. *Biometrika* **59**, 303–307.

Olkin, I., and Sobel, M. (1965). Integral expressions for tail probabilities of the multinomial and the negative multinomial distributions. *Biometrika* **52**, 167–179.

Olkin, I., Sobel, M., and Tong, Y. L. (1976). Estimating the true probability of correct selection for location and scale parameter families. Tech. Rep. No. 110, Department of Statistics, Stanford Univ., Stanford, California.

Olkin, I., Sobel, M., and Tong, Y. L. (1979). Bounds for a k-fold integral for location and scale parameter models with applications to statistical ranking and selection problems. Tech. Rep. No. 141, Department of Statistics, Stanford Univ., Stanford, California.

Pledger, G., and Proschan, F. (1971). Comparisons of order statistics and of spacings from heterogeneous distributions. In *Optimizing Methods in Statistics* (J. S. Rustagi, ed.), pp. 89–113. Academic Press, New York.

Pledger, G., and Proschan, F. (1973). Stochastic comparisons of random processes, with applications in reliability. *J. Appl. Probab.* **10**, 572–585.

Proschan, F. (1965). Peakedness of distributions of convex combinations. *Ann. Math. Statist.* **36**, 1703–1706.

Proschan, F. (1975). Applications of majorization and Schur functions in reliability and life testing. In *Reliability and Fault Tree Analysis* (R. E. Barlow, J. B. Fussell, and N. D. Singpurwalla, eds.), pp. 237–258. SIAM, Philadelphia, Pennsylvania.

Proschan, F., and Sethuraman, J. (1976). Stochastic comparisons of order statistics from heterogeneous populations, with applications in reliability. *J. Multivariate Anal.* **6**, 608–616.

Proschan, F., and Sethuraman, J. (1977). Schur functions in Statistics. I. The preservation theorem. *Ann. Statist.* **5**, 256–262.

Rinott, Y. (1973). Multivariate majorization and rearrangement inequalities with some applications to probability and statistics. *Israel J. Math.* **15**, 60–67.

Rinott, Y., and Santner, T. J. (1977). An inequality for multivariate normal probabilities with application to a design problem. *Ann. Statist.* **5**, 1228–1234.

Samuels, S. M. (1965). On the number of successes of independent trials. *Ann. Math. Statist.* **36**, 1272–1278.

Sen, P. K. (1970). A note on order statistics for heterogeneous distributions. *Ann. Math. Statist.* **41**, 2137–2139.

Shaked, M. (1977). A concept of positive dependence for exchangeable random variables. *Ann. Statist.* **5**, 505–515.

Tong, Y. L. (1977). An ordering theorem for conditionally independent and identically distributed random variables. *Ann. Statist.* **5**, 274–277.

Wong, C. K., and Yue, P. C. (1973). A majorization theorem for the number of distinct outcomes in N independent trials. *Discrete Math.* **6**, 391–398.

G. DISTRIBUTION-FREE INEQUALITIES†

Berge, P. O. (1937). A note on a form of Tchebycheff's theorem for two variables. *Biometrika* **29**, 405–406.

Birnbaum, Z. W., and Marshall, A. W. (1961). Some multivariate Chebyshev inequalities with extensions to continuous parameter processes. *Ann. Math. Statist.* **32**, 687–703.

Birnbaum, Z. W., Raymond, J., and Zuckerman, H. S. (1947). A generalization of Tshebyshev's inequality to two dimensions. *Ann. Math. Statist.* **18**, 70–79.

Camp, B. H. (1948). Generalization to *N* dimensions of inequalities of the Tchebycheff-type. *Ann. Math. Statist.* **19**, 568–574.

Chung, K. L., and Erdős, P. (1952). On the application of the Borel–Cantelli lemma. *Trans. Amer. Math. Soc.* **72**, 179–186.

David, H. A. (1970). *Order Statistics*. Wiley, New York.

Dawson, D. A., and Sankoff, D. (1967). An inequality for probabilities. *Proc. Amer. Math. Soc.* **18**, 504–507.

Fréchet, M. (1940, 1943). Les probabilités associées à un système d'événements compatibles et dépendents, I, II. *Actualités Scientifiques et Industrielles*, No. 859 and 942. Hermann, Paris.

Galambos, J. (1975). Methods for proving Bonferroni type inequalities. *J. London Math. Soc.* (2) **9**, 561–564.

Gallot, S. (1966). A bound for the maximum of a number of random variables. *J. Appl. Probab.* **3**, 556–558.

Godwin, H. J. (1955). On generalizations of Tchebycheff's inequality. *J. Amer. Statist. Assoc.* **50**, 923–945.

Godwin, H. J. (1964). *Inequalities on Distribution Functions*. Charles Griffin, London.

Hájek, J., and Rényi, A. (1955). Generalization of an inequality of Kolmogorov. *Acta Math. Acad. Sci. Hungar.* **6**, 281–283.

Hoeffding, W. (1963). Probability inequalities for sums of bounded random variables. *J. Amer. Statist. Assoc.* **58**, 13–20.

Hunter, D. (1976). An upper bound for the probability of a union. *J. Appl. Probab.* **13**, 597–603.

Karlin, S. (1974). Inequalities for symmetric sampling plans I. *Ann. Statist.* **2**, 1065–1094.

Karlin, S., and Studden, W. J. (1966). *Tchebycheff Systems with Applications in Analysis and Statistics*. Wiley (Interscience), New York.

Kounias, E. G. (1968). Bounds for the probability of a union, with applications. *Ann. Math. Statist.* **39**, 2154–2158.

Kounias, S., and Martin, J. (1976). Best linear Bonferroni bounds. *SIAM J. Appl. Math.* **30**, 307–323.

Lal, D. N. (1955). A note on a form of Tchebycheff's inequality for two or more variables. *Sankhyā* **15**, 317–320.

Leser, C. E. V. (1942). Inequalities for multivariate frequency distributions. *Biometrika* **32**, 284–293.

Marshall, A. W. (1960). A one-sided analog of Kolmogoroff's inequality. *Ann. Math. Statist.* **31**, 483–487.

†See also the references listed in Godwin (1955) and Savage (1961).

Marshall, A. W., and Olkin, I. (1960). A one-sided inequality of the Chebyshev type. *Ann. Math. Statist.* **31**, 488–491.

Marshall, A. W., and Olkin, I. (1960). Multivariate Chebyshev inequalities. *Ann. Math. Statist.* **31**, 1001–1014.

Marshall, A. W., and Olkin, I. (1960). A bivariate Chebyshev inequality for symmetric convex polygons. In *Contributions to Probability and Statistics* (I. Olkin, S. G. Ghurye, W. Hoeffding, W. G. Madow, and H. B. Mann, eds.), pp. 299–308. Stanford Univ. Press, Stanford, California.

Marshall, A. W., and Olkin, I. (1961). Game theoretic proof that Chebyshev inequalities are sharp. *Pacific J. Math.* **11**, 1421–1429.

Meyer, R. M. (1969). Note on a "multivariate" form of Bonferroni's inequalities. *Ann. Math. Statist.* **40**, 692–693.

Mudholkar, G. S. (1970). Some Tchebycheff type inequalities for matrix valued random variables. In *Essays in Probability and Statistics* (R. C. Bose, I. M. Chakravarti, P. C. Mahalanobis, C. R. Rao, and K. J. C. Smith, eds.), pp. 489–494. Univ. of North Carolina Press, Chapel Hill, North Carolina.

Mudholkar, G. S., and Rao, P. S. R. S. (1967). Some sharp multivariate Tchebycheff inequalities. *Ann. Math. Statist.* **38**, 393–400.

Olkin, I., and Pratt, J. W. (1958). A multivariate Tchebycheff inequality. *Ann. Math. Statist.* **29**, 226–234.

Rényi, A. (1962). *Wahrscheinlichkeitsrechnung.* Veb Deutscher Verlag der Wissenschaften, Berlin.

Savage, I. R. (1961). Probability inequalities of the Tchebycheff-type. *J. Res. Nat. Bur. Standards Ser. B* **65**, 211–222.

Sen, P. K. (1971). A Hájek–Rényi type inequality for stochastic vectors with applications to simultaneous confidence regions. *Ann. Math. Statist.* **42**, 1132–1134.

Serfling, R. J. (1974). Probability inequalities for the sum in sampling without replacement. *Ann. Statist.* **2**, 39–48.

Sobel, M., and Uppuluri, V. R. R. (1972). On Bonferroni-type inequalities of the same degree for the probability of unions and intersections. *Ann. Math. Statist.* **43**, 1549–1558.

Whittle, P. (1958). A multivariate generalization of Tchebychev's inequality. *Quart. J. Math. Oxford Ser.* (2) **9**, 232–240.

Whittle, P. (1959). Sur la distribution du maximum d'un polynôme trigonométrique à coefficients aléatoires. *Le Calcul des Probabilités et ses Applications*, p. 173. CNRS, Paris.

H. APPLICATIONS

a. Simultaneous Confidence Regions

Broemeling, L. D. (1969). Confidence regions for variance ratios of random models. *J. Amer. Statist. Assoc.* **64**, 660–664.

Broemeling, L. D. (1978). Simultaneous inferences for variance ratios of some mixed linear models. *Comm. Statist. A—Theory Methods* **7**, 297–305.

Broemeling, L. D., and Bee, D. E. (1976). Simultaneous confidence intervals for parameters of a balanced incomplete block. *J. Amer. Statist. Assoc.* **71**, 425–428.

Chew, V. (1968). Simultaneous prediction intervals. *Technometrics* **10**, 323–331.

Dunn, O. J. (1958). Estimation of the means of dependent variables. *Ann. Math. Statist.* **29**, 1095–1111.

Dunn, O. J. (1959). Confidence interva s for the means of dependent, normally distributed variables. *J. Amer. Statist. Assoc.* **54**, 613–621.

Dunn, O. J. (1959). Estimation of the median for dependent variables. *Ann. Math. Statist.* **30**, 192–197.

Dykstra, R. L. (1979). Product inequalities involving the multivariate normal distribution. Tech. Rep. No. 85, Department of Statistics, Univ. of Missouri, Columbia, Missouri.

Folks, J. L., and Antle, C. E. (1967). Straight line confidence regions for linear models. *J. Amer. Statist. Assoc.* **62**, 1365–1374.

Gabriel, K. R. (1969). Simultaneous test procedures—some theory of multiple comparisons. *Ann. Math. Statist.* **40**, 224–250.

Gafarian, A. V. (1964). Confidence bands in straight line regression. *J. Amer. Statist. Assoc.* **59**, 182–213.

Graybill, F. A., and Bowden, D. C. (1967). Linear segment confidence bands for simple linear models. *J. Amer. Statist. Assoc.* **62**, 403–408.

Hahn, G. J. (1972). Simultaneous prediction intervals to contain the variability of future samples from a normal distribution. *J. Amer. Statist. Assoc.* **68**, 938–942.

Hahn, G. J. (1975). A simultaneous prediction limit on the means of future samples from an exponential distribution. *Technometrics* **17**, 341–345.

Hewett, J. E., and Moeschberger, M. L. (1976). Some approximate simultaneous prediction intervals for reliability analysis. *Technometrics* **18**, 227–229.

Hoel, P. G. (1951). Confidence regions for linear regression. *Proc. Second Berkeley Symp. Math. Statist. Probab.* (J. Neyman, ed.), pp. 75–81. Univ. of California Press, Berkeley, California.

Jensen, D. R., and Jones, M. Q. (1969). Simultaneous confidence intervals for variances. *J. Amer. Statist. Assoc.* **64**, 324–332.

Kadiyala, K. R. (1968). An inequality for the ratio of two quadratic forms in normal variates. *Ann. Math. Statist.* **39**, 1762–1763.

Khatri, C. G. (1965). A note on the confidence bounds for the characteristic roots of dispersion matrices of normal variates. *Ann. Inst. Statist. Math.* **17**, 175–183.

Khatri, C. G. (1967). On certain inequalities for normal distributions and their applications to simultaneous confidence bounds. *Ann. Math. Statist.* **38**, 1853–1867.

Khatri, C. G. (1970). Further contributions to some inequalities for normal distributions and their applications to simultaneous confidence bounds. *Ann. Inst. Statist. Math.* **22**, 451–458.

Low, L. (1970). An application of majorization to comparison of variances. *Technometrics* **12**, 141–145.

Miller, R. G., Jr. (1966). *Simultaneous Statistical Inference*. McGraw-Hill, New York.

Mudholkar, G. S. (1966). On confidence bounds associated with multivariate analysis of variance and nonindependence between two sets of variates. *Ann. Math. Statist.* **37**, 1736–1746.

Roy, S. N. (1954). Some further results in simultaneous confidence interval estimation. *Ann. Math. Statist.* **25**, 752–761.

Roy, S. N., and Bargmann, R. E. (1958). Test of multivariate independence and the associated confidence bounds. *Ann. Math. Statist.* **29**, 491–503.

Roy, S. N., and Bose, R. C. (1953). Simultaneous confidence interval estimation. *Ann. Math. Statist.* **24**, 513–536.

Roy, S. N., and Gnanadesikan, R. (1958). A note on "further contributions to multivariate confidence bounds." *Biometrika* **45**, 581.

Roy, S. N., and Potthoff, R. F. (1958). Confidence bounds on vector analogues of the "ratios of means" and the "ratios of variances" for two correlated normal variates and some associated tests. *Ann. Math. Statist.* **27**, 829–841.

Sahai, H. (1974). Simultaneous confidence intervals for variance components in some balanced random effects models. *Sankhyā Ser. B* **36**, 278–287.

Sahai, H., and Anderson, R. L. (1973). Confidence regions for variance ratios of random models for balanced data. *J. Amer. Statist. Assoc.* **68**, 951–952.

Scott, A. (1967). A note on conservative confidence regions for the mean of a multivariate normal. *Ann. Math. Statist.* **38**, 278–280. [Corrigenda (1968). *Ann. Math. Statist.* **39**, 2161.]

Sen, P. K. (1971). A Hájek–Rényi type inequality for stochastic vectors with applications to simultaneous confidence regions. *Ann. Math. Statist.* **42**, 1132–1134.

Šidák, Z. (1967). Rectangular confidence regions for the means of multivariate normal distributions. *J. Amer. Statist. Assoc.* **62**, 626–633.

Šidák, Z. (1968). On multivariate normal probabilities of rectangles: Their dependence on correlations. *Ann. Math. Statist.* **39**, 1425–1434.

Šidák, Z. (1971). On probabilities of rectangles in multivariate Student distributions: Their dependence on correlations. *Ann. Math. Statist.* **42**, 169–175.

Tong, Y. L. (1979). Counterexamples to a result of Broemeling on simultaneous inferences for variance ratios of some mixed linear models. *Comm. Statist. A—Theory Methods* **8**, 1197–1204.

Wynn, H. P., and Bohrer, R. (1978). Inequalities for simultaneous confidence levels. Tech. Rep., Department of Mathematics, Univ. of Illinois, Urbana, Illinois.

b. Hypothesis-Testing and Simultaneous Comparisons

Alt, F., and Spruill, C. (1977). A comparison of confidence intervals generated by the Scheffé and Bonferroni methods. *Comm. Statist. A—Theory Methods* **6**, 1503–1510.

Anderson, T. W., and Das Gupta, S. (1964). Monotonicity of the power functions of some tests of independence between two sets of variates. *Ann. Math. Statist.* **35**, 206–208.

Bandyopadhyay, S. (1977). Probability inequalities involving estimates of probability of correct classification using dependent samples. *Sankhyā Ser. B* **39**, 145–150.

Bechhofer, R. E. (1968). Multiple comparisons with a control for multiple-classified variances of normal populations. *Technometrics* **10**, 715–718.

Bechhofer, R. E. (1969). Optimal allocation of observations when comparing several treatments with a control. In *Multivariate Analysis* (P. R. Krishnaiah, ed.), Vol. II, pp. 463–473. Academic Press, New York.

Bechhofer, R. E., and Nocturne, D. J.-M. (1972). Optimal allocation of observations when comparing several treatments with a control (II): 2-sided comparisons. *Technometrics* **14**, 423–436.

Bechhofer, R. E., and Turnbull, B. W. (1971). Optimal allocation of observations when comparing several treatments with a control (III): globally best one-sided intervals for unequal variances. In *Statistical Decision Theory and Related Topics* (S. S. Gupta and J. Yackel, eds.), pp. 41–78. Academic Press, New York.

Bradu, D., and Gabriel, K. R. (1974). Simultaneous statistical inference on interactions in two-way analysis of variance. *J. Amer. Statist. Assoc.* **69**, 428–436.

Cohen, A., and Strawderman, W. E. (1971). Unbiasedness of tests for homogeneity of variances. *Ann. Math. Statist.* **42**, 355–360.

Cox, C. M., Krishnaiah, P. R., Lee, J. C., Reising, J., and Schuurmann, F. J. (1979). A study on finite intersection tests for multiple comparisons of means. In *Multivariate Analysis* (P. R. Krishnaiah, ed.), Vol. V. North-Holland Publ., Amsterdam (to appear).

Das Gupta, S. (1969). Properties of power functions of some tests concerning dispersion matrices of multivariate normal distributions. *Ann. Math. Statist.* **40**, 697–701.

Das Gupta, S. (1974). Probability inequalities and errors in classification. *Ann. Statist.* **2**, 751–762.

Das Gupta, S., Anderson, T. W., and Mudholkar, G. S. (1964). Monotonicity of the power functions of some tests of the multivariate linear hypothesis. *Ann. Math. Statist.* **35**, 200–205.

Dunn, O. J., and Massey, F. J. (1965). Estimation of multiple contrasts using *t*-distributions. *J. Amer. Statist. Assoc.* **60**, 573–583.

Dunnett, C. W. (1955). A multiple comparison procedure for comparing several treatments with a control. *J. Amer. Statist. Assoc.* **50**, 1096–1121.

Dunnett, C. W. (1964). New tables for multiple comparisons with a control. *Biometrics* **20**, 482–491.

Dykstra, R. L. (1979). On dependent tests of significance in the multivariate analysis of variance. *Ann. Statist.* **7**, 459–461.

Eaton, M. L., and Perlman, M. D. (1974). A monotonicity property of the power functions of some invariant tests for MANOVA. *Ann. Statist.* **2**, 1022–1028.

Fujikoshi, Y. (1973). Monotonicity of the power functions of some tests in general MANOVA models. *Ann. Statist.* **1**, 388–391.

Gabriel, K. R. (1969). Simultaneous test procedures—some theory of multiple comparisons. *Ann. Math. Statist.* **40**, 224–250.

Gabriel, K. R., and Sen, P. K. (1968). Simultaneous test procedures for one-way ANOVA and MANOVA based on rank scores. *Sankhyā Ser. A* **30**, 303–312.

Ghosh, M. N. (1955). Simultaneous tests of linear hypothesis. *Biometrika* **42**, 441–449.

Ghosh, M. N. (1963). Hotelling's generalized T^2 in the multivariate analysis of variance. *J. Roy. Statist. Soc. Ser. B* **25**, 358–367.

Gupta, S. S., Nagel, K., and Panchapakesan, S. (1973). On the order statistics from equally correlated normal random variables. *Biometrika* **60**, 403–413.

Hochberg, Y., and Marcus, R. (1978). On partitioning successive increments in means or ratios of variances in a chain of normal populations. *Comm. Statist. A—Theory Methods* **7**, 1501–1513.

Jensen, D. R. (1976). The comparison of several response functions with a standard. *Biometrics* **32**, 51–59.

Kimball, A. W. (1951). On dependent tests of significance in the analysis of variance. *Ann. Math. Statist.* **22**, 600–602.

Krishnaiah, P. R. (1965). On the simultaneous ANOVA and MANOVA tests. *Ann. Inst. Statist. Math.* **17**, 35–53.

Krishnaiah, P. R. (1968). Simultaneous tests for the equality of covariance matrices against certain alternatives. *Ann. Math. Statist.* **39**, 1303–1309.

Krishnaiah, P. R. (1969). Simultaneous test procedures under general MANOVA models. In *Multivariate Analysis* (P. R. Krishnaiah, ed.), Vol. II, pp. 121–143. Academic Press, New York.

Krishnaiah, P. R. (1979). Some developments on simultaneous test procedures. In *Developments in Statistics* (P. R. Krishnaiah, ed.), Vol. 2, pp. 157–201. Academic Press, New York.

Lehmann, E. L. (1964). Asymptotically nonparametric inference in some linear models with one observation per cell. *Ann. Math. Statist.* **35**, 726–734.

Lehmann, E. L. (1966). Some concepts of dependence. *Ann. Math. Statist.* **37**, 1137–1153.

Marshall, A. W., and Olkin, I. (1974). Majorization in multivariate distributions. *Ann. Statist.* **2**, 1189–1200.

Miller, R. G., Jr. (1966). *Simultaneous Statistical Inference*. McGraw-Hill, New York.

Miller, R. G., Jr. (1977). Developments in multiple comparisons 1966–1976. *J. Amer. Statist. Assoc.* **72**, 779–788.

Mudholkar, G. S. (1965). A class of tests with monotone power functions for two problems in multivariate statistical analysis. *Ann. Math. Statist.* **36**, 1794–1801.

Mudholkar, G. S., Davidson, M. L., and Subbaiah, P. (1974). Extended linear hypotheses and simultaneous tests in multivariate analysis of variance. *Biometrika* **61**, 467–477.

Olkin, I. (1972). Monotonicity properties of Dirichlet integrals with applications to the multinomial distribution and the analysis of variance. *Biometrika* **59**, 303–307.

O'Neil, R., and Wetherill, G. B. (1971). The present state of multiple comparison methods (with discussion). *J. Roy. Statist. Soc. Ser. B* **33**, 218–250.

Paulson, E. (1952). On the comparison of several experimental categories with a control. *Ann. Math. Statist.* **23**, 239–246.

Perlman, M. D. (1972). Monotonicity of the power function of Pillai's trace test. Tech. Rep. No. 194, School of Statistics, Univ. of Minnesota, Minneapolis, Minnesota.

Perlman, M. D. (1974). On the monotonicity of the power functions of tests based on traces of multivariate beta matrices. *J. Multivariate Anal.* **4**, 22–30.

Perlman, M. D. (1979). Unbiasedness of multivariate tests: Recent results. In *Multivariate Analysis* (P. R. Krishnaiah, ed.), Vol. V. North-Holland Publ., Amsterdam (to appear).

Perlman, M. D. (1980). Unbiasedness of the likelihood ratio tests for equality of several covariance matrices and equality of several multivariate normal populations. *Ann. Statist.* (to appear).

Perlman, M. D., and Olkin, I. (1978). Unbiasedness of invariant tests for MANOVA and other multivariate problems. Tech Rep. No. 70, Department of Statistics, Univ. of Chicago, Chicago, Illinois.

Ramachandran, K. V. (1956). On the simultaneous analysis of variance test. *Ann. Math. Statist.* **27**, 521–528.

Roy, S. N. (1953). On a heuristic method of test and its use in multivariate analysis. *Ann. Math. Statist.* **24**, 220–238.
Roy, S. N., and Bargmann, R. E. (1958). Test of multivariate independence and the associated confidence bounds. *Ann. Math. Statist.* **29**, 491–503.
Schafer, W. D., and Macready, G. B. (1975). A modification of the Bonferroni procedure on contrasts which are grouped into internally independent sets. *Biometrics* **31**, 227–228.
Shaffer, J. P. (1977). Multiple comparisons emphasizing selected contrasts: An extension and generalization of Dunnett's procedure. *Biometrics* **33**, 293–304.
Sobel, M., and Tong, Y. L. (1971). Optimal allocation of observations for partitioning a set of normal populations in comparison with a control. *Biometrika* **58**, 177–181.
Spjøtvoll, E. (1972). Multiple comparison of regression functions. *Ann. Math. Statist.* **43**, 1076–1088.
Steel, R. G. D. (1959). A multiple comparison rank sum test: Treatments versus control. *Biometrics* **15**, 560–572.
Steel, R. G. D. (1959). A multiple comparison sign test: Treatments versus control. *J. Amer. Statist. Assoc.* **54**, 767–775.
Sugiura, N., and Nagao, N. (1968). Unbiasedness of some test criteria for the equality of one or two covariance matrices. *Ann. Math. Statist.* **39**, 1686–1692.
Tong, Y. L. (1969). On partitioning a set of normal populations by their locations with respect to a control. *Ann. Math. Statist.* **40**, 1300–1324.
Tong, Y. L. (1970). Some probability inequalities of multivariate normal and multivariate *t*. *J. Amer. Statist. Assoc.* **65**, 1243–1247.

c. Ranking and Selection Problems

Bechhofer, R. E. (1954). A single-stage multiple decision procedure for ranking means of normal populations with known variances. *Ann. Math. Statist.* **25**, 16–39.
Bechhofer, R. E., Santner, T. J., and Turnbull, B. W. (1977). Selecting the largest intersection in a two-factor experiment. In *Statistical Decision Theory and Related Topics* (S. S. Gupta and D. S. Moore, eds.), Vol. II, pp. 1–18. Academic Press, New York.
Carroll, R. J., and Gupta, S. S. (1977). On the probabilities of rankings of *k* populations with applications. *J. Statist. Comput. Simulation* **5**, 145–157.
Dudewicz, E. J. (1969). An approximation to the sample size in selection problems. *Ann. Math. Statist.* **40**, 492–497.
Dudewicz, E. J., and Dalal, S. R. (1975). Allocation of observations in ranking and selection with unequal variances. *Sankhyā Ser. B* **37**, 28–78.
Dudewicz, E. J., and Zaino, N. A., Jr. (1971). Sample size for selection. In *Statistical Decision Theory and Related Topics* (S. S. Gupta and J. Yackel, eds.), pp. 347–362. Academic Press, New York.
Gibbons, J. D., Olkin, I., and Sobel, M. (1977). *Selecting and Ordering Populations: A New Statistical Methodology*. Wiley, New York.
Gupta, S. S. (1963). Probability integrals of multivariate normal and multivariate *t*. *Ann. Math. Statist.* **34**, 792–828.
Gupta, S. S. (1966). On some selection and ranking procedures for multivariate normal

populations using distance functions. In *Multivariate Analysis* (P. R. Krishnaiah, ed.), pp. 457–475. Academic Press, New York.

Gupta, S. S., and Huang, D. Y. (1976). Subset Selection procedures for the means and variances of normal populations: Unequal sample size case. *Sankhyā Ser. B* **38**, 112–118.

Gupta, S. S., and Huang, D. Y. (1977). Some multiple decision problems in analysis of variance. *Comm. Statist. A—Theory Methods* **6**, 1035–1054.

Gupta, S. S., and Panchapakesan, S. (1969). Some selection and ranking procedures for multivariate normal populations. In *Multivariate Analysis* (P. R. Krishnaiah, ed.), Vol. II, pp. 475–505. Academic Press, New York.

Gupta, S. S., and Panchapakesan, S. (1972). On a class of subset selection procedures. *Ann. Math. Statist.* **43**, 814–822.

Gupta, S. S., and Panchapakesan, S. (1974). Inference for restricted families: (A) Multiple decision procedures; (B) Order statistics inequalities. In *Probability and Biometry* (F. Proschan and R. J. Serfling, eds.), pp. 503–596. SIAM, Philadelphia, Pennsylvania.

Gupta, S. S., and Panchapakesan, S. (1979). *Multiple Decision Procedures: Theory and Methodology of Selecting and Ranking Populations*. Wiley, New York.

Gupta, S. S., and Sobel, M. (1957). On a statistic which arises in selection and ranking problems. *Ann. Math. Statist.* **28**, 957–967.

Gupta, S. S., and Studden, W. J. (1970). On some selection and ranking procedures with applications to multivariate normal populations. In *Essays in Probability and Statistics* (R. C. Bose, I. M. Chakravarti, P. C. Mahalanobis, C. R. Rao, and K. J. C. Smith, eds.), pp. 327–338. Univ. of North Carolina Press, Chapel Hill, North Carolina.

Gupta, S. S., Nagel, K., and Panchapakesan, S. (1973). On the order statistics from equally correlated normal random variables. *Biometrika* **60**, 403–413.

Hooper, J. H., and Santner, T. J. (1979). Design of experiments for selection from ordered families of distributions. *Ann. Statist.* **7**, 615–643.

McDonald, G. C. (1971). On approximating constants required to implement a selection procedure based on ranks. In *Statistical Decision Theory and Related Topics* (S. S. Gupta and J. Yackel, eds.), pp. 299–312. Academic Press, New York.

Olkin, I., Sobel, M., and Tong, Y. L. (1976). Estimating the true probability of correct selection for location and scale parameter families. Tech. Rep. No. 110, Department of Statistics, Stanford Univ., Stanford, California.

Olkin, I., Sobel, M., and Tong, Y. L. (1979). Bounds for a k-fold integral for location and scale parameter models with applications to statistical ranking and selection problems. Tech. Rep. No. 141, Department of Statistics, Stanford Univ., Stanford, California.

Paulson, E. (1952). An optimum solution to the k-sample slippage problem for the normal distribution. *Ann. Math. Statist.* **23**, 610–616.

Ramberg, J. S. (1972). Selection sample size approximations. *Ann. Math. Statist.* **43**, 1977–1980.

Rinott, Y. (1978). On two-stage selection procedures and related probability inequalities. *Comm. Statist. A—Theory Methods* **7**, 799–811.

Tamhane, A. C., and Bechhofer, R. E. (1977). A two-stage minimax procedure with screening for selecting the largest normal mean. *Comm. Statist. A—Theory Methods* **6**, 1003–1033.

Tamhane, A. C., and Bechhofer, R. E. (1979). A two-stage minimax procedure with screening for selecting the largest normal mean (II): An improved PCS lower bound and associated tables. *Comm. Statist. A—Theory Methods* **8**, 337–358.

Tong, Y. L. (1977). An ordering theorem for conditionally independent and identically distributed random variables. *Ann. Statist.* **5**, 274–277.

Tong, Y. L. (1977). Applications of a probability inequality to ranking and selection and other related problems. *Comm. Statist. A—Theory Methods* **6**, 1105–1120.

Tong, Y. L. (1978). An adaptive solution to ranking and selection problems. *Ann. Statist.* **6**, 658–672.

Tong, Y. L., and Wetzell, D. E. (1979). On the behavior of the probability function for selecting the best normal population. *Biometrika* **66**, 174–176.

Wetzell, D. E. (1979). Allocation of observations in ranking and selection problems via majorization and other related inequalities. Ph.D. thesis, Department of Mathematics and Statistics, Univ. of Nebraska, Lincoln, Nebraska.

d. Reliability and Life Testing

Ahmed, A. H. N., Langberg, N. A., León, R. V., and Proschan, F. (1979). Two concepts of positive dependence, with applications in multivariate analysis. Tech. Rep. No. M486, Department of Statistics, Florida State Univ., Tallahassee, Florida.

Barlow, R. E. (1965). Bounds on integrals with applications to reliability problems. *Ann. Math. Statist.* **36**, 565–574.

Barlow, R. E., and Proschan, F. (1966). Inequalities for linear combinations of order statistics from restricted families. *Ann. Math. Statist.* **37**, 1574–1592.

Barlow, R. E., and Proschan, F. (1975). *Statistical Theory of Reliability and Life Testing*. Holt, New York.

Barlow, R. E., Marshall, A. W., and Proschan, F. (1963). Properties of probability distributions with monotone hazard rate. *Ann. Math. Statist.* **34**, 375–389.

Birnbaum, Z. W. (1969). On the importance of different components in a multicomponent system. In *Multivariate Analysis* (P. R. Krishnaiah, ed.), Vol. II, pp. 581–592. Academic Press, New York.

Birnbaum, Z. W., Esary, J. D., and Saunders, S. C. (1961). Multi-component systems and structures and their reliability. *Technometrics* **3**, 55–77.

Birnbaum, Z. W., Esary, J. D., and Marshall, A. W. (1966). Stochastic characterization of wearout for components and systems. *Ann. Math. Statist.* **37**, 816–825.

Brindley, E. C., Jr., and Thompson, W. A., Jr. (1972). Dependence and aging aspects of multivariate survival. *J. Amer. Statist. Assoc.* **67**, 822–830.

Downton, F. (1970). Bivariate exponential distributions in reliability theory. *J. Roy. Statist. Soc. Ser. B* **32**, 408–417.

Esary, J. D., and Marshall, A. W. (1964). System structure and the existence of a system life. *Technometrics* **6**, 459–462.

Esary, J. D., and Marshall, A. W. (1970). Coherent life functions. *SIAM J. Appl. Math.* **18**, 810–814.

Esary, J. D., and Marshall, A. W. (1974). Multivariate distributions with exponential minimums. *Ann. Statist.* **1**, 84–98.

Esary, J. D., and Marshall, A. W. (1974). Families of components and systems, exposed to a compound Poisson damage process. In *Reliability and Biometry* (F. Proschan and R. J. Serfling, eds.), pp. 31–46. SIAM, Philadelphia, Pennsylvania.

Esary, J. D., and Marshall, A. W. (1979). Multivariate distributions with increasing hazard rate average. *Ann. Probab.* **7**, 359–370.

Esary, J. D., and Proschan, F. (1963). Coherent structures of non-identical components. *Technometrics* **5**, 191–209.

Esary, J. D., and Proschan, F. (1970). A reliability bound for systems of maintained, interdependent components. *J. Amer. Statist. Assoc.* **65**, 329–338.

Harris, R. (1970). A multivariate definition for increasing hazard rate distribution functions. *Ann. Math. Statist.* **41**, 713–717.

Hollander, M., Proschan, F., and Sethuraman, J. (1979). Functions decreasing in transposition with applications to shock, damage and down time. Tech. Rep. No. M494, Department of Statistics, Florida State Univ., Tallahassee, Florida.

Langberg, N., and Proschan, F. (1977). Converting dependent models into independent ones, with applications in reliability. In *The Theory and Applications of Reliability, with Emphasis on Bayesian and Nonparametric Methods* (C. P. Tsokos and I. N. Shimi, eds.), Vol. I, pp. 259–275. Academic Press, New York.

Langberg, N., Proschan, F., and Quinzi, A. J. (1977). Transformations yielding reliability models based on independent random variables: A survey. In *Applications of Statistics* (P. R. Krishnaiah, ed.), pp. 323–337. North-Holland Publ., Amsterdam.

Langberg, N., Proschan, F., and Quinzi, A. J. (1977). Estimating dependent life lengths, with applications to the theory of competing risks. Tech. Rep. No. M438, Department of Statistics, Florida State Univ., Tallahassee, Florida.

Langberg, N., Proschan, F., and Quinzi, A. J. (1978). Converting dependent models into independent ones, preserving essential features. *Ann. Probab.* **6**, 174–181.

Marshall, A. W. (1975). Multivariate distributions with monotone hazard rate. In *Reliability and Fault Tree Analysis* (R. E. Barlow, J. B. Fussell, and N. D. Singpurwalla, eds.), pp. 259–284. SIAM, Philadelphia, Pennsylvania.

Marshall, A. W., and Olkin, I. (1967). A multivariate exponential distribution. *J. Amer. Statist. Assoc.* **62**, 30–44.

Marshall, A. W., and Olkin, I. (1967). A generalized bivariate exponential distribution. *J. Appl. Probab.* **4**, 291–302.

Marshall, A. W., and Proschan, F. (1970). Mean life of series and parallel systems. *J. Appl. Probab.* **7**, 165–174.

Marshall, A. W., and Proschan, F. (1972). Classes of distributions applicable in replacement with renewal theory implications. *Proc. Sixth Berkeley Symp. Math. Statist. Probab.* **1** (L. M. LeCam, J. Neyman, and E. L. Scott, eds.), 395–415. Univ. of California Press, Berkeley, California.

Marshall, A. W., and Shaked, M. (1979). Multivariate shock models for distributions with increasing hazard rate average. *Ann. Probab.* **7**, 343–358.

Pledger, G., and Proschan, F. (1971). Comparisons of order statistics and of spacings from heterogeneous distributions. In *Optimizing Methods in Statistics* (J. S. Rustagi, ed.), pp. 89–113. Academic Press, New York.

Pledger, G., and Proschan, F. (1973). Stochastic comparisons of random processes, with applications in reliability. *J. Appl. Probab.* **10**, 572–585.

Proschan, F. (1975). Applications of majorization and Schur functions in reliability and life testing. In *Reliability and Fault Tree Analysis* (R. E. Barlow, J. B. Fussell, and N. D. Singpurwalla, eds.), pp. 237–258. SIAM, Philadelphia, Pennsylvania.

Proschan, F., and Sethuraman, J. (1976). Stochastic comparisons of order statistics from heterogeneous populations, with applications in reliability. *J. Multivariate Anal.* **6**, 608–616.

Sen, P. K. (1970). A note on order statistics for heterogeneous distributions. *Ann. Math. Statist.* **41**, 2137–2139.

Shaked, M. (1977). A family of concepts of dependence for bivariate distributions. *J. Amer. Statist. Assoc.* **72**, 642–650.

Shaked, M. (1977). A concept of positive dependence for exchangeable random variables. *Ann. Statist.* **5**, 505–515.

I. STATISTICAL TABLES IN MULTIVARIATE DISTRIBUTIONS

Armitage, J. V., and Krishnaiah, P. R. (1964). Tables for the studentized largest chi-square distribution and their applications. Tech. Rep. No. ARL 64-188, Aerospace Research Laboratories, Wright-Patterson AFB, Ohio.

Bechhofer, R. E. (1954). A single-stage multiple decision procedure for ranking means of normal populations with known variances. *Ann. Math. Statist.* **25**, 16–39.

Beckman, R. J., and Tietjen, G. L. (1973). Upper 10% and 25% points of the maximum *F* ratio. *Biometrika* **60**, 213–214.

Chambers, C. (1967). Extension of tables of percentage points of the largest variance ratio, $s^2_{\max}/s^2_0$. *Biometrika* **54**, 225–227.

Chen, H. J. (1974). Percentage points of multivariate *t* distribution and their application. Tech. Rep. No. 74-14, Department of Mathematical Sciences, Memphis State Univ., Memphis, Tennessee.

David, H. A. (1952). Upper 5 and 1% points of the maximum *F*-ratio. *Biometrika* **39**, 422–424.

Dayton, C. M., and Schafer, W. D. (1973). Extended tables of *t* and chi-square for Bonferroni tests with unequal error allocation. *J. Amer. Statist. Assoc.* **68**, 78–83.

Dudewicz, E. J., and Dalal, S. R. (1975). Allocation of observations in ranking and selection with unequal variances. *Sankhyā Ser. B* **37**, 28–78.

Dudewicz, E. J., Ramberg, J. S., and Chen, H. J. (1975). New tables for multiple comparisons with a control (unknown variances). *Biomet. Z.* **17**, 13–26.

Dunn, O. J., and Massey, F. J. (1965). Estimation of multiple contrasts using *t*-distributions. *J. Amer. Statist. Assoc.* **60**, 573–583.

Dunn, O. J., Kronmal, R. A., and Yee, W. J. (1968). Tables of the multivariate *t*-distribution. Tech. Rep., School of Public Health, UCLA, Los Angeles, California.

Dunnett, C. W. (1964). New tables for multiple comparisons with a control. *Biometrics* **20**, 482–491.

Dunnett, C. W., and Sobel, M. (1954). A bivariate generalization of Student's *t*-distribution with tables for certain special cases. *Biometrika* **41**, 153–169.

Foster, F. G., and Rees, D. H. (1957). Upper percentage points of the generalized Beta distribution. *Biometrika* **44**, 237–247.

Gibbons, J. D., Olkin, I., and Sobel, M. (1977). *Selecting and Ordering Populations: A New Statistical Methodology*. Wiley, New York.

Graybill, F. A., and Bowden, D. C. (1967). Linear segment confidence bands for simple linear models. *J. Amer. Statist. Assoc.* **62**, 403–408.

Gupta, S. S. (1963). Probability integrals of multivariate normal and multivariate *t*. *Ann. Math. Statist.* **34**, 792–828.

Gupta, S. S. (1963). On a selection and ranking procedure for gamma populations. *Ann. Inst. Statist. Math.* **14**, 199–216.

Gupta, S. S., and Sobel, M. (1957). On a statistic which arises in selection and ranking problems. *Ann. Math. Statist.* **28**, 957–967.

Gupta, S. S., and Sobel, M. (1962). On the smallest of several correlated *F* statistics. *Biometrika* **49**, 509–523.

Gupta, S. S., Nagel, K., and Panchapakesan, S. (1973). On the order statistics from equally correlated normal random variables. *Biometrika* **60**, 403–413.

Hahn, G. J., and Hendrickson, R. W. (1971). A table of percentage points of the distribution of the largest absolute value of *k* Student *t* variates and its applications. *Biometrika* **58**, 323–332.

Halperin, M., Greenhouse, S. W., Cornfield, J., and Zalokar, J. (1955). Tables of percentage points for the studentized maximum absolute deviation in normal samples. *J. Amer. Statist. Assoc.* **50**, 185–195.

Hanumara, R. C., and Thompson, W. A., Jr. (1968). Percentage points of the extreme roots of a Wishart matrix. *Biometrika* **55**, 505–512.

Harter, H. L. (1960). Tables of range and studentized range. *Ann. Math. Statist.* **31**, 1122–1147.

Harter, H. L. (1963). Percentage points of the ratio of two ranges and power of the associated test. *Biometrika* **50**, 187–194.

Heck, D. L. (1960). Charts on some upper percentage points of the distribution of the largest characteristic root. *Ann. Math. Statist.* **31**, 625–641.

Krishnaiah, P. R., and Armitage, J. V. (1964). Distribution of the studentized smallest chi-square, with tables and applications. Tech. Rep. No. ARL 64-218, Aerospace Research Laboratories, Wright-Patterson AFB, Ohio.

Krishnaiah, P. R., and Armitage, J. V. (1965). Tables for the distribution of the maximum of correlated chi-square variates with one degree of freedom. Tech. Rep. No. ARL 65-136, Aerospace Research Laboratories, Wright-Patterson AFB, Ohio.

Krishnaiah, P. R., and Armitage, J. V. (1965). Probability integrals of the multivariate *F* distribution, with tables and applications. Tech. Rep. No. ARL 65-236, Aerospace Research Laboratories, Wright-Patterson AFB, Ohio.

Krishnaiah, P. R., and Armitage, J. V. (1966). Tables for multivariate *t* distribution. *Sankhyā Ser. B* **28**, 31–56.

Krishnaiah, P. R., Armitage, J. V., and Breiter, M. C. (1969). Tables for the probability integrals of the bivariate *t* distribution. Tech. Rep. No. ARL 69-0060, Aerospace Research Laboratories, Wright-Patterson AFB, Ohio.

Krishnaiah, P. R., Armitage, J. V., and Breiter, M. C. (1969). Tables for the bivariate $|t|$ distribution. Tech. Rep. No. ARL 69-0210, Aerospace Research Laboratories, Wright-Patterson AFB, Ohio.

Miller, R. G., Jr. (1966). *Simultaneous Statistical Inference*. McGraw-Hill, New York.

Milton, R. C. (1963). Tables of the equally correlated multivariate normal probability integral. Tech. Rep. No. 27, Department of Statistics, Univ. of Minnesota, Minneapolis, Minnesota.

Milton, R. C. (1970). *Rank Order Probabilities: Two-Sample Normal Shift Alternatives*. Wiley, New York.

Olkin, I., Sobel, M., and Tong, Y. L. (1976). Estimating the true probability of correct selection for location and scale parameter families. Tech. Rep. No. 110, Department of Statistics, Stanford Univ., Stanford, California.

Owen, D. B. (1956). The bivariate normal probability distribution. Tech. Rep. No. SC-3831, Sandia Corp., Albuquerque, New Mexico.

Owen, D. B. (1962). *Handbook of Statistical Tables*. Addison-Wesley, Reading, Massachusetts.

Pearson, E. S., and Hartley, H. O. (1966). *Biometrika Tables for Statisticians*, 3rd ed. Cambridge Univ. Press, London and New York.

Pillai, K. C. S. (1967). Upper percentage points of the largest root of a matrix in multivariate analysis. *Biometrika* **54**, 189–194.

Pillai, K. C. S., and Buenaventura, A. R. (1961). Upper percentage points of a substitute F-ratio using ranges. *Biometrika* **48**, 195–196.

Pillai, K. C. S., and Ramachandran, K. V. (1954). On the distribution of the ith observation in an ordered sample from a normal population to an independent estimate of the standard deviation. *Ann. Math. Statist.* **25**, 565–572.

Schuurmann, F. J., Krishnaiah, P. R., and Chattopadhyay, A. K. (1975). Tables for a multivariate F distribution. *Sankhyā Ser. B* **37**, 308–331.

Sobel, M., Uppuluri, V. R. R., and Frankowski, K. (1976). *Dirichlet Distributions, Type 1*, Selected Tables in Mathematical Statistics, Vol. 4. Amer. Math. Soc., Providence, Rhode Island.

Steck, G. P. (1958). A table for computing trivariate normal probabilities. *Ann. Math. Statist.* **29**, 780–800.

Steck, G. P. (1962). Orthant probabilities for the equicorrelated multivariate normal distribution. *Biometrika* **49**, 433–445.

Steffens, F. E. (1969). Critical values for bivariate Student t-tests. *J. Amer. Statist. Assoc.* **64**, 637–646.

Tietjen, G. L., and Beckman, R. J. (1972). Tables for using the maximum F-ratio in multiple comparison procedures. *J. Amer. Statist. Assoc.* **67**, 581–583.

Tong, Y. L. (1969). On partitioning a set of normal populations by their locations with respect to a control. *Ann. Math. Statist.* **40**, 1300–1324.

Trout, J. R., and Chow, B. (1972). Table of percentage points of the trivariate t-distribution with an application to uniform confidence bands. *Technometrics* **14**, 855–879.

Zelen, M., and Severo, N. C. (1964). Probability functions. In *Handbook of Mathematical Functions* (M. Abramowitz and I. A. Stegun, eds.) 5th printing, pp. 925–995. Dover, New York.

Author Index

Subject Index

Probability and Mathematical Statistics

A Series of Monographs and Textbooks

Thomas Ferguson. Mathematical Statistics: A Decision Theoretic Approach. 1967

Howard Tucker. A Graduate Course in Probability. 1967

K. R. Parthasarathy. Probability Measures on Metric Spaces. 1967

P. Révész. The Laws of Large Numbers. 1968

H. P. McKean, Jr. Stochastic Integrals. 1969

B. V. Gnedenko, Yu. K. Belyayev, and A. D. Solovyev. Mathematical Methods of Reliability Theory. 1969

Demetrios A. Kappos. Probability Algebras and Stochastic Spaces. 1969

Ivan N. Pesin. Classical and Modern Integration Theories. 1970

S. Vajda. Probabilistic Programming. 1972

Sheldon M. Ross. Introduction to Probability Models. 1972

Robert B. Ash. Real Analysis and Probability. 1972

V. V. Fedorov. Theory of Optimal Experiments. 1972

K. V. Mardia. Statistics of Directional Data. 1972

H. Dym and H. P. McKean. Fourier Series and Integrals. 1972

Tatsuo Kawata. Fourier Analysis in Probability Theory. 1972

Fritz Oberhettinger. Fourier Transforms of Distributions and Their Inverses: A Collection of Tables. 1973

Paul Erdös and Joel Spencer. Probabilistic Methods in Combinatorics. 1973

K. Sarkadi and I. Vincze. Mathematical Methods of Statistical Quality Control. 1973

Michael R. Anderberg. Cluster Analysis for Applications. 1973

W. Hengartner and R. Theodorescu. Concentration Functions. 1973

Kai Lai Chung. A Course in Probability Theory, Second Edition. 1974

L. H. Koopmans. The Spectral Analysis of Time Series. 1974

L. E. Maistrov. Probability Theory: A Historical Sketch. 1974

William F. Stout. Almost Sure Convergence. 1974

E. J. McShane. Stochastic Calculus and Stochastic Models. 1974

Robert B. Ash and Melvin F. Gardner. Topics in Stochastic Processes. 1975

Avner Friedman. Stochastic Differential Equations and Applications, Volume 1. 1975; Volume 2. 1975

Roger Cuppens. Decomposition of Multivariate Probabilities. 1975

Eugene Lukacs. Stochastic Convergence, Second Edition. 1975

H. Dym and H. P. McKean. Gaussian Processes, Function Theory, and the Inverse Spectral Problem. 1976

N. C. Giri. Multivariate Statistical Inference. 1977

Lloyd Fisher and John McDonald. Fixed Effects Analysis of Variance. 1978

Sidney C. Port and Charles J. Stone. Brownian Motion and Classical Potential Theory. 1978

Konrad Jacobs. Measure and Integral. 1978

K. V. Mardia, J. T. Kent, and J. M. Biddy. Multivariate Analysis. 1979

Sri Gopal Mohanty. Lattice Path Counting and Applications. 1979

Y. L. Tong. Probability Inequalities in Multivariate Distributions. 1980

in preparation

Michel Metivier and J. Pellaumail. Stochastic Integration. 1980

M. B. Priestly, Spectral Analysis and Time Series. 1980

Ishwar V. Basawa and B. L. S. Prakasa Rao, Statistical Inference for Stochastic Processes. 1980

M. Csörgö and P. Révész. Strong Approximations in Probability and Statistics. 1980